W0112134

Springer Tracts in Modern Physics 107

Springer Tracts in Modern Physics

---

* denotes a volume which contains a Classified Index starting from Volume 36

N. Schwentner
E.-E. Koch · J. Jortner

Electronic Excitations in Condensed Rare Gases

With 131 Figures

Springer-Verlag Berlin Heidelberg GmbH

Professor Dr. Nikolaus Schwentner

Institut für Atom- und Festkörperphysik, Freie Universität Berlin,
D-1000 Berlin 33, Germany

Professor Dr. Ernst-Eckhard Koch

Fritz-Haber-Institut der Max-Planck-Gesellschaft, Faradayweg 4–6,
D-1000 Berlin 33, Germany

Professor Dr. Joshua Jortner

Department of Chemistry, Tel Aviv University, 69978 Tel Aviv, Israel

Manuscripts for publication should be addressed to:
Gerhard Höhler

Institut für Theoretische Kernphysik der Universität Karlsruhe
Postfach 6380, D-7500 Karlsruhe 1, Fed. Rep. of Germany

*Proofs and all correspondence concerning papers in the process of publication
should be addressed to:*

Ernst A. Niekisch

Haubourdinstrasse 6, D-5170 Jülich 1, Fed. Rep. of Germany

ISBN 978-3-662-15221-8   ISBN 978-3-540-39423-5 (eBook)
DOI 10.1007/978-3-540-39423-5

Library of Congress Cataloging in Publication Data. Schwentner, N. (Nikolaus), 1945–.Electronic excitations in condensed rare gases. (Springer tracts in modern physics; 107). Bibliography: p. Includes index. 1. Electronic excitation. 2. Relaxation phenomena (Physics). 3. Solid rare gases. 4. Liquids. 5. Condensed matter, I. Koch, E.-E. (Ernst-Eckhard). II. Jortner, Joshua. III. Title. IV. Series. QC1.S797   vol.107   [QC176.8.E9]   539 s [530.4]   85-9950

© Springer-Verlag Berlin Heidelberg 1985
Originally published by Springer-Verlag Berlin Heidelberg New York Tokyo in 1985
Softcover reprint of the hardcover 1st edition 1985

Offset printing and bookbinding: Brühlsche Universitätsdruckerei, Giessen
2153/3130-543210

Preface

This book is concerned with the electronic structure and excited-state nonradiative relaxation phenomena in pure rare-gas solids (RGS), alloys of rare gases, solid two-component mixtures, metals in rare-gas solids, pure rare-gas liquids and liquid mixtures. Starting from the experimental point of view the techniques utilized to probe the optical constants, transient absorption, energy- and time-resolved luminescence, as well as photoelectron yield and energy distribution are considered. In the survey of the electronic structure of valence bands, conduction bands and excitonic states in pure RGS, recent new experimental and theoretical results for bulk excitations and for surface excitons are emphasized. This is followed by a review of electronic excitations in rare-gas alloys and liquid rare gases. Next, basic properties of metal rare-gas solid mixtures are considered and information is gained concerning electronic structure, transport properties and metal-nonmetal transitions.

Knowledge of the electronic structure provides the basic input data required for the elucidation of excited-state dynamics in condensed rare gases. We discuss the experimental information obtained from luminescence and photoemission studies and the available theoretical framework pertaining to the microscopic relaxation processes in solid and liquid rare gases. A variety of dynamic processes occur, including exciton trapping, vibrational relaxation, electronic relaxation, electronic energy migration via exciton states, electronic energy transfer between localized states, autoionization and electron-hole recombination. Finally, electron transport properties and electron-hole pair creation processes for electrons with kinetic energies of a few eV up to the MeV region are treated and discussed in terms of the electronic structure and the dynamical processes involved. We emphasize the effects of structural and compositional disorder on the electronic properties and are concerned with the microscopic aspects of excited-state energy conversion, storage and disposal in these materials.

We wish to express our sincere thanks and gratitude to our colleagues from the Institute for Experimental Physics of the University of Kiel, the Hamburg Synchrotron Radiation Laboratory HASYLAB at DESY and the Department of Chemistry of Tel Aviv

University, for their contributions to this research area. We are especially in-
debted and grateful to O. Cheshnovsky, U. Even, A. Gedanken, R. Haensel, U. Hahn,
I. Messing, Z. Ophir, P. Rabe, B. Raz, V. Saile, M. Skibowski, B. Sonntag, I. Steinberger, W. Steinmann and G. Zimmerer, with whom we have enjoyed many stimulating,
interesting and rewarding discussions.

Berlin, Tel Aviv,   *N. Schwentner · E.-E. Koch · J. Jortner*
April 1985

Contents

1. Introduction

Rare-gas solids (RGS) constitute the simplest molecular crystals, which are held together by van der Waals forces. Many properties of rare-gas crystals can well be accounted for in terms of a superposition of two-body potentials, each consisting of a short-range repulsion and a long-range London attraction. By adjusting the parameters of a Lennard-Jones spherically symmetric pair potential, $V(r) = 4\varepsilon[(\sigma/r)^{12} - (\sigma/r)^{6}]$, where ε is a depth of the potential well and σ is the distance r at which repulsion and attraction cancel out each other, a rather good account of ground-state energetics data was provided /Dobbs and Jones, 1957; Cook, 1961; Pollak, 1964/. The Lennard-Jones theory predicts the lattice energy within an accuracy of a few percent. A general conclusion concerning the structure of RGS, which stems from this theory, pertains to the energetic stability of a closed-packed lattice. However, very small energy differences can go a long way in determining structural stability. The energetic difference between the two closed-packed structures, i.e., the hexagonal closed-packed lattice and the face-centre cubic lattice, is below 1% of the total lattice energy /Meyer, 1969/. The simple theory predicts the energetic stability of the hexagonal closed-packed structure relative to the face-centre cubic structure, which is in contrast with the real life situation. All RGS crystallize in a face-centre cubic structure with the exception of solid He, which constitutes a quantum crystal for the nuclear motion. Fascinating structural modifications of RGS, which involve the existence of a metastable hexagonal closed-packed structure and a face-centre cubic to a hexagonal closed-packed phase transition in solid Ar, have been documented /Meyer, 1969/. These extraordinary structural features of RGS require a sophisticated theoretical description to account for a modification of about (approximately) 1% of the total lattice energy. In this context, the role of three-body electron exchange interactions /Janson, 1964/, electronic overlap effects /Meyer, 1969/, zero-point nuclear motion /Gillis et al., 1968/ and anharmonicity effects /Meyer, 1969/ were invoked. A complete self-consistent determination of the minor energy changes, which govern the structural stability of rare-gas lattices, has not yet been provided. In an analogy to a famous discussion of the chemical bond /Coulson, 1951/, it may be stated that the determination of the stabilization energy of one closed-packed RGS structure relative to another is like "weighing the captain of a ship by determining the difference in the displacement of the ship when he is or is not on board". The intensive and extensive effort aimed towards the elucidation of the structure and the dynamics of nuclear motion of RGS has been documented in several good books /Cook, 1961; Klein and Venables, 1967,1977/.

The electronic states of RGS constitute a prototype for insulators, serving as a testing ground for novel theoretical ideas and experimental techniques. RGS are large-gap insulators, being optically transparent into the vacuum ultraviolet (VUV) spectral region. The onset of optical absorption is dominated by the excitation of intense bound exciton states. At higher energies interband excitations from the valence to the conduction band are exhibited, while at even higher energies core excitations prevail. Our current understanding of the electronic structure and the electronic excitations of pure RGS is rather complete. An extensive review of the experimental work, which led to the elucidation of the dielectric and optical properties of RGS with an emphasis on the spectra in the VUV and soft X-ray region has been provided by Sonntag /1977/, excitons in RGS have been discussed by Fugol /1978/, while a review of the theoretical calculations of electronic states in pure RGS has been presented by Rössler /1976/. The mechanisms of nonradiative relaxation of bound electronic excitations in pure RGS are dominated by intrastate (adiabatic) nuclear deformations, resulting in the partial radiationless dissipation of the electronic energy. Adopting, for a moment, a "molecular" point of view, we can assert that an electronically excited rare-gas atom is highly chemically reactive in contrast to an inert ground state atom because the excited electron resembles the lone electron of an alkali atom, and the remaining hole in the valence shell corresponds to the partly filled shell of a halogen atom. In the gas phase strongly bound rare-gas excimers between an excited atom and a ground state atom are formed. The reactive excited atoms in the RGS induce local structural changes which are similar to the form of either excimers or local density dilations, i.e., "bubbles". The emission bands of pure RGS are nearly identical with the gas phase excimer bands, which are used in excimer lasers for the VUV. Such an admittedly over-simplified molecular approach for relaxation in RGS has to be supplemented by solid-state concepts. In this context, it should be emphasized that the electronic excitation of rare gases can be interrogated over a broad density range by continuously varying the density from the gas phase through the liquid phase to the solid. The effects of density change, as well as of structural disorder and order on the electronic level structure and radiationless relaxation of rare gases, can be explored in a systematic manner.

The exploration of diverse facets pertaining to the structure and to the electronic level structure of RGS is by no means restricted to chemically neat one-component systems. RGS are easily doped with guest atoms or molecules by codeposi-

tion on a cold substrate. This approach provides the basis for the celebrated
matrix isolation spectroscopy /Pimentel, 1958; McCarty and Robinson, 1959/. The
characteristics of the vibrational and electronic states of the guest are often
only weakly perturbed by the RGS. This is the case in certain intravalence excitations of the guest, when the guest energy levels are only slightly modified by
the host and when the coupling of the guest's electronic and/or vibrational excitations with the host phonons is weak. Under these circumstances, the radiationless depopulation of excited vibrational and electronic guest states may be com-
paratively slow on the time scale of the radiative decay. By an appropriate choice
of the guest and the host matrix, photoselective studies can be conducted eluci-
dating the dependence of the excited-state dynamics on the strength of the guest-
host coupling and on the phonon spectrum of the host. An extensive survey of experimental and theoretical studies of impurity states has been provided in several
books (see for example /Meyer, 1971; Moskowitz and Ozin, 1976; Dash, 1975; Barnes
et al., 1981/).

During the past two decades the experimental and theoretical effort has been
extended to investigate the mutual interdependence of structural and electronic
properties. Furthermore, attempts have been made to understand the excited-state
relaxation phenomena in condensed rare gases. Pure RGS, solid alloys, condensed
rare gases including solid two-component mixtures, pure liquids and liquid alloys
provide useful information concerning the effects of the state of aggregation and
disorder on the electronic structure and, in particular, establish the influence
of structural and compositional changes on the electronic properties. Structural,
positional-type disorder is exhibited in liquids, in glasses and in structurally
deformed solids. Compositional disorder prevails in binary mixtures. The effects
of disorder on the electronic structure are characterized by some universal features /Economou et al., 1974; Kramer, 1976/ which are drastically different from
the electronic states of ideally ordered solids /Mott and Davis, 1971/. It turns
out that rare-gas solids, liquids and alloys can be considered as prototype materials to explore electronic structure on a much deeper level than originally an-
ticipated. For an investigation of the static electronic properties and the influence of structural changes on the electronic states, the closed-shell electron
configuration and the weak dispersion forces in the ground state are of major importance. Comparative studies on rare-gas alloys in the gaseous, liquid and solid
phase provide a wealth of information concerning electronic states of disordered
insulators. In addition, metal rare-gas solid mixtures yield insight into the
challenging problems of transport properties and the metal-nonmetal transition in
disordered materials /Mott, 1974/.

One major theme of this monograph addresses the relation between the electronic
level structure and the geometrical studies of condensed phases of rare gases,
with an emphasis on the effects of the state of aggregation and compositional

disorder on the electronic properties. The diverse information concerning the static features of electronic structure provides the basic input information for the understanding of dynamic processes in ordered and disordered condensed rare gases. In fact, the static information and the dynamic processes are complementary and cannot be disentangled, as the understanding of the electronic structure provides a prerequisite for the elucidation of dynamic relaxation phenomena. The second major theme of this monograph is concerned with the diverse and interesting problems of excited-state dynamics in condensed rare gases, considering the microscopic aspects of energy conversion, storage and disposal in these materials. The fate of electronically excited states in ordered and disordered condensed rare gases involves a variety of nonradiative channels, such as exciton self-trapping, electronic energy transfer between localized states and auto-ionization, just to mention a few examples. There has been remarkable progress in the experimental investigation of these various pathways of nonradiative energy dissipation in condensed rare gases. This experimental information established general trends and rules which can be used as a testing ground for theoretical ideas.

Apart from the basic intrinsic interest in the electronic structure and in the general concepts and mechanisms governing radiationless transitions and dynamic processes, rare-gas solids, liquids and alloys have attracted interest in other areas of science and technology. There has been important progress in matrix isolation spectroscopy /Pimentel, 1958; McCarty and Robinson, 1959; Meyer, 1971; Barnes et al., 1981/, and cryochemistry in rare-gas matrices /Moskowitz and Ozin, 1976/. As a scientific and technological application, we mention the progress in high power VUV excimer lasers based on excited rare-gas molecules and rare-gas halogen compounds /Rhodes, 1984/. To provide another example of technical relevance, we recall that doped rare-gas liquids play an important role in the recent development of new high energy particle detectors /Weber, 1979/. For all these developments a detailed understanding of the electronic structure and dynamic processes in RGS is required. The scientific and technological aspects of structural features and of electronic excitations should not be limited to the bulk. The structural and electronic phenomena observed for two-dimensional physisorbed and adsorbed rare gases provide a clue for a variety of surface phenomena. The remarkable progress in the area of supersonic jets and cluster beams /Hagena and Obert, 1972/ makes it possible to explore clusters of rare gases, elucidate macroscopic phenomena, such as nucleation, adsorption and catalysis from the microscopic point of view.

What are the relevant excited states and what are the important decay mechanisms in RGS? To answer this question we have sketched in Fig. 1.1 an overview of the energy regions for basic excitations and radiationless decay channels in molecular solids. Phonons, librational and vibrational excitations mark the lower energy limit of 0.01 eV. The elementary electronic excitations at higher energies involve intravalence and low intervalence electronic excitations, which in RGS can be

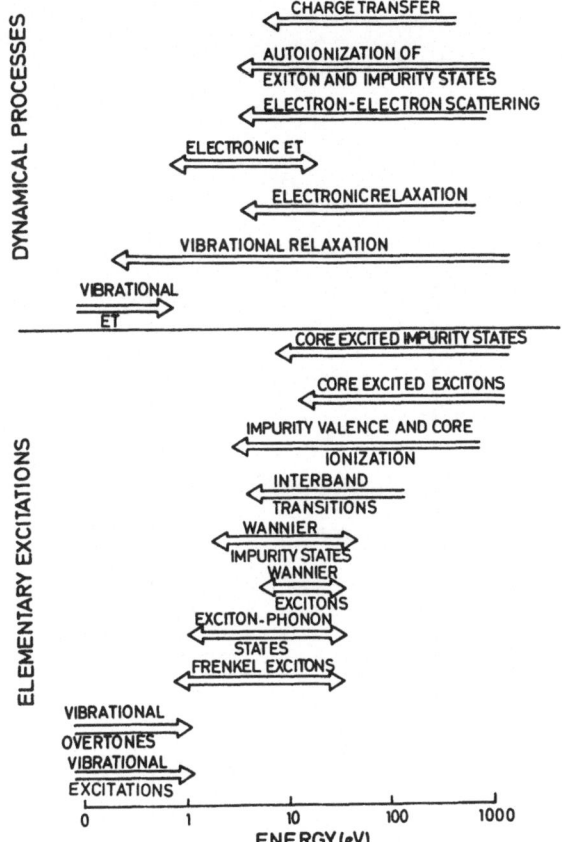

Fig. 1.1. Energy regions for excited-state photophysical and chemical processes in molecular crystals (after Jortner and Leach /1980/)

described in terms of the tight-binding Frenkel scheme, as well as Wannier exciton and impurity states for higher-energy excitations. At even higher energies one encounters interband transitions, metastable exciton and impurity auto-ionizing states. At very high energies core excitons and core impurity excitations are exhibited. The dynamic processes involve electronic energy transfer between the exciton states, exciton diffusion, electronic relaxation of exciton states, exciton trapping with excimer formation, all occurring in the energy range 5 - 20 eV, electron-hole recombination in the energy range above the band gap energy E_G, as well as auto-ionization processes of core excited states and charge transfer processes in the energy range up to several thousand eV's. In order to provide an estimate for the orders of magnitude of the energies involved, we have collected in Table 1.1 the electron-binding energies for all rare gases. To be more specific, we have sketched in Figs. 1.2,3,4 the valence states and the decay processes in some detail. The optical absorption spectra of pure and lightly doped solid and liquid rare gases can be analysed in terms of stable exciton or impurity states. A schematic diagram of the

5

Table 1.1. Binding energies of the occupied levels of the rare-gas atoms obtained from photoemission spectroscopy. The binding energies are given in eV relative to the vacuum level. Some optical data as given by Moore a) and Lotz c) are also included

	$1s_{1/2}$ K	$2s_{1/2}$ L_I	$2p_{1/2}$ L_{II}	$2p_{3/2}$ L_{III}	$3s_{1/2}$ M_I	$3p_{1/2}$ M_{II}	$3p_{3/2}$ M_{III}	$3d_{3/2}$ M_{IV}	$3d_{5/2}$ M_V	$4s_{1/2}$ N_I
$_1$He	24.580 a)									
$_{10}$Ne	870.2 b)	48.42 b) / 48.47 c)	21.661 c) / 21.656 a)	21.564 c) / 21.559 a)						
$_{18}$Ar	3205.9 b)	326.3 b)	250.56 b)	248.45 b)	29.3 b)	15.94 c) / 15.933 a)	15.56 c) / 15.755 a)			
$_{36}$Kr	14326 b)	1921 b)	1730.9 b)	1678.4 b)	292.8 b)	222.2 b)	214.4 b)	94.9 b) / 95.04 a)	93.7 b) / 93.83 a)	
$_{54}$Xe	34561 b)	5453 b)	5104 b)	4782 b)	1148.7 b)	1002.1 b)	940.6 b)	689.0 b)	676.4 b)	213.2 b)
$_{86}$Rn										1208 d)

	$4p_{1/2}$ N_{II}	$4p_{3/2}$ N_{III}	$4d_{3/2}$ N_{IV}	$4d_{5/2}$ N_V	$4f_{5/2}$ N_{VI}	$4f_{7/2}$ N_{VII}	$5s$ O_I	$5p_{1/2}$ O_{II}	$5p_{3/2}$ O_{III}	$5d_{3/2}$ O_{IV}	$5d_{5/2}$ O_V
$_1$He											
$_{10}$Ne											
$_{18}$Ar											
$_{36}$Kr	14.662 a) / 14.08 b)	13.996 b)									
$_{54}$Xe	155 c)	145.5 b)	69.5 b)	67.5 b)		-	23.3 b)	13.4 b) / 13.433 a)	12.13 b) / 12.127 a)		
$_{86}$Rn	1958 d)	879 d)	636 d)	603 d)		299 d)	254 d)	200 d)	153 d)		80 d)

a) Moore /1949, 1952, 1958/
b) Siegbahn et al. /1969/
c) Lotz /1967, 1968/ from optical data
d) Siegbahn et al. /1967/

energy levels appears in Fig. 1.2 for the case of an Xe impurity in an Ar matrix.
The relative simplicity of these excitons and impurity states makes it possible
to gain detailed information concerning the energy levels which is of prime importance for a subsequent discussion of the dynamical processes. The sources of
experimental information concerning the energy levels are as follows:

i) Absorption spectroscopy. Wannier exciton series are identified in pure materials and of excitonic atomic or molecular impurity states in doped insulators by
absorption spectroscopy. The Wannier series converge to the bottom of the conduction band. The lowest exciton ($n = 1$) is of the intermediate type between the Wannier
and the Frenkel schemes. From the convergence limit of the Wannier series studied
by absorption spectroscopy, one can obtain the band gap, E_G, in the pure substance
or the impurity ionization potential, E_G^i, of the impurity in the medium. The energetics of the Wannier states, monitored by absorption spectroscopy results in basic
information regarding the characteristics of the conduction band. In particular,
from the effective Rydberg constant of the Wannier series, one can deduce the effective mass of the electron near the minimum of the conduction band.

ii) Photodonductivity. The threshold for photoconductivity in the pure solid or
liquid results in a direct measurement of E_G.

iii) Photoelectron spectroscopy. The threshold for external photoemission from
the pure material E_{Th}, or from the doped insulator, E_{Th}^i, results in the external

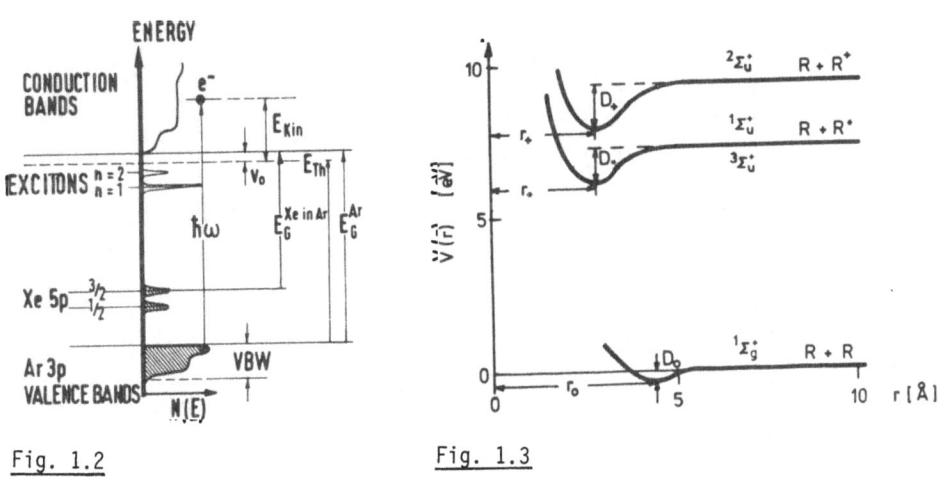

Fig. 1.2   Fig. 1.3

Fig. 1.2. Schematic scheme of the energy levels of an atomic impurity state in a
solid rare-gas matrix depicted for the case of Xe in Ar. The energies given are
discussed in the text

Fig. 1.3. Potential curves for a diatomic rare-gas molecule. The nomenclature is
explained in the text

ionization potentials. From the combination of optical spectra and photoemission yields in pure or in doped insulators, one can determine the energy V_0 of the bottom of the conduction band relative to the vacuum level. From the photoelectron energy distribution curves (EDC's), one obtains a detailed picture of the initial states. Furthermore, the dependence of the EDC's on the exciting photon energy gives some information about final states. Angle resolved photoemission is well suited to measure the energy and momentum of electrons in a solid. Thus a determination of the $E = E(\underline{k})$ relation describing an energy band becomes possible.

Table 1.2. Energies and structural parameters of RGS and rare-gas molecules. T_0: melting point at 760 Torr; $\hbar\omega_p$: largest energies of transversal-acoustic phonons TA(X) and longitudinal-acoustic phonons LA(X); D_K: binding energy per atom in a crystal; D_0, D_*, D_+: dissociation energies of rare-gas molecules in the ground state, the excited state and of the molecular ions respectively (see Fig. 1.3); r_K: nearest neighbour separation in a crystal; r_0, r_*, r_+: internuclear distance of rare-gas molecules in the ground state, the excited state and of the molecular ions, respectively (see Fig. 1.3)

		He	Ne	Ar	Kr	Xe
T_0 (K) [a]			24.6	83.7	115.8	161.3
$\hbar\omega_p$ (eV)[a]	TA(X)		0.0046	0.0059	0.0043	0.0038
	LA(X)		0.0068	0.0086	0.0062	0.0054
D_K (eV/Atom) [d]			0.02	0.08	0.116	0.17
D_0 (eV)[c]		0.00095	0.0036	0.0012	0.0017	0.0024
D_* (eV)		2.5 [d]	0.5 [e]	0.68[f]		0.79[g]
D_+ (eV)		2.67[h]	1.2 [e]	1.25[h]	1.15[i]	0.99[g]
r_K (Å) [k]			3.156	3.755	3.992	4.335
r_0 (Å) [c]		2.96	3.102	3.761	4.006	4.361
r_* (Å)		1.04[d]	1.79[e]	2.42[f]		3.04[g]
r_+ (Å)		1.06[h]	1.75[e]	2.43[h]	2.79[k]	3.04[g]

[a] Powell and Dolling /1977/
[b] Kittel /1971/
[c] Barker /1976/
[d] Sando /1971/
[e] Cohen and Schneider /1974/
[f] Saxon and Liu /1976/
[g] Ermler, Lee, Pitzer and Winter /1978/
[h] Gilbert and Wahl /1971/
[i] Ng, Trevor, Mahan and Lee /1977/
[k] Wadt /1979/

iv) Luminescence spectroscopy. The potential curves of the ground and excited states in different stages of structural rearrangement can be probed by time- and energy-resolved luminescence spectroscopy. For a cursory discussion of the dynamical processes in the excited states, it is important to notice that the properties of a rare-gas atom change dramatically upon electronic excitation and ionization (Fig. 1.3). In the ground state we have the typical weakly bound van der Waals potential curve (Fig. 1.3) with a shallow minimum and a strongly repulsive part for smaller distances. The equilibrium distance r_0 of rare-gas molecules practically coincides with the near neighbour distance r_K in the crystal. The small depth D_0 of the molecular potential curve corresponds to the weak bonding energy D_K of the atoms in the crystal (see Table 1.2). In the excited state a chemically active radical is formed. Thus, strongly bound rare-gas molecular ions, R_2^* with binding energies, D_+, ranging from 1 to 2 eV, can be formed. These are characterized by a considerably smaller equilibrium distance, r_t, as compared to the ground state. Excited stable neutral states, R_2^* (excimer states) also exist /Mulliken, 1970/. The additional electron on a large orbit around the R_2^* centre only weakly disturbs the R^*-R bond. The excited R_2^* molecules can decay radiatively into a pair of atoms in the ground state (Fig. 1.4). Since the two atoms are on the strongly repulsive part of the potential curve, they rapidly fly apart. An excited rare-gas atom in the crystal will exert strong attractive forces upon the neighbouring atoms. Experiments show

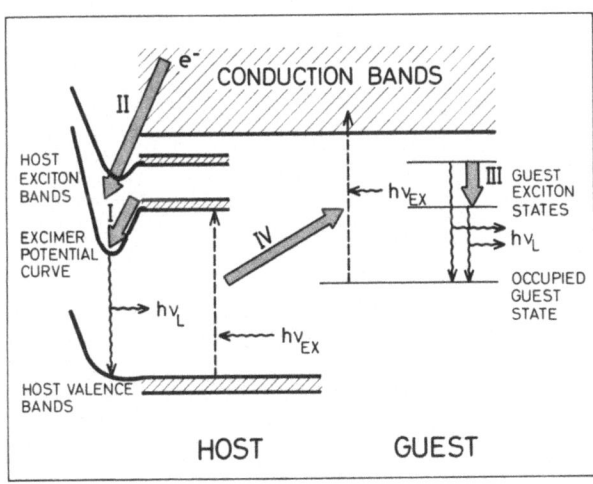

Fig. 1.4. Schematic scheme for various relaxation and energy transfer processes discussed in the text. (*Case I*) selftrapping of excitons and nonradiative relaxation; (*Case II*) recapture of free electrons and formation of luminescence centres; (*Case III*) competition of nonradiative relaxation and radiative decay in exciton states of guest atoms; (*Case IV*) energy transfer of free and selftrapped host excitons to guest atoms and boundaries

that molecular centres of the R_2^* type are strongly favoured in the solid. In this case, the nearest neighbour distance is reduced by about 30% from 4 Å to about 3 Å. These large structural changes upon excitation are favoured by the weak bonding forces in the ground state resulting in only weak forces opposing the rearrangement, and by the strong bonding in the excited state. The electronic excitation energy will be converted by a number of competing processes, such as non-radiative electronic relaxation, auto-ionization, recapture of free electrons, formation of luminescence centres, energy transfer processes to guest atoms and boundaries and radiative decay. Some of these decay processes are schematically portrayed in Fig. 1.4 and will be discussed in detail in Chap. 6 of this book.

From the examination of Fig. 1.1 and Table 1.1, it is immediately apparent that the utilization of VUV and soft X-ray radiation is essential for a detailed study of the electronic structure and dynamics in condensed rare gases. Aside from the broad band excitation by electron or particle excitation, the use of synchrotron radiation for selective excitation brought about a major experimental breakthrough in these studies. Multiphoton excitation by lasers has also recently been invoked. The combination of basic spectroscopic techniques, such as energy- and time-resolved luminescence and photoelectron spectroscopy, has been of prime importance for the investigation of decay processes.

2. Experimental Aspects

The electronic structure of pure RGS has been studied mainly by absorption-, re-
flection-, electron energy loss- and photoelectron-spectroscopy. In the visible and
long wavelength region measurements of the index of refraction by the minimum de-
viation method, of the Faraday effect and of the capacity have also been carried
out. These techniques and experimental results are well documented and discussed
by Sonntag /1977/. Here, only a brief summary of the earlier experiments is given
with emphasis on the progress obtained in the intervening time. Especially the extension of these techniques to study rare-gas alloys and liquids will be described.

Major advances have been achieved in luminescence experiments. Furthermore, magnetic circular dichroism experiments, photoconductivity measurements, transient
absorption spectroscopy and two photon ionization spectroscopy have successfully
been applied. From extended X-ray absorption fine structure (EXAFS) measurements
structural information is obtained using the optical spectra.

A stage has now been reached when careful attention has to be devoted to the
characterization of the samples. Improvements concerning these problems are just
emerging. For instance, we compare data obtained for thin polycrystalline films
to those obtained for crystals; surface states and surface quenching processes are
investigated, the influence of phase transitions is studied, and more experience
has been gained concerning annealing effects and sample temperature. In what follows
we illustrate a few typical examples for sample preparation techniques (for more
references, see, for example /Meyer, 1971/ and the many spectroscopic techniques.
For additional references see /Samson, 1967; Cardona and Ley, 1978 , 1979; Lumb,
1978/).

2.1 Sample Preparation and Structure

A general review of crystal growth and crystal defects has been given by Venables
and Smith /1977/. The wealth of details and references is considered as a background material, which will not be cited here again. Also, we will not repeat the

discussion of problems of crystal structure relating to optical investigations, as discussed by Sonntag /1977/.

Liquid rare gases are prepared in cells. An example for a high pressure absorption cell is shown in Fig. 2.1. On a larger scale, liquid rare gases are used in ionization chambers for high energy physics. The proportional increase of the amount of charge with particle energy yields a high energy resolution which has been applied in liquid rare-gas particle detectors. The high electron mobility assures a fast response. As an example of such a device, a shower counter filled with argon for high energy particle detection at the e^+e^- storage ring PETRA in Hamburg is shown in Fig. 2.2. For these applications the purity of the gases is crucial, since

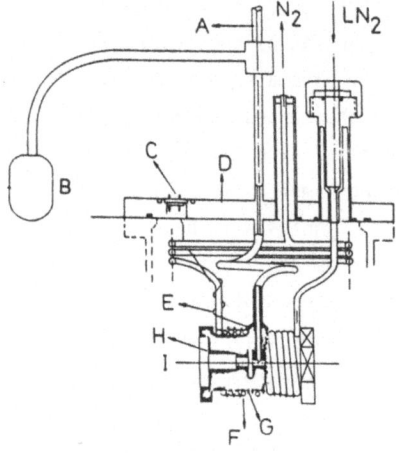

Fig. 2.1   Fig. 2.2

Fig. 2.1. Absorption cell for high pressure vacuum ultraviolet optical experiments (from Messing et al. /1977e/). (A) pipe leading to the gas handling system; (B) sample preparation chamber; (C) electrical feed-through; (D) vacuum chamber cover flange; (E) thermocouple; (F) cooling coil; (G) heating element; (H) closing nut; (I) optical axis

Fig. 2.2. Liquid Ar barrel shower counter for the CELLO high energy particle detector used at the Deutsche Elektronen-Synchrotron (DESY)

the smallest admixture of O_2, even below the ppm level, captures electrons quite effectively.

Commonly, for spectroscopic investigations, RGS's are condensed as thin films onto a cooled substrate. Rare-gas alloys are usually prepared by mixing the constituents in the gas phase with appropriate partial pressures and depositing them in situ on a cryostat in the form of a thin film. In this kind of preparation, care has to be taken to work at extreme low base pressures in the gas handling system

and the ultra-high-vacuum experimental vessel (in the range of 10^{-10} Torr or better).
Only UHV bakeable stainless steel systems satisfy these conditions (see for example,
/Harmsen et al., 1974; Saile, 1978/). The concentration in the sample may differ
from the gas phase composition due to demixing during the transfer from the storage
vessel to the substrate (see, for example, /Mann and Behrens, 1978/) and due to different sticking coefficients for both components on the cold substrate.

The thickness of the film, typically some tenths of a nm up to some hundred μm,
can be determined by monitoring the sequence of interference fringes observed for
the light reflected from the sample during evaporation /Baldini, 1965/. Either laser
light or, in order to get higher contrast and higher accuracy for film thicknesses
in the Å region, VUV light in the transparent region of the sample is used /Harmsen,
1975/. We mention that rare-gas metal mixtures have to be prepared in a different
way. Rare gas and metal vapour are deposited simultaneously from a gas nozzle and
from a metal furnace on a substrate, which has to be cold enough to avoid clustering
of the metal atoms. The concentration is derived by measuring both deposition rates
independently. Clustering is a main difficulty for the preparation of these mixtures
/Kolb, 1981/. Coufal et al. /1978/ demonstrated for Ar^{40} : K^{41} mixtures a new possibility by converting in an Ar^{40} sample some atoms to K^{41} by γ activation.

Vapour deposition of thin films is used because of its simplicity and because
some spectroscopic techniques require thin films as samples. In absorption measurements a sufficiently high transmission of light is needed; in photoelectron emission
experiments charging has to be avoided and in the study of surface states the bulk
background has to be reduced. Further, thin films are required for the investigation
of those transport properties where small penetration depths are involved, such as
the electron mean free path and the range of energy transfer of excitons to substrate
and surface layers. In these experiments the thickness dependence of the signal
gives additional information. Finally, for experiments which are sensitive to trapping sites, thin films are in some cases favourable. The structure of films and its
dependence on the preparation conditions has been investigated and the epitaxial
growth of single-crystalline films has been demonstrated (see, for example, /Venables
and Smith, 1977/). Usually thin films have a polycrystalline structure. For pure
polycrystalline rare-gas films an admixture of hcp structure has been observed in
addition to the expected fcc structure /Sonnenblick et al., 1977/.

We note that in spectroscopic investigations on thin films a definite correlation
with the structure of the samples is generally missing. In a few cases the influence
of preparation conditions and of annealing has been discussed. Schulze and Kolb
/1974a/ and Schulze et al. /1974b/ have studied changes of the density ρ and refrac-
tive index n of solid films of rare gases on a metal mirror substrate as a function
of the condensation conditions such as temperature, growth rate and layer thickness.
For condensation temperatures above a value characteristic for each RGS these authors
obtained results for ρ and n in agreement with those given in the literature. Below

this temperature an approximately linear decrease of ρ and n with temperature was found.

For doped samples, structural investigations are available showing, for example, that Ar/Kr and Kr/Xe can be prepared as homogeneous polycrystalline films with fcc structure up to large concentrations /Curzon and Mascal, 1969; Kovalenko et al., 1972; Venables and Smith, 1977/. The concentration dependence of the lattice parameter in substitutional solid solutions of RGS is in good agreement with the mean potential model /Prigogine, 1957/. In most of the optical investigations the actual structure of the samples is not reported. This lack of information is even more severe for doped than for pure samples because here local distortions due to the guest atoms are a further point of concern. The size of the dopand determines if one or more matrix atoms in the lattice are replaced. Further, it determines also the degree of rearrangement in the surrounding matrix atoms. In addition, clustering and accumulation of crystal defects at the site of a dopand may be important for the explanation of spectroscopic results. Some aspects like changes in selection rules and splittings are dominant in spectra of metal atoms and can help to classify at least the local symmetry. The rate constants and pathways of relaxation processes will be influenced by local phonons as well /Luchner and Micklitz, 1978/. EXAFS /Stern and Heald, 1983/ provides an excellent tool for the investigation of local structure. First results for RGS concerning the internal structure of metal clusters in matrices /Purdum et al., 1982/ and the structure around a rare-gas guest atom /Malzfeldt et al., 1983/ are very promising.

The growth of large RGS single crystals by modified Bridgeman methods is well known /Hingsammer and Lüscher, 1968/. The dominating stable structure of RGS crystals is the fcc lattice. For different growth conditions the coexistence of the fcc and hcp phases and the stabilization of the hcp phase by stress and impurities (O_2, N_2, CO) has been analysed /Meyer, 1969; Kovalenko et al., 1975/. In Fig. 2.3 a device to grow free standing RGS polycrystals for luminescence experiments /Schuberth and Creuzburg, 1975/ is shown.

A clear relation between crystal structure and spectroscopic data is evident in X-ray induced VUV luminescence /Schuberth et al., 1976/. A strong increase in luminescence efficiency of Ne crystals at 10.5 K is caused by a partial phase transition from fcc to hcp in polycrystalline samples. This phase transition has been identified by Raman scattering experiments. The optical phonons in the hcp phase will enhance the formation of the relaxed emission centres after X-ray excitation by dissipating more energy per phonon for small wave vectors than the acoustical phonons in the fcc phase /Schuberth and Creuzburg, 1975/. A similar explanation for intensity changes in Ar, Kr and Xe polycrystals has been given by Heumüller /1978/ and has been supported by the observation of correlated changes in birefringence.

Large polycrystals ($2 \times 1 \times 1$ cm^3) of excellent optical quality have been grown in an attempt to exploit the intrinsic VUV luminescence of rare-gas crystals for

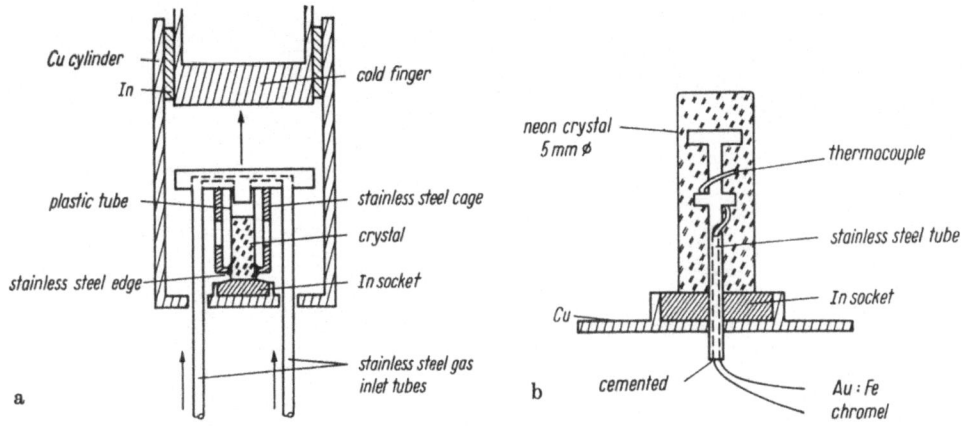

Fig. 2.3a,b. Crystal growing device for the preparation of neon crystals. (a) Crucible removable for observation in the vacuum ultraviolet; (b) Temperature measurement inside the crystal by a thermocouple led through a stainless steel tube which also supports the crystal (from Schuberth and Creuzburg /1975/)

a VUV solid state laser /Schwentner et al., 1982/. The light output depends on the preparation conditions which influence the size of the microcrystallites. An increase of the crystallite size from 0.05 mm to 0.7 mm doubles the luminescence intesnity /Rudnick, 1983/.

2.2 Spectroscopic Techniques

In this section we first deal with the various sources for excitation. Here we can distinguish between broad band or energy selective excitation and pulsed or continuous excitation. High energy particles, which are frequently used as excitation sources, deposit their energy via a broad spectrum of energy loss processes. The following sources have been used to study luminescence emission: (i) α particles /Jortner et al., 1965; Brodmann et al., 1976/, (ii) electrons /Basov et al., 1970; Coletti and Bonnot, 1978; Huber et al., 1974; Packard et al., 1970; Fugol et al., 1974; Stockton et al., 1970; Surko et al., 1970; Keto et al., 1974; Dössel et al., 1983/ and (iii) X rays /Schuberth et al., 1975; Nanba et al., 1974/. An example of a set-up for excitation of luminescence in liquid He by fast electrons is shown in Fig. 2.4 /Fitzsimmons, 1973/. For time-resolved luminescence spectroscopy electron beams can be pulsed in a convenient way /Hahn et al., 1977; Kink et al., 1977; Coletti and Hanus, 1977; Dössel et al., 1983/. Pulsed electric field discharge in RGS crystals has also been used /Schörner, 1977/. The statistical time structure of electrons and α particles of ^{207}Bi, ^{210}Po, ^{90}Sr, ^{90}Y sources have been exploited

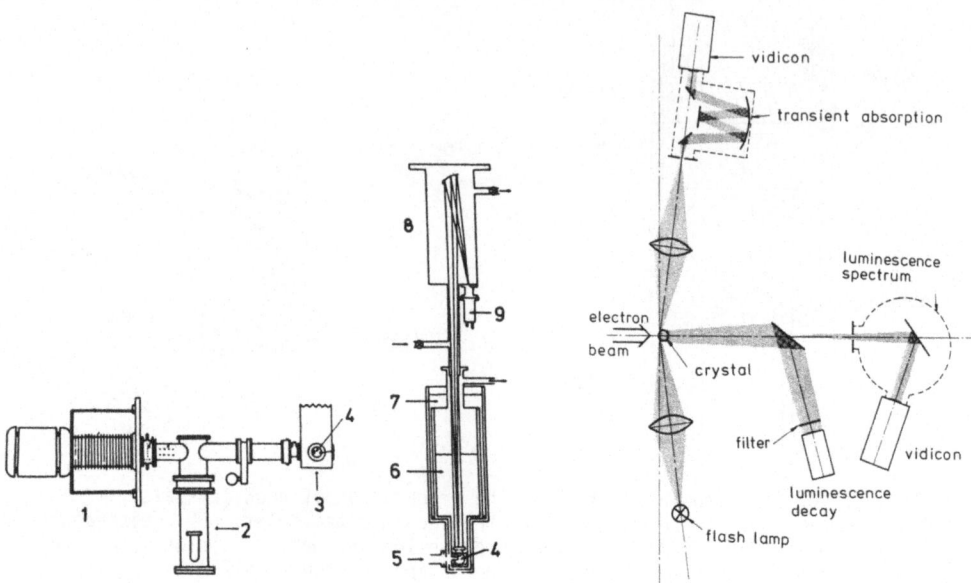

Fig. 2.4   Fig. 2.5

Fig. 2.4. Device for luminescence experiments in liquid He by excitation with high energetic electrons (after Fitzsimmons /1973/). (*1*) electron gun; (*2*) vacuum pumps; (*3*) sample chamber; (*4*) main cryostat; (*5*) electron beam; (*6*) cryogenic helium; (*7*) liquid nitrogen; (*8*) monochromator; (*9*) sodium salicylate and photo-multiplier

Fig. 2.5. Experimental set up for simultaneous measurement of transient absorption spectra after excitation with a $5 \cdot 10^{-9}$ s pulse of 500 keV electrons and registration of luminescence emission spectra and decay curves (after Wilcke /1979/)

for static and time dependent luminescence emission spectroscopy /Kubota et al., 1978; Carvalho and Klein, 1978/. In the same way electron-hole pair production rates have been analysed in liquid rare gases /Takahashi et al., 1975/.

The high intensities, which are rather easily obtained, are the major advantage of broad band excitation with high energy particles or X rays compared to selective excitation with light. Therefore, high resolution luminescence studies and the search for weak emission lines have been the domain of broad band excitation.

Also transient absorption spectroscopy /Suemoto and Kanzaki, 1979; Keto et al., 1974; Dössel et al., 1983/ requires the high currents which can be delivered from pulsed electron beams. In these experiments a sufficient high density of excited states is prepared by a high energy ($\approx$ 500 keV) and short duration ($\approx 5 \cdot 10^{-9}$ s) electron pulse. The absorption spectrum for transitions from e-beam excited states to higher-lying excited states is measured using a light flash which is correlated in time to the excitation pulse (Fig. 2.5).

In general, energy selective excitation is more desirable than broad band excitation, because the initially excited state can be specified. Energy selective excitation needs light sources in the VUV, The source should be strong and also tuneable to allow a free choice of the primary excited state.

Absorption, photoelectron yield /Sonntag; 1977/ and luminescence studies have used monochromatic light from conventional discharge and flash lamps /Danilychev et al., 1970; Nagasawa and Nanba, 1974/. The experiments are hampered by the weak intensity in the continuum, the superimposed lines and by problems with suitable window materials. Synchrotron radiation (see, for example, /Kunz, 1979; Winick and Doniach, 1980; Koch, 1983/) brought a major breakthrough with tremendous improvements in absorption, reflection and photoelectron yield spectroscopy. It made possible a new generation of photoelectron energy distribution experiments and luminescence emission experiments from selected primary states. The following advantages of synchrotron radiation have been exploited for the investigation of RGS, (i) the intense continuum from the visible to the X-ray region, (ii) the time structure providing short light pulses of the order of 100 ps with high repetition rates of 1 MHz up to 500 MHz, and (iii) the low pressure in the source (10^{-9} Torr) which facilitates the operation of the experiment under UHV conditions.

The intensity of the monochromatic ($\Delta\lambda = 0.007$ nm) VUV light impinging onto the sample is shown in Fig. 2.6 /Wilcke et al., 1983/. The time distribution of the light at the storage ring DORIS is given in the lower panel of Fig. 2.6. The repeti-

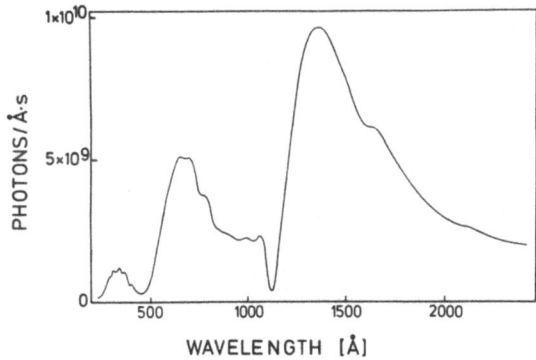

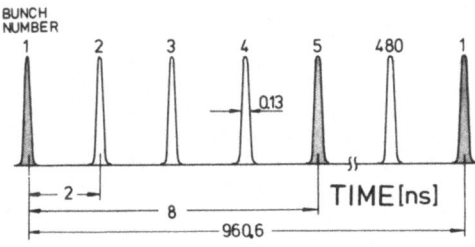

Fig. 2.6. Typical photon flux of monochromatized synchrotron radiation at the sample (*upper panel*). Time structure of the synchrotron light pulses provided by the storage ring DORIS in Hamburg (after Hahn et al. /1978/) (*lower panel*)

tion rate of the pulses depends on the mode of operation of the storage ring. Due to the detectors and electronics, the measured pulse width in actual set ups is broadened to ≈ 0.4 ns. With dedicated storage rings and more sophisticated timing techniques a time resolution of some ps has been obtained and time resolutions in the subpicosecond regime are possible /Rhen, 1980; Munro and Schwentner, 1983/.

Several experimental set-ups at synchrotron radiation sources have been realized to exploit these properties and to provide as detailed information as possible by specifying the primary excited states and by analysing secondary processes. For this purpose optical spectroscopy has been combined with photoelectron and/or luminescence spectroscopy.

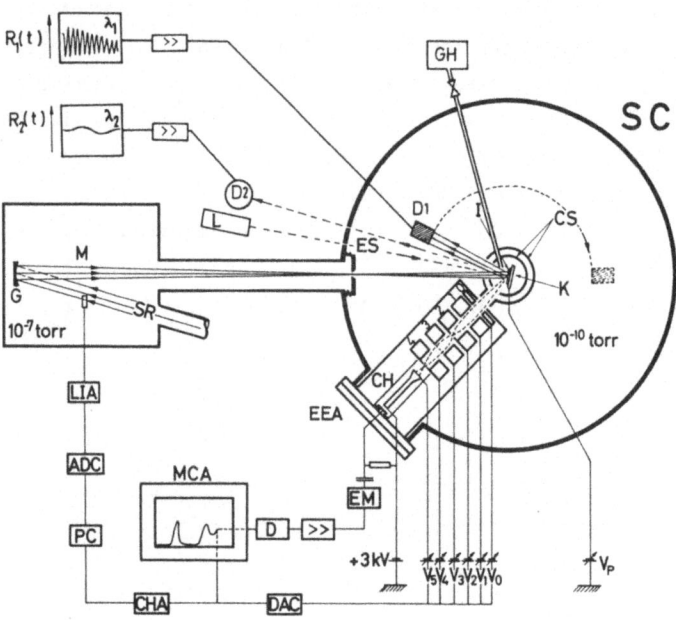

Fig. 2.7. Set-up for simultaneous VUV-reflection and photoemission experiments on condensed gases. Synchrotron light (SR) enters the sample chamber (SC) from the monochromator (M) with concave grating (G) via the excit slit (ES). A cryostat (K) with two cryoshields (CS) and an insulated sample substrate (I), an open electrostatic photomultiplier (D1), a gashandling system (GH), and a photoelectron energy analyser (EEA) with a channeltron as detector (CH) are incorporated into the sample chamber. Photoelectron analysis: V_0 to V_5, lens voltages; V_p: sample voltage; EM: emitter follower; D: discriminator; MCA: multichannel analyser; DAC: digital-analog converter. Channel advance (CHA) is triggered by a reference signal via lock-in amplifier (LIA), analog-digital converter (ADC) and a preset counter (PC). The reflectance as a function of wavelength is measured by D1. Film thickness is determined by comparing the reflectance $R_1(t)$ at λ_1(VUV) and $R_2(t)$ at λ_2 (laser wavelength, laser (L) via detector D2) simultaneously during evaporation time t (from Schwentner et al. /1974/)

Absorption and reflection experiments under extreme UHV condition have reached a level of sophistication which allows lineshape analysis of exciton bands with high resolution /Saile, 1976; Saile et al., 1976/. In addition, surface and bulk exciton states in thin films of variable thicknesses have been investigated with a set up for simultaneous reflection and transmission measurements which is attached to a high resolution ($\Delta\lambda$ = 0.003 nm) 3 m normal incidence monochromator at the storage ring DORIS /Saile et al., 1976/. Because of the high and stable light flux of this instrument it is possible to separate the weak absorption of surface states from the bulk absorption background.

Luminescence spectroscopy in RGS, exploiting synchrotron radiation, has been performed at the synchrotron DESY /Brodmann et al., 1974/, at the storage ring DORIS /Brodmann et al., 1976; Hahn et al., 1978; Wilcke et al., 1983; Gürtler et al., 1983/, at the storage ring SPEAR /Monahan et al., 1976 , 1978/ and at the synchrotron NINA /Hasnain et al., 1977/. Photoelectron yields from RGS have been studied at the synchrotron DESY /Schwentner et al., 1973/ and at the synchrotron NINA /Hasnain et al., 1977/. Photoelectron energy distribution measurements from RGS and matrices have been carried out at DESY by Schwentner /1974/ and Schwentner et al. /1974/. In Fig. 2.7 the set-up using a combination of retarding field and electrostatic lenses as an electron energy analyser in connection with a 1.0 m

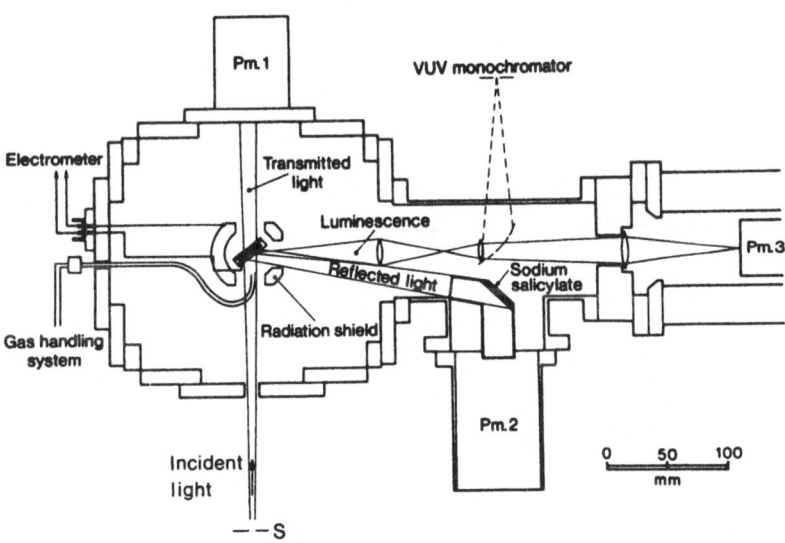

Fig. 2.8. Experimental arrangement for simultaneous measurement of the VUV reflection, luminescence excitation and transmission spectra. The radiation shield is used as a collecting electrode for photoemission yield measurements. Synchrotron radiation is dispersed with a 1.5 m horizontal Wadsworth monochromator. S is the exit slit of the monochromator, Pm.1,2,3 are photomultipliers (from Hasnain et al. /1977/)

vertical Wadsworth monochromator is shown. Simultaneously, the absolute sample reflectivity and the absolute photoelectric yield can be recorded with this instrument.

Photoelectron EDC's from physisorbed monolayers and multilayers on various metal substrates have been carried out by several groups with HeI sources (see, for example, /Unwin et al., 1980; Hulse et al., 1980; Mandel et al., 1982/) and with synchrotron radiation from the TANTALUS storage ring (e.g., /Kaindl et al., 1980; Chiang et al., 1980/) and from the storage ring DORIS /Mandel et al., 1983/.

The experimental arrangement for simultaneous measurements of the reflection, transmission, luminescence excitation and photoemission yield spectra at the Synchrotron Radiation Facility at Daresbury Laboratory is shown in Fig. 2.8 /Hasnain et al., 1977/.

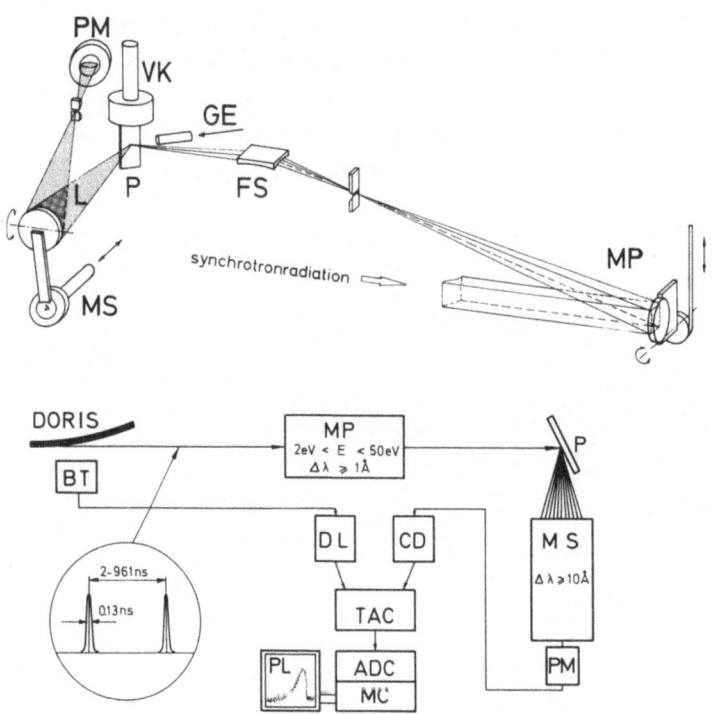

Fig. 2.9. Experimental arrangement for time-resolved luminescence experiments with selective excitation. *Upper panel*: experimental arrangement consisting of: (MP) primary monochromator; (FS) focusing mirror; (P) sample holder; (L) luminescence light; (MS) secondary monochromator; (PM) photomultiplier; (VK) cryostat; (GE) gas inlet tube. *Lower panel*: electronics for time-resolved luminescence spectroscopy consisting of: (DORIS) storage ring; (BT) bunch trigger; (DL) delay line; (CD) constant fraction discriminator; (TAC) time to amplitude converter; (ADC) analog to digital converter; (MC) multichannel analyser; (PL) plotter (from Hahn and Schwentner /1980/)

Photoluminescence experiments, in which the primary excitation energy can be
selected and where the emission spectra are dispersed by a second monochromator,
have been carried out at the storage ring DORIS (Fig. 1.3) /Hahn et al., 1978/.
Further, this set-up allows the analysis of decay curves of the emission bands in
a single photon counting mode (Fig. 2.9). For long lifetimes and for alignment purposes a pulsed electron gun (not shown in Fig. 2.9) has been integrated.

A promising two-photon ionization experiment combining synchrotron radiation for
exciton excitation and a laser pulse to ionize these excitons has been developed
by Saile and collaborators /Saile, 1980; Saile et al., 1980/. In another experimental approach multiphoton absorption in a focused high-power laser beam is used.
This technique was first applied to pure and doped liquid rare gases /Müller et al.,
1982/ and recently to pure and doped RGS /Kessler et al., 1985/.

3. Electronic Structure of Valence and Conduction Bands and Excitonic States

A description of the physical properties of a solid starts with the total Hamiltonian for all nuclei electrons and their mutual interactions. The electronic states follow from one-electron band structure calculations, which yield the electron energy E(k̲) versus the wave vector k̲ of the electron. An example for Xe (Fig. 3.1) shows the valence bands around -10 eV, originating from the outer atomic p levels. The upper two levels with a total angular momentum j = 3/2 of the hole are degenerated at the centre of the Brillouin zone Γ. At Γ the j = 3/2 and the lower-lying

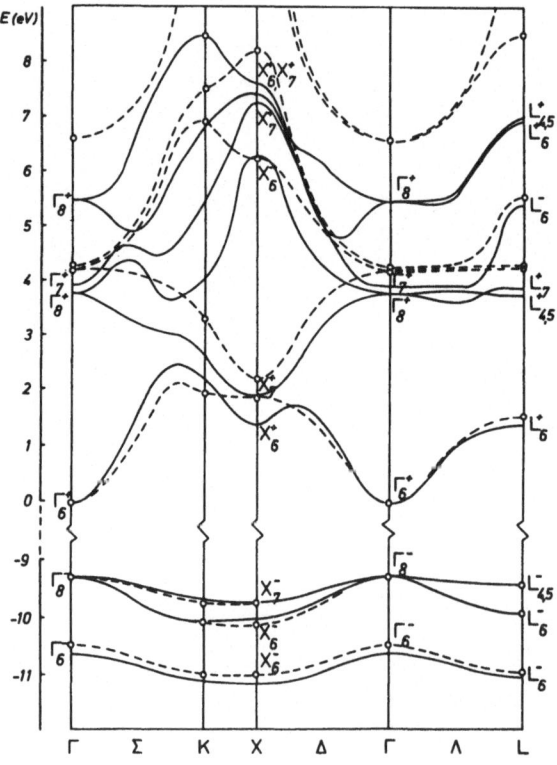

Fig. 3.1. Electronic energy bands of xenon. (————): KKR calculations /Rössler, 1970/; (– – –): OPW (tight-binding) calculation /Reilly, 1967/ (from Rössler /1976/)

j = 1/2 bands are separated by a spin-orbit splitting of ≈ 1.3 eV. The considerable dispersion, i.e. width, of the individual valence bands of the order of 1 eV indicates a significant overlap of the p-wave functions of neighbouring atoms in the crystal and establishes a rather high group velocity of these resonant hole states. The lowest empty conduction-band states are separated from the occupied valence bands by a large band gap E_G of the order of 10 eV. Near Γ the lowest conduction band is free electron-like (Fig. 3.1) and has s symmetry. The contributions of additional bands complicate the picture at higher energies.

The partly screened Coulomb interaction between an electron in the conduction band and a hole in the valence band leads to a hydrogen-like series of bound states within the band gap. Two such exciton series converge to the bottom of the conduction band.

Extensive information on the band structure of RGS has been inferred from optical data and photoemission experiments. The interest in the band structure is largely motivated by two aspects. First, RGS are prototype materials for the large class of van der Waals crystals. In this respect comparison of experimental data with state-of-the-art ab initio band-structure calculations plays a major role. Second, an accurate description of the band structure and of the excitonic states of pure RGS is basic for the discussion of impurity states, the more complex electronic structure of metal rare-gas mixtures and the dynamical processes. Pertinent information about the dielectric and optical properties of pure RGS, as obtained by optical absorption or reflection and energy loss spectroscopy as well as by photoemission, has been collected and discussed by Sonntag /1977/. The theoretical concepts including band structure calculations and the theory of excitons in RGS have been summarized by Rössler /1976/. The main emphasis in our discussion will be on some new experimental and theoretical developments concerning the band structure (Sect. 3.1) and the excitonic states (Sect. 3.2). Within Sect. 3.2 we shall also discuss the recent observations of surface excitons and our current understanding of this phenomenon. Finally, we shall focus our discussion on results for dilute and concentrated rare-gas alloys (Sect. 3.3).

3.1 Band Structure of Rare-Gas Solids, Ordered Layers, and Clusters

Over the past few years there has been a steady increase in our knowledge of the band structure (e.g., determination of band gaps and spin-orbit splittings) mainly based on the analysis of optical spectra. Since the optical spectra are dominated by excitonic excitations little information was available about the dispersion of the bands, their widths and the absolute position of the energy levels. Thus, a detailed comparison with the many, partly conflicting, band structure calculations was largely an indirect procedure.

Photoelectron spectroscopy (see, for example, /Cardona and Ley, 1978,1979/) has become the most powerful technique to probe for the density and dispersion of the valence bands and core levels. In measurements of the photoelectron energy distribution curves (EDC's) the sample is illuminated with monochromatic light and those excited electrons which are emitted into vacuum are analysed according to their energy and/or angle. In photoemission yield experiments all electrons are collected by an appropriate grid and the photocurrent is measured as a function of the energy of the exciting light.

For the following discussion, in particular the comparison of band structure calculations with experimentally determined EDC's, the widely used phenomenological three-step model /Berglund and Spicer, 1964/ is used. Within this model the three main steps are (i) absorption of photons and excitation above the band gap, (ii) transport of electrons from the point of excitation including possible scattering events within the sample, and (iii) escape of electrons into vacuum and detection.

Rare-gas solids have the advantage that the escape depth of photoelectrons is rather large for a wide range of energies below the electron exciton scattering onset (see Chap. 7). Consequently, the EDC's are governed largely by the energy distribution at the site of excitation. The absorbed photons excite electrons from occupied states of the valence bands into empty conduction band states. Assuming vertical transitions with $\Delta k = 0$ the energy distribution $N(E,h\nu)$ is given by

$$N(E,h\nu) = \text{const} \sum_{V,C} \int_{BZ} d^3\underline{k} \ |M_{VC}|^2 \ \delta[E_C(\underline{k}) - E_V(\underline{k}) - h\nu] \ \delta[E_C(\underline{k}) - E] \ . \tag{3.1}$$

Here M_{VC} is the matrix element of the transition probability, E_V is the energy of the valence band states, E_C is the energy of the final states and E is the energy of the excited electrons. The first δ function guarantees energy conservation taking into account the momentum conservation. The second δ function reflects the choice of the energy of the electrons considered. The summation is over all valence and conduction band states and the volume of integration is the Brillouin zone (BZ).

For a comparison with experiment the matrix elements are assumed to be constant leading to the simplified expression

$$N(E,h\nu) = \text{const} \sum_{V,C} \int_{BZ} d^3\underline{k} \ \delta[E_C(\underline{k}) - E_V(\underline{k}) - h\nu] \ \delta[E_C(\underline{k}) - E] \ . \tag{3.2}$$

This equation neglecting the influence of electron scattering is frequently used for a discussion of the EDC's.

Due to the experimental difficulties encountered in photoemission experiments, when applied to insulators, and due to the general difficulty of working at very low temperatures, only a few photoemission experiments have been performed on RGS. In most of these experiments synchrotron radiation has been used as a source of excitation. There have been a couple of photoemission yield experiments reported for pure RGS /O'Brion and Teegarden, 1966; Schwentner et al., 1973; Ophir et al., 1974; Koch et al., 1974a; Koch et al., 1974b; Ophir et al., 1975; Steinberger et al., 1974; Pudewill et al., 1976/. Photoelectron energy distribution studies have been carried out at Deutsches Elektronen-Synchrotron (DESY) since 1974. In a series of papers Schwentner and his colleagues /Schwentner, 1974; Schwentner et al., 1974; Schwentner et al., 1975; Schwentner and Koch, 1976; Schwentner, 1976; Nürnberger et al., 1977; Koch et al., 1978/ have measured EDC's for Ne, Ar, Kr and Xe for a number of photon energies. The experimental set-up used in these experiments was discussed in Chap. 2 (see Fig. 2.7). On the basis of these experiments detailed comparisons with predictions of band structure calculations for the various parameters have become possible.

It turned out that although the general features of the valence bands are predicted correctly by the calculations, almost all available band-structure calculations failed to predict quantitative features other than the spin-orbit splitting (ΔE_{SO}). In order to illustrate this situation a comparison of some representative EDC's taken from Schwentner's work /1974/ with calculated valence bands are shown in Fig. 3.2. Theoretical results for the Ne 2p-derived valence bands are taken from the work of Rössler /1970/ using the KKR method of Kunz and Mickish /1973/ and Dagens and Perrot /1972/. For the 3p-derived valence bands of solid Ar results from Lipari and Fowler /1970/, Lipari /1972/, Knox and Bassani /1961/ and Mattheis /1964/ are displayed. In the heavier RGS the width of the valence bands (W_{VB}) increases considerably. For Kr calculated valence bands, according to Rössler /1970/, Kunz and Mickish /1973/, Fowler /1963/ and Lipari /1970/, are shown together with the experimentally determined EDC. Finally, for the Xe 5p-derived bands comparison is made with the calculated valence bands according to Rössler /1970/ and Reilly /1967/. In the uppermost panel for Xe in Fig. 3.2 a quantitative estimate based on the experimentally observed valence features /Schwentner, 1974/ is included. Some of the relevant parameters characterizing the band structure have been compiled in Table 3.1. Results for slightly doped RGS, which we shall discuss below, are also given. Inspection of Fig. 3.2 shows that the general features of the valence bands common for all rare-gas solids are (i) a spin-orbit splitting of the np 3/2 and np 1/2 states in the centre of the Brillouin zone, (ii) a k dependent dispersion to lower energies going from Γ to the border of the Brillouin zone, and (iii) a splitting of the np 3/2 bands, which are degenerate at Γ, into two subbands. A graphical summary of the parameters deduced from experimental data, as displayed in Fig. 3.2, appears in Fig. 3.3. They are plotted versus the energy gaps E_G. It is interesting to note that there is a nice correlation and linear dependence of all parameters on this scale. A more extensive and quantitative comparison with the predictions for the

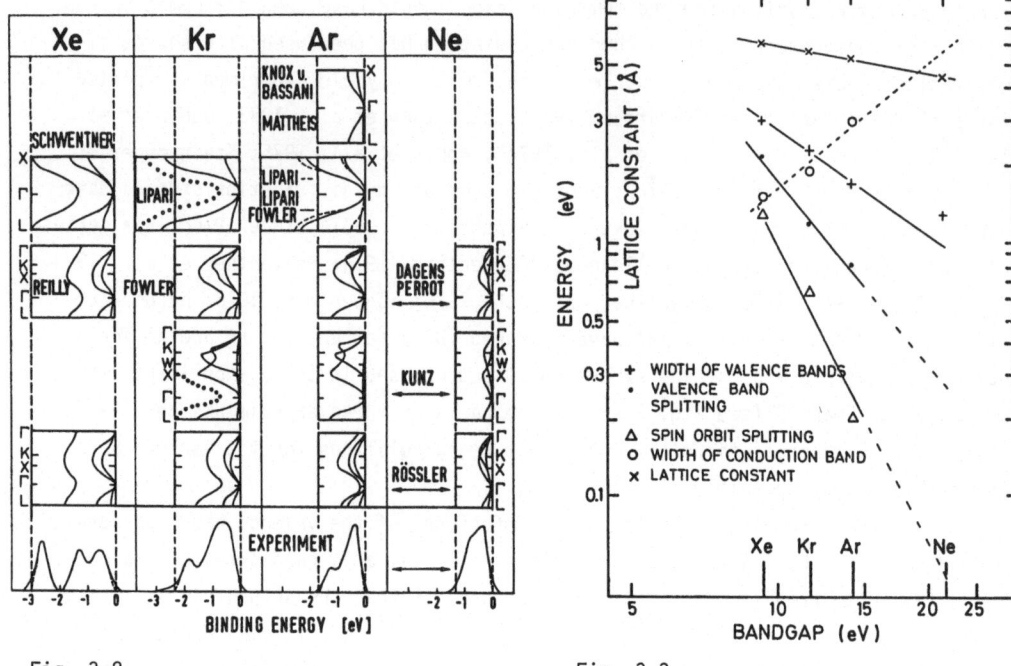

Fig. 3.2

Fig. 3.3

Fig. 3.2. Comparison of some representative photoelectron energy distribution curves for solid Xe (at hν = 13.78 eV), Kr (at hν = 19.84 eV), Ar (at hν = 19.07 eV) and Ne (at hν = 26.72 eV) with calculated valence bands (from Schwentner /1974/)

Fig. 3.3. The plot of characteristic parameters of the valence and conduction bands (+ width of valence bands; • valence band splitting; Δ spin-orbit splitting; ○ width of conduction band; × lattice constant) on a linear band-gap scale shows a continuous change of these parameters from Xe to Ne (from Schwentner /1974/)

total width and the structure of the valence bands has been given by Schwentner /1974/, Schwentner et al. /1975/ and Kunz et al. /1975/. These comparisons have been significant in that they support the idea that the valence states of the RGS are band-like with a k dependence of the bands stronger than predicted by most calculations (see also the discussion in Sect. 3.3).

At this stage it is appropriate to discuss some limits of the band model. At first sight the fact that most of the band-structure calculations for RGS are only moderately successful in yielding a quantitative description is surprising, particularly in view of the success of band-structure calculations in interpreting EDC's from metals and semiconductors (see, for example, /Grobman, 1975; Cardona and Ley, 1978,1979/). We note that the above comparison involves experimental excited-state energies versus calculated ground state single-particle energies; such comparisons involve as yet unresolved issues concerning electron-phonon coupling and relaxation effects which accompany electronic excitations. Thus, the local symmetry

Table 3.1. Parameters for the band structure and for excitons of pure RGS and for impurity states in rare-gas matrices. All energies in eV. A superscript "i" denotes values for the impurity states. E_G, E_G^i: band gap energy; E_{Th}, E_{Th}^i: threshold energy for photoelectron emission = vacuum level; $V_0 \equiv E_G - E_{Th}$: electron affinity; ΔE_{SO}, ΔE_{SO}^i: spin-orbit splitting of the valence bands and of impurity states with j = 3/2 and j = 1/2; W_{VB}: total width of the valence bands; B, B^i: binding energy of excitons; Δ, Δ^i: central cell correction of n = 1 excitons and δ, δ^i: quantum defect. We have attempted to give a consistent set of values from the presumably most reliable data. The binding energies B, B^i depend on the applied model (see, for example: B in Ne corresponds to 5.00 eV in the Wannier model and 6.93 eV in the quantum defect model[a])

	E_G, E_G^i	E_{Th}, E_{Th}^i	V_0	ΔE_{SO}, ΔE_{SO}^i	W_{VB}	B, B^i	Δ, Δ^i	δ, δ^i
Ne	21.58[a]	20.3 [b]	+1.3	0.09[a]	1.3[b]	5.00 - 6.93[a]	0.86[a]	0.28[a]
Ar in Ne	16.1 [d][e]	15.05[f]	+1.1	0.34[g]		"	1.39[e]	0.40[e][l]
Kr in Ne	14.6 [d][e]	13.48[f]	+1.1	0.61[g]		"	1.03[e]	0.31[e][l]
Xe in Ne	12.6 [d][e]	11.60[f]	+1.0	1.27[g]		"	1.53[e]	0.43[e][l]
Ar	14.16[h]	13.9 [b][c]	+0.3	0.18[h]	1.7[b]	2.36[h]	0.26[h]	0.21-0.28[k]
Kr in Ar	12.5 [i]	12.2 [i]	+0.3				0.5	
Xe in Ar	10.97[o]	10.7 [o]	+0.3	0.75[o]		1.76[o]	0.0 [o]	
Kr	11.61[h]	11.9 [b][c]	-0.3	0.69[h]	2.3[b]	1.53[h]	0.09[h]	0.08-0.17[k]
Xe in Kr	10.1 [i]	10.30[i]	-0.2				0.4	
Xe	9.33 [h]	9.7 [b][c]	-0.4	1.3	3.0[b]	1.02[h]	0.06[h][m]	-0.03-0.06[k]

a) Saile and Koch /1979/
b) Schwentner, Himpsel, Saile, Skibowski, Steinmann and Koch /1975/
c) Schwentner, Skibowski and Steinmann /1973/
d) Hahn and Schwentner /1980/
e) Böhmer, Haensel and Schwentner /1980/
f) Himpsel, Saile, Schwentner, Skibowski, Koch and Jortner /1976/
g) Pudewill, Himpsel, Saile, Schwentner, Skibowski and Koch /1976/
h) Saile, Steinmann and Koch /1977/
i) Ophir, Raz, Jortner, Saile, Schwentner, Koch, Skibowski and Steinmann /1975/
j) Schwentner and Koch /1976/
k) Resca, Resta and Rodriguez /1978/
l) Resca, Resta /1979/
m) Gedanken, Raz and Jortner /1973/
n) Baldini /1965/
o) Reininger, Bernstorff, Laporte, Saile and Steinberger /1984/

of the hole in the valence band in RGS might be important. One may also raise the iconoclastic question of whether, in view of strong electron-phonon coupling which results in the narrowing of the valence band /Holstein, 1959/, an atomic-like approach is not more appropriate for the description of these narrow np bands in RGS. In this context, support for a band-structure model was obtained from photoemission spectroscopy of rare-gas alloys. Nürnberger et al. /1977/ have studied the gradual band formation for Xe-Ar alloys for concentrations ranging from 0 to 100 per cent by photoemission spectroscopy. These authors have been able to describe their results by concentration dependent tight-binding band structure. Rather high concentrations of Xe or Ar have been necessary in order to reach the fully developed Xe or Ar bands. We shall discuss these experiments in more detail in Sect. 3.3.

The studies of the valence states of clusters provide central evidence for a gradual transition from an atomic to a solid-state level structure. Such information emerges from recent photoemission experiments on Xe, Kr and Ar clusters by Dehmer and Dehmer /1978a,b/. In these experiments an HeI line source, together with a hemispherical photoelectron analyser, gave a resolution of 20 meV. Dimers and larger clusters were produced by a supersonic molecular beam source. The results for a nozzle stagnation pressure of 6.76 atm appear in Fig. 3.4 together with the EDC's of atomic Xe /Turner et al., 1970/ and Xe in Ne matrix, respectively, /Schwentner et al., 1975/. Although the exact mass distribution of the clusters is unknown, the supersonic beam experiments showed qualitatively how the photo-

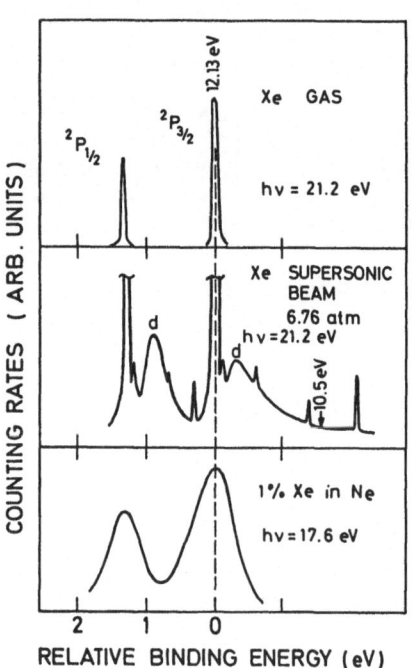

Fig. 3.4. Comparison of photoelectron spectra for the Xe 5p levels of atomic Xe /Turner et al., 1970/, Xe clusters in the gas phase /Dehmer and Dehmer, 1978/ and Xe embedded in an Ar and Ne matrix /Schwentner et al., 1975/. The binding energy scales have been aligned at the Xe $5p_{3/2}$ maximum

electron spectrum changes as the average cluster size is increased. In addition to the $^2P_{1/2}$ and $^2P_{3/2}$ bands observed at low pressures and in the matrix spectra, several peaks due to the dimers (marked by d in Fig. 3.4) are clearly visible and very broad emission features have been assigned to larger clusters; they increase with pressure relative to the dimer peaks. More detailed studies of this kind promise to provide a wealth of information on the energy level structure of clusters, bridging the gap between atomic and solid-state physics.

Interesting experimental information has emerged for the nature of two-dimensional rare-gas films. Photoemission experiments of Xe overlayers on metal surfaces provided evidence for ligand field splitting and strong dispersion effects. Surface crystal field effects have been obtained by Waclawski and Herbst /1975/ in photoemission experiments for Xe physisorbed on a W(100) surface. The EDC for $h\nu = 21.2$ eV revealed two 5p levels. The spin-orbit splitting of the two levels and their intensity ratio was close to the gas phase. A significant broadening of the 5p 3/2 peak with respect to the 5p 1/2 was observed. It has been interpreted as an unresolved doublet resulting from a splitting of the Xe 5p 3/2 states in the tungsten-surface crystal field. Herbst /1977/, Matthew and Devey /1976/ as well as Antoniewicz /1977/ have pointed out, however, that this model requires an unreasonable large positive charge on the metal surface atoms.

More stringent tests of the electronic band structure require angle resolved photoemission on single crystals in order to determine the $E(\underline{k})$ relation experimentally (see e.g. /Smith, 1978; Himpsel, 1983/). Such experiments still need to be done for bulk RGS where they are hampered by charging of the samples and by difficulties in producing and handling single crystals.

Thin films of condensed rare gases are however ideal candidates for investigating the electronic structure of two-dimensional ordered monolayers and multilayers. Usually adsorbate orbitals admix some substrate character due to the bonding interaction with the substrate. The degree of mixing depends on the orbital and the mechanism and strength of the substrate-adsorbate bonding. At higher coverages the lateral adsorbate-adsorbate interaction leads in many cases to the formation of two-dimensional ordered overlayers. The electronic states of the two-dimensional adsorbate layer are then Bloch-type functions characterized by a two-dimensional wave vector $\underline{k}_{\parallel}$ in the surface Brillouin zone (SBZ) and the energy levels are bands in a two dimensional band structure $E(\underline{k})$. They can be measured directly by angle-resolved photoelectron spectroscopy due to the fact that electron momentum parallel to the surface is conserved in the photoemission process, whereas the momentum component $\underline{k}_{\perp}$ is changed because the electron has to cross the potential step at the surface.

Influences of the substrate are particularly weak in the case of physisorption where only van der Waals interactions occur. Thus it is possible to investigate the type and strength of the lateral interaction between the adsorbed particles and to compare experiment with band-structure calculations. This idea has now been successfully applied to several rare gas / substrate combinations (see e.g. /Scheffler et al., 1978,1979; Horn et al., 1978,1979; Hermann et al., 1980; Mariani et al., 1982/).

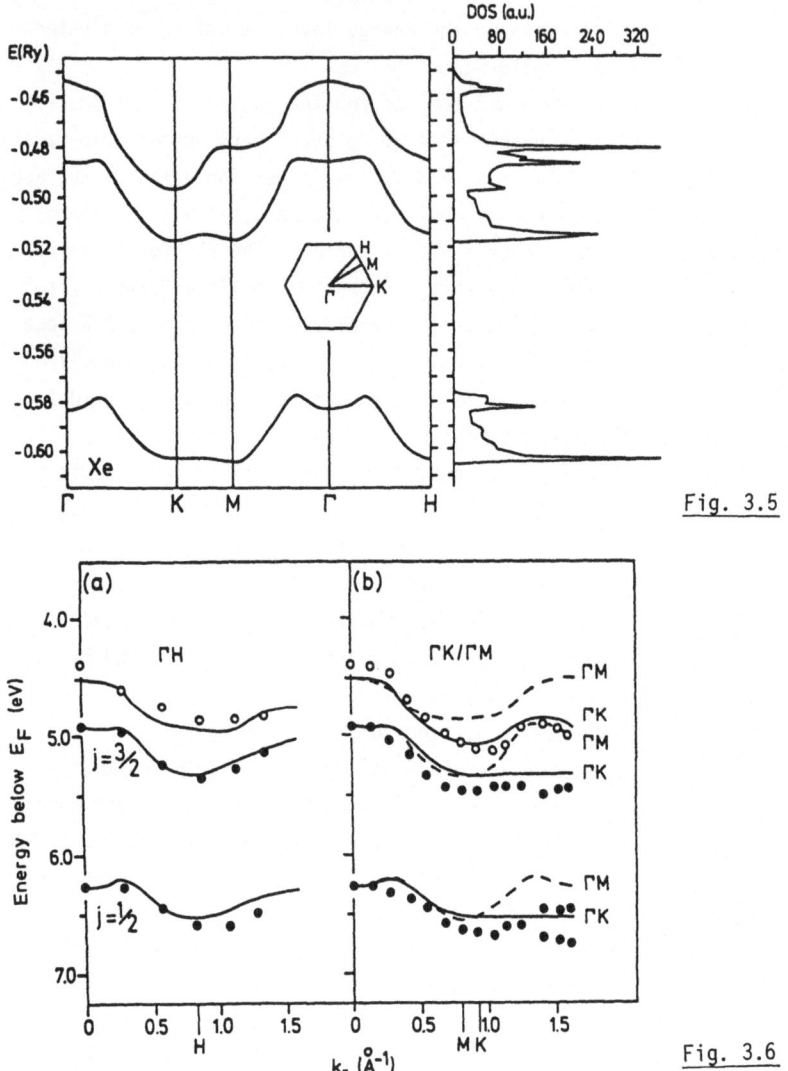

Fig. 3.5

Fig. 3.6

Fig. 3.5. Calculated band structure of an isolated hexagonal xenon monolayer along the main symmetry directions, as well as the direction ΓH. The surface Brillouin zone is shown in the insert. In the right part the computed density of states is shown (from Hermann et al. /1980/)

Fig. 3.6a,b. Comparison between theoretical and experimental band structure for xenon along the ΓH direction (*Panel a*) and the superimposed ΓK/ΓM directions (*Panel b*). The appropriate quantum numbers for the Xe 5p states at Γ are shown in (a). The computed band scheme has been rigidly shifted such that the lowest band at Γ coincides with the experimental value (from Hermann et al. /1980/)

In Fig. 3.5 the calculated energy band structure of an isolated hexagonal xenon monolayer is shown /Hermann et al., 1980/. Both the upper part of the valence bands derived from the $5p_{3/2}$ states and the $5p_{1/2}$ derived lower band show considerable dispersion. In Fig. 3.6 a comparison is shown of these calculated energy bands with data for xenon on Pd(100) obtained by Horn et al. /1980/. Xenon adsorbs on Pd(100) in a hexagonal layer with two domains orthogonal to each other. Thus, photoelectron spectra measured in any one azimuthal direction will be the sum of emission from the two domains, except in the unique direction labeled H (corresponding to the [011] Pd azimuth) which is common to both domains but does not represent a high symmetry direction. The computed band scheme has been rigidly shifted in such a way that the value of the lower band (j = 1/2) coincides with the respective experimental value at Γ. This shift is introduced in order to account for local relaxation effects in the photoemission process that are not contained in the calculations. For ΓH, good agreement of the theoretical dispersion curves with experiment for the whole range of k vectors can be noted.

In the right part of Fig. 3.6 the calculated band dispersion along ΓK and ΓM with experimental data for the [001] azimuth for the Xe/Pd(100) system are shown. In this case ΓM and ΓK are superimposed due to the presence of the two orthogonal domains. For k vectors inside the first surface Brillouin zone where the computed dispersion along ΓM and ΓK are almost identical, good agreement with the experimental data is obtained. Outside the first surface Brillouin zone one would expect a splitting of the bands, due to the superposition of ΓM and ΓK which is experimentally only resolved for the low-lying j = 1/2 band.

Similarly the theoretical dispersion of krypton monolayers and argon physisorbed on Pd(100) agree well with the experimental band structure derived from angle resolved photoemission /Hermann et al., 1980/. Mariani et al. /1982/ have carried these experiments a step further when they studied the formation of two-dimensional electronic bands in ordered layers of Xe adsorbed on Cu(110). Again they found the general trends in the band structure to be in agreement with expectations for quasi-hexagonal layers. Furthermore, at higher coverage they were able to study the commensurate-incommensurate transition, which leads to a substantial increase in bandwidths and causes new photoemission peaks to occur due to multiple scattering at the substrate surface ("substrate umklapp").

These studies emphasize the importance of a simultaneous study of structural and electronic properties of these systems. In fact, Kaindl and coworkers /Kaindl et al., 1980; Chiang et al., 1980,1982/ have shown in a series of papers, that photo- and Auger-electrons from rare-gas atoms physisorbed in form of monolayers, bilayers, and multilayers on metal surfaces exhibit a well-resolved increase in kinetic energy with decreasing distance of the rare gas layers from the surface. The observed binding energy shifts have been described as being mainly due to extraatomic hole-relaxation final state effects, using a point charge image-potential model. An important aspect is the fact that they allow a direct labeling of the first few layers of an adsorbed multilayer configuration. For instance, these binding energy shifts have been used to obtain information for a roughening transition in adsorbed xenon multilayers on palladium /Miranda et al., 1983/ and to study the dispersion of the Xe valence-band states in a layer by layer way for Xe on Al(111) /Mandel et al., 1982/.

In Fig. 3.7 angle resolved photoemission spectra at normal exit angle obtained with HeI radiation (21.2 eV) from a monolayer, a bilayer, and a trilayer Xe/Al (111) are shown /Mandel et al., 1982/. In the monolayer spectrum five peaks can be clearly

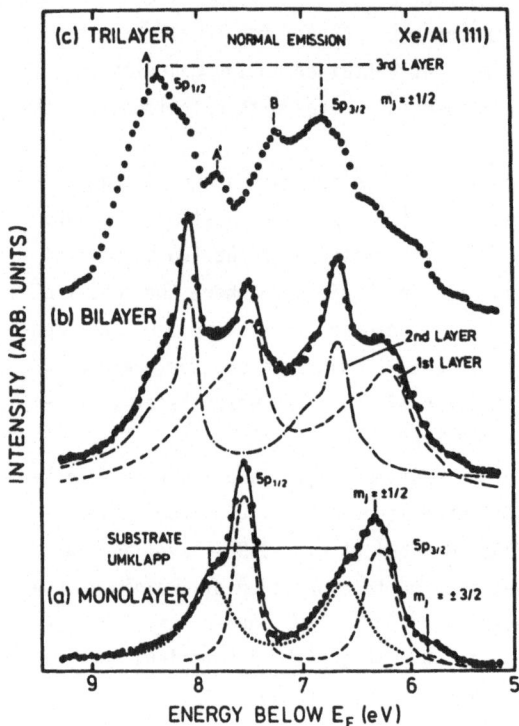

Fig. 3.7a-c. Angle resolved normal emission photoelectron spectra of (a) a monolayer, (b) a bilayer, and (c) a trilayer of Xe on Al (111). The solid lines represent the results of least-square fits. In (a) the peaks due to substrate-umklapp processes are given by the dotted curves; in (b) the total contributions from the first (second) layer are indicated by dashed (dashed-dotted) lines. In the trilayer spectrum (c) the peaks labelled A, A', and B are identified as bulk-Xe features (from Mandel et al. /1983/)

identified: three of them are due to direct photoemission from Xe $5p_{1/2}$ - and Xe $5p_{3/2}$ - derived valence band states at the Γ point (dashed curves), while two of them are caused by photoelectrons backscattered from the incommensurate Al (111) substrate (substrate-umklapp peaks, dotted). The solid line for the monolayer spectrum represents the result of a least-squares fit of a superposition of five Lorentzian lines folded by a Gaussian to the experimental points. The assignment of the direct photoemission lines follows from theoretical bandstructure calculations for ordered Xe monolayers (see Fig. 3.5) /Hermann et al., 1980/.

The photoemission spectrum of the Xe bilayer (Fig. 3.7b) is decomposed into two almost identical single-layer contributions from the first (dashed) and second (dashed-dotted) Xe layer, respectively, separated by a binding energy shift of 0.5 eV. In the trilayer Xe photoemission spectrum (Fig. 3.7c) peaks in addition to the single layer contributions from the three separate Xe layers are observed. The new features at binding energies of 7.8 eV (A') and 8.5 eV (A) and peak B are assigned to emission from critical points of the 3-dimensional *bulk* Xe band structure (see also /Horn and Bradshaw, 1979/). On the other hand, the peaks at binding energies of 8.4 eV and 6.8 eV, respectively, are due to emission from $5p_{1/2}$ and $5p_{3/2}$ single-layer valence-band states of 3rd layer Xe atoms.

Mandel et al. /1982/ have also studied the photoemission of all three Xe-configurations as a function of the polar angle and found a clear dispersion of all spectral features.

These experimental investigations have clearly shown the enormous potential of angle resolved photoemission for mapping out band dispersions. On the other hand, they have also demonstrated the effects of the final state photo hole on experimental data. It is clear from these studies that such effects should be taken into account when comparing experimental photoemission band dispersion curves with the results of band-structure calculations.

3.2 Excitons in Pure Rare-Gas Solids

The optical spectra of pure and slightly doped RGS, in particular the rich exciton structure at the onset of interband transitions (see Figs. 3.8 - 11), have attracted much interest from both theoreticians and experimentalists. Since the dielectric screening in these large band gap insulators is weak, one can expect to study the limits of the single particle model. The dominance of exciton states in the optical spectra already signals the importance of electron-hole correlation effects. Concerning the experimental development we may distinguish several stages: (i) The first systematic experiments in the VUV by Schnepp and Dressler /1960/ and the following detailed investigations by Baldini /1962/, Baldini and Knox /1963/ and Baldini /1965/ and others have led to a fairly detailed description of the excitons in Xe, Kr and Ar. (ii) With the possibilities offered by synchrotron radiation, detailed results including Ne became available for the valence band excitations /Haensel et al., 1969c; Haensel et al., 1970c,d; Boursey et al., 1970; Scharber and Webber, 1971; Steinberger and Asaf, 1973; Pudewill et al., 1976a,b/. Of particular importance during this stage of development was the elucidation of core level excitons (Fig.3.8) for Ne, Ar, Kr and Xe /Haensel et al., 1969a , 1969b , 1970a , 1970b , 1971 , 1973/. The spectra available up to early 1977 have been critically reviewed

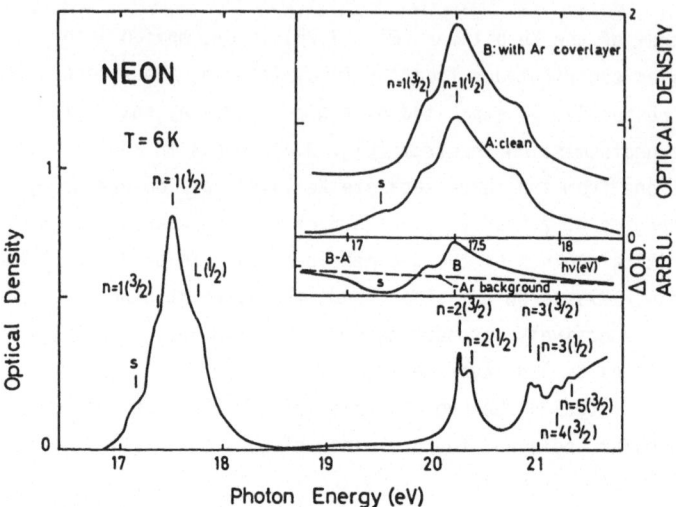

Fig. 3.8. Optical density (– ln I/I$_0$) of solid neon at 6 K in the excitonic range of the spectrum. The main features can be grouped into two series split by spin-orbit interaction which converge to the band gap. For the assignment see text. In the insert, results of a surface coverage experiment for solid neon in the range of the n = 1 exciton are shown. Curve A is the optical density of a clean neon sample. Upon evaporation of a thin Ar coverlayer the structure denoted by S disappears whereas the band shape of the remaining peak remains unchanged. In the lower part of the insert, the difference spectrum B-A is shown and the smooth and structureless Ar background is indicated by the broken line (from Saile and Koch /1979/)

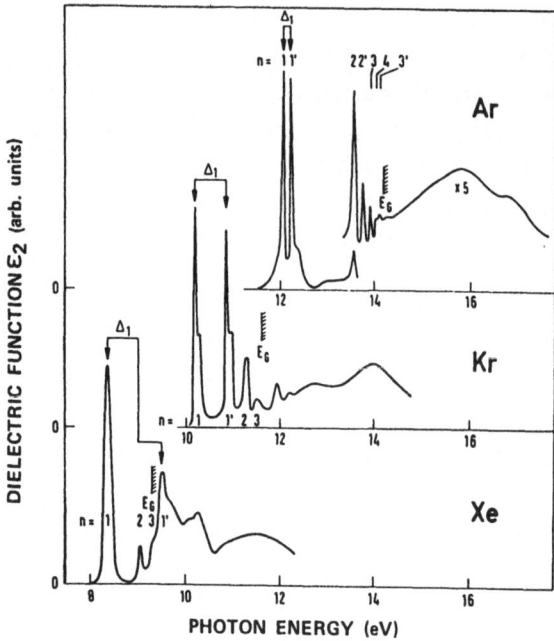

Fig. 3.9. The spectra of the imaginary part of the dielectric constant ε_2 for solid Ar, Kr and Xe in the region of the valence excitons. The spin-orbit splitting in the n = 1 excitons Δ_1, the assignments of the higher members of the exciton series and the band gap E$_G$ are indicated. For the Ne spectrum see Fig. 3.8 (from Saile /1976/)

and summarized by Sonntag /1977/. (iii) Presently there is further progress in the experimental development associated with the use of high resolution spectroscopy in the VUV which exploits the possibilities of a storage ring as an extremely stable and intense radiation source /Saile, 1976; Saile et al., 1976b; Saile et al., 1977a; Saile and Koch, 1978a,1980/. A notable recent accomplishment of these experiments was the unambiguous identification of surface excitons in RGS. Concomitant with these developments remarkable theoretical progress has been made in order to provide an adequate description of exciton states in RGS.

3.2.1 Brief Remarks on Recent Calculations

The theory of excitons dates back to the thirties when the basic concepts were formulated by Frenkel /1931a,1931b,1936/, Peierls /1932/ and Wannier /1937/. Today excitons play a major role in the description of excited electronic states of almost every type of non metallic solids. Excellent treatments of the general theory of excitons are available (see, for example, /Davydov, 1971; Knox, 1963; Rice and Jortner, 1967/).

For RGS the excitation energies E_n of the exciton bands have been described in terms of a hydrogenic Wannier-Mott exciton model based on the effective mass approximation and expressed in the well-known form

$$E_n = E_G - \frac{B}{n^2} \tag{3.3}$$

with E_G being the gap energy, B the binding energy of the first exciton and n the principal quantum number. B is given by

$$B = \frac{\mu}{\varepsilon_0^2} \cdot \frac{e^4}{2\hbar^2} \tag{3.4}$$

where ε_0 is the static dielectric constant and μ is the reduced effective mass of the exciton

$$\frac{1}{\mu} = \frac{1}{m_e} + \frac{1}{m_h} \tag{3.5}$$

with m_e and m_h as effective mass of the electron and hole, respectively. This simple model has been applied successfully to describe the $n \geq 2$ states in pure RGS and for the description of excitations from deep impurity levels in RGS. Equation (3.3) is also frequently used to obtain the energy gap E_G and the binding energy B of the first exciton and hence the effective mass μ.

For the first exciton one observes an energy defect Δ defined as the difference between the experimental and the effective mass approximation (EMA) values:

$$\Delta = E_{expt}^{n=1} - E_{EMA}^{n=1} \; . \tag{3.6}$$

This energy defect Δ reaches values of up to 1 eV. Considerable effort has been exerted to understand the nature of the intermediate excitons, where neither the Wannier model nor the Frenkel picture (appropriate for tightly bound excitons) are strictly applicable. These intermediate excitons can be described either in terms of Frenkel excitons modified by large non-orthogonality corrections or, alternatively, in terms of n = 1 Wannier states subjected to large central cell corrections. These large central cell corrections for n = 1 Wannier excitons, reflected in the large value of Δ, manifest the failure of the simple Wannier model due to the spatial localization of the electron and hole for the n = 1 state. This is evident from Table 3.2, where we compare the radii (r_{Bohr} = 0.529 Å) calculated from

$$r_n = n^2 \cdot \frac{\varepsilon_0}{\mu} \cdot r_{Bohr} \tag{3.7}$$

with the nearest neighbour distances r_K in RGS's. Since in all cases $r_K > r_{n=1}$ the description using a screening of the Coulomb forces between electron and hole by a static dielectric constant becomes problematic. Different theoretical approaches have been persued in this context.

Table 3.2. Radii of the n = 1 excitons for Ne, Ar, Kr and Xe in comparison with the nearest neighbour distances r_K in a fcc-lattice (see Table 1.2)

	Ne	Ar	Kr	Xe
$r^{n=1}$ (Å) a)		1.8	2.5	3.2
r_K (Å) b)	3.156	3.755	3.992	4.335

a) after Saile /1976/ calculated with equation (3.7)
b) Horton /1968/

i) *Pseudopotential theory.* Phillips /1966/ considered contributions to central cell corrections for the energies of deep Wannier excitons, which originate from the following sources: First, spatial dispersion of the effective mass, second, spatial dispersion of the dielectric constant, and third, core non-orthogonality corrections. This approach suggests the use of a model potential, which is Coulombic at large distances and constant for small distances from the positive ion core.

ii) *The integral equation method.* Using a band structure model and assuming a localization of the electron and hole in the same unit cell, Andreoni et al. /1975, 1976/ applied an integral equation method to calculate the energy positions, oscil-

lator strengths and the longitudinal (LO)- transverse (TO) splittings of the first
excitons in solid Ar /Andreoni et al., 1975/ and Ne /Andreoni et al., 1976/. The
main drawback of this theory is its limitation to states with electron and hole
confined to the same unit cell. An extension to states with n > 1 seems to be pro-
hibited. Subsequently, Resca et al. /1978b/ described the energy positions of the
n = 1 excitons for all four rare-gas solids within the framework of an integral
equation approach. In this approach the excitation energy is fixed between the
atomic value and the EMA result by a parameter ρ/ρ_{cc} with an effective exciton
radius ρ and a central cell radius ρ_{cc}.

iii) *Molecular tight-binding calculations*, starting from the corresponding atomic
transitions $2p^6 \rightarrow 2p^5$ ns, ns' in Ne. Boursey et al. /1977/ calculated the energies
of n = 1 excitons in solid Ne on the basis of the potential curves of the molecular
excited states. This theory works very well for the n = 1 states but is inadequate
for the higher transitions.

iv) *The quantum defect method.* This approach /Resca et al., 1978a,c/ is of considerable interest as it bridges the gap between atomic (molecular) excitations and
solid-state large-radius exciton states for all values of n. Starting from the one-
electron picture for deep Wannier excitons the energy levels can be written in terms
of a quantum defect expression

$$E_n = E_G - \frac{B}{(n + \delta_n)^2} \tag{3.8}$$

where n is the principal quantum number, while δ_n is the quantum defect which depends on n. A heuristic approach to the quantum defect method rests on the use of
a model potential $V_n(r)$, which is different for each principal quantum number n,
whose explicit form is /Resca et al., 1978a,c/

$$V_n(r) = -C_n \quad ; \quad r < \rho_n$$

$$\tag{3.9}$$

$$V_n(r) = -\frac{e^2}{\varepsilon r} \quad ; \quad r > \rho_n \quad ,$$

being constant for small distances and Coulombic for large r. The advancement of
the model potential (3.9) is in the spirit of the pseudopotential theory (i). The
Schrödinger equation is now

$$\left(-\frac{\hbar^2}{2\mu} \nabla^2 + V_n(r) \right) \psi_n = (E_G - E_n) \psi_n \tag{3.10}$$

which makes it possible to solve this equation by equating the logarithmic deriva-
tives of ψ_n at r = ρ_n. The input data for the solution of the quantum defect equations (3.10) for RGS are of two types: (a) Atomic parameters. The energies for the

excitations of the rare-gas atoms can be utilized for the calculation of the potential parameters C_n and ρ_n. (b) Solid-state parameters μ, ε and E_G. The quantum
defect equation is then solved to yield the quantum defects δ_n, the energies E_n
(3.8) and the relative intensities $I_n \propto (n + \delta_n)^{-3}$. Two practical advantages of this
scheme should be noted. First, this scheme is applicable for valence excitations
and for core excitations (see Sect. 3.2.3) in RGS, as well as for atomic impurity
states. Second, the energies of deep Wannier exciton states, including the problematic $n = 1$ state, have been properly taken into account. The methodological dis-
advantage of this scheme is the proliferation of parameters involved in the calculation. Nevertheless, the quantum defect method is of considerable interest from the
viewpoint of general methodology because of two reasons. It establishes the atomic
parentage of Wannier states with high n, and it is capable of dealing, in principle,
with the transition from the Frenkel (atomic) to the Wannier (solid-state) picture.

Progress in band structure calculations of RGS were also recently accomplished.
In this context, Baroni et al. /1980,1981/ have computed the energy bands of solid
Ar and Ne using a modified orthogonal plane-wave method and the Gaussian represen-
tation of occupied orbitals. A good feature of such a procedure is that the exchange potential in its non-local form can be included and that the matrix elements
are still in analytic form. Thus, a numerically simple, and still fairly accurate
description of the electronic structure is obtained.

With the recent observation of surface exciton states in Ar, Kr and Xe and new
structures in some of the bulk exciton bands, a new challenge has been put forward
for a further theoretical investigation of these states. Several models have been
developed to interpret surface excitons in rare-gas solids. The energy shifts and
splittings in environments with different symmetry have been treated by Wolff /1977/
starting from the corresponding atomic excitations. The calculated splittings for
localized excitations at the surface of Ar and Kr compare favourably with the experimental results /Saile and Wolff, 1977b/. In the same spirit Chandrasekharan
and Boursey /1978/ extended their picture, described above, to excitations in the
(100)- and (111)-surface planes and obtained good agreement for the excitation
energies of surface excitons in all rare-gas solids. Ueba and Ichimura /1976/ established conditions for the energies of surface excitons relative to the bulk states
by a localized perturbation method. In this approach the excitation energies, as
well as the Davydov splitting, are determined by two energetic parameters, the
environmental shift term and the exciton transfer term. In an application to Ar,
Kr and Xe, Ueba /1977/ related the observed splittings of the surface excitons to
the spin-orbit splitting of the valence bands.

3.2.2 Volume and Surface Excitons – Experimental Results

A comprehensive survey of the optical constants of RGS obtained until 1977 was provided by Sonntag /1977/. In this section we want to summarize the recent new experimental results for volume and surface excitons obtained by Saile et al. /1976, 1978a,b/ and their connection to the recent theoretical developments, as summarized in the previous section. Transmission and reflection spectra of solid Ne in the valence-exciton range ($16 \leq h\nu \leq 22$ eV) have carefully been reinvestigated by Saile and Koch /1978/. An overview of the absorption spectrum for thin films of solid Ne appears in Fig. 3.8. The results of a coverage experiment are shown for the range of the n = 1 excitons in the insert. Using highly monochromatized ($\Delta E = 4$ meV) synchrotron radiation with a resolution of about thirty times better than the half-width of the sharpest spectral features, these experiments revealed a number of new features. Consequently, information from the exciton lineshapes could be extracted and more precise energy values than previously possible could also be obtained. The results of this study may be summarized as follows (see also Tables 3.3 and 3.4):

1) For mono- or submonolayers of Ne evidence for an adsorbate induced resonance was found. The excitation energy of 16.91 eV of this resonance, which disappears for thicker samples, is close to the value for the $2p^6 \rightarrow 2p^5$ 3s, j = 1/2 state in the gas phase at 16.85 eV.

2) For Ar, Kr and Xe surface excitons were observed. The behaviour of the surface exciton state at 17.15 eV was found to be different from that of the resonance mentioned above in that it can be observed for a full sequence of film thicknesses. Evidence that this exciton state is confined to the Ne surface is based on the observation that the surface exciton is only observed under ultrahigh-vacuum conditions. Also, upon coating with a different rare-gas film the surface exciton disappears (see the insert in Fig. 3.8). Finally, from the fact that the contribution of the surface exciton remains almost constant, whereas the bulk peaks increase with increasing film thickness, it was estimated that the absorption due to the surface exciton is confined to about one layer at the vacuum-sample boundary.

A full width at half maximum (FWHM) of about 80 meV to 300 meV was found for the surface states depending on sample preparation and background subtraction for deconvolution. These halfwidths exceed considerably those for surface states in other RGS where FWHM = 20 to 30 meV have been observed /Saile et al., 1976b/. Different reasons can be responsible for such broadening: either the surface exciton is broadened by the same mechanism which causes a broadening of the bulk n = 1 excitons in solid Ne (see below) or this surface peak consists of several overlapping excitations. This latter explanation was found to be valid. It has been shown both experimentally /Saile et al., 1976b/ and theoretically /Saile and Wolff, 1977; Chandrasekharan and Boursey, 1978/ that three transitions to surface states cor-

Table 3.3. Comparison between experimental transition energies for bulk excitons in solid neon and other parameters with theoretical calculations $B_1 = E_G - E(n=1)$ being the real binding energies of the $n = 1$ excitons and μ is the effective mass. The other notations are the same as in Table 3.1. All energies are in eV

	Experiment				Theory							
	Ref. a) 3/2	Ref. a) 1/2	Ref. b)	Ref. b)	Ref. c)	Ref. c)	Ref. d)	Ref. d)	Ref. e)	Ref. e)	Ref. f)	Ref. f)
j												
$n = 1$	17.36	17.50			17.50	17.63	17.65	17.75	17.75	17.85	17.58	17.79
2	20.25	20.36	19.98	20.08					20.32	20.42	20.24	20.35
3	20.94	21.02	20.93	21.13					20.94	21.04	20.91	21.03
4	21.19	21.29	21.25	21.34					21.19	21.29	21.19	21.31
5	21.32		21.40	21.50					21.31	21.41	21.33	21.45
E_G	21.58	21.62	21.67	21.77					21.55	21.65	21.61	21.73
B_1	4.22	4.12	4.16	3.81								
Δ_{SO}	0.09		0.10						0.10		0.12	
δ	0.28	0.24							0.35	0.35	~0.5	~0.5
μ	0.8	0.7	0.8						0.8		0.97	

a) Saile and Koch /1979/
b) Andreoni, Perrot and Bassani /1976/
c) Boursey, Castex and Chandrasekharan /1977/
d) Resca and Rodriguez /1978b/
e) Resta /1978/; gap energies are obtained by a fit of the experimental data and Δ_{SO} is the atomic value. From these gap energies the theoretical exciton binding energies have been subtracted.
f) Resca, Resta and Rodriguez /1978,1978a/

respond to the two $n = 1$ bulk excitons in RGS. Due to the small spin-orbit inter-action in Ne, we may extrapolate the splittings of surface excitons in Ar and Kr to the case of Ne. The resulting three states below the $n = 1$ (3/2), bulk exciton for solid Ne are then confined to an energy interval of less than 200 meV. In the mole-cular-type calculation by Chandrasekharan and Boursey /1978/ surface excitons are introduced by the different surface symmetry C_{4v} or C_{3v}, as compared to the bulk

Table 3.4. Exciton energies of bulk and surface excitons. Exciton energies at 20 K (Xe, Kr, Ar) and 8 K (Ne) in rare-gas solids in the Wannier notation n = 1, 2, ... j = 3/2 and 1/2 indicates the spin-orbit of the series and L is the longitudinal exciton. For Ne an adsorbate state A at 16.91 eV has been observed. All values are in eV. The accuracy is better or equal to 0.01 eV for peak positions

		Xe 5p Ref. a)		Kr 4p Ref. a)		Ar 3p Ref. a)		Ne 2p Ref. b)	
j =		3/2	1/2	3/2	1/2	3/2	1/2	3/2	1/2
bulk excitons	n = 1	8.37	9.51	10.17	10.86	12.06	12.24	17.36	17.50
	2	9.07		11.23	11.92	13.57	13.75	20.25	20.36
	3	9.21		11.44	(12.21)	13.87	14.07	20.94	21.02
	4			11.52		13.97		21.19	21.29
	5							21.32	
	L	8.43		10.29	10.95		12.50		17.75
surface excitons								A 16.91	
	1	8.21		9.95	10.68	11.71	11.93		17.15
				10.02		11.81			
	2			11.03			12.99		
							13.07		

a) Saile, Steinmann and Koch /1977/
b) Saile and Koch /1979/

O_h symmetry. These ab initio calculations yield three states at about 17.30 eV with a splitting of about 150 meV between the first and third peak. Taking into account a natural half width of 80 meV for the surface excitons (which correspond to 33% of the bulk widths for the lighter rare-gas solids), this theoretical result is consistent with the experimental observation of one rather broad structure at 17.1 eV.

3) Extensive information has been obtained for the bulk exciton states in Ne. For the first time the spin-orbit splitting of the j = 3/2 and j = 1/2 states was resolved and thus the main features of the spectrum can be grouped into two series split by spin-orbit interaction which converge to the band gap. The prescise values obtained, as well as derived quantities, appear in Table 3.3. The value of 90 meV for the spin-orbit splitting of the valence band maxima at the Γ point in the Brillouin zone is close to the gas phase result. Further, using the quantum defect formalism, according to (3.8), Resca et al. /1978a,b,c/ have determined a new value for the band gap of 21.58 eV for solid Ne.

In Table 3.3 the experimental results are compared with theoretical predictions. Generally, there is good agreement between theory and experiment. Since the higher numbers of the series are rather well described in a hydrogenic-like picture the good agreement for $n \geq 2$ is not surprising once the proper value for E_G has been obtained. This holds also for a quantum defect theory as long as the quantum defect is small. The tightly bound lowest $n = 1$ levels are more difficult to describe and pose more serious problems.

A complete theoretical analysis of the exciton states of solid Ne has been presented by Andreoni et al. /1976/ (Table 3.3). For the $n = 1$ exciton the main approximation was to assume that the electron and the hole are confined to the same unit cell (one-site approximation). This seems justified in view of the small Bohr radii of the $n = 1$ excitons (Table 3.2). While these theoretical results for the $n = 1$ binding energies are in fair agreement with experiment, the calculated intensity ratio $I(n = 1, 3/2) / I(n = 1, 1/2) = 1 : 50$ is considerably smaller than the experimental one (between $1 : 5$ and $1 : 10$). Using the experimental data for an estimate of the exchange interaction one arrives at smaller values for this quantity than predicted /Saile and Koch, 1979/. For the states $n \geq 2$, Andreoni et al. /1976/ also gave values for the excitation energies and intensity ratios on the basis of the atomic data and the effective mass approximation with a given B and corrections for electron-hole exchange and spin-orbit splitting. As for the $n = 1$ excitons one observes experimentally in this case quite different intensity ratios. Obviously the exchange interaction is considerably overestimated in this theory, an observation which is also supported by the analysis of the solid Ar data /Saile, 1976/.

4) Finally for the $n = 1$ excitons of solid Ne the excitation of longitudinal modes has been observed by Saile and Koch /1978/ in both transmission and reflection experiments (shoulder denoted by L(1/2) in Fig. 3.8). This assignment is supported by the results of electron energy loss experiments (see /Sonntag, 1977/) and by the observation of the broad "quasi"-stop band between 17.5 eV and 18.0 eV in the reflectance from thick Ne films /Pudewill et al., 1976a/ which yields $h\nu_{TO} \approx 17.5$ eV and $h\nu_{LO} \approx 17.8$ eV. The numerical value of 0.25 eV which was deduced for the LO-TO splitting by Saile and Koch /1978/ for solid Ne is in very good agreement with the values calculated by Andreoni et al. /1976/ of 0.232 eV and by Chandrasekharan and Boursey /1979/ who calculated 0.252 eV.

Now we turn to the discussion of volume and surface excitons in the heavier RGS. The excitonic range of the absorption spectra (see Fig. 3.9) with onsets above 8 eV (Xe), 10 eV (Kr) and 11.5 eV (Ar) are characterized by sharp intense exciton bands which, when studied with sufficient resolution /Saile, 1976; Saile et al., 1976, 1977,1978a,b/, reveal considerable structure and a wealth of new information. For thin films of Ar, Kr and Xe, spectra covering the first exciton states are displayed in Fig. 3.10. Energy positions for all structures including the results for Ne discussed above are compiled in Table 3.4. Derived parameters such as the band gaps, spin-orbit splittings and binding energies have been included in Table 3.1. The additional sharp peaks and shoulders 200 - 500 meV below the bulk exciton peaks with half widths of the order of 30 meV have been interpreted as surface excitons by experiments with overlayers and by their thickness dependence in transmission /Saile, 1976; Saile et al., 1976a,b; Saile et al., 1977a/. In the upper part of Fig. 3.10 the results for thin films of Ar are presented. For the clean sample the absorption spectrum shows three additional features at energies roughly 100 - 600 meV below the $n = 1$ volume excitons and two additional peaks at ≈ 500 meV below the $n = 2$ exciton which disappear upon coverage of the Ar film with a Kr overlayer.

For Kr the results shown in the middle part of Fig. 3.10 are analogous; with an Ar overlayer the additional maxima below the $n = 1$ and $n = 1'$ bulk excitons disappear. Removal of the overlayer by gently heating the sample makes the surface excitons reappear again. The difference spectrum of the absorption for a clean sample minus the absorption of the same Kr film with an Ar overlayer (also shown in Fig. 3.10) demonstrates that roughly the same amount by oscillator strength associated with the surface excitons is gained by the volume excitons upon coverage. It is also interesting to point out that the transmission in the surface excitons remains constant with thickness, as has been shown by careful studies of the thickness dependence of the optical absorption of Ar and Kr /Saile, 1976/.

Finally, in the lower part of Fig. 3.10 the results for solid Xe in the excitonic part are displayed. For this sample with a film thickness of ≈ 38 Å the $n = 3$ volume exciton is missing since its diameter $d_{n=3} = 57$ Å already exceeds the sample thickness. The position of this exciton found for thicker samples however, is marked in the figure. The most important point is the double structure of the $n = 1$ exciton which is displayed in the insert on an expanded scale for a film with $d = 27$ Å. On the low energy tail of the known $n = 1$ volume excitons a sharp (FWHM 20 - 30 meV) maximum is found at 8.2 eV which, again, has been identified as a surface exciton.

Saile et al. /1977/ have performed a lineshape analysis for the $n = 1$ bulk and surface exciton for solid Xe. Their studies have shown that reasonable fits can only be obtained with Lorentzians. For the Xe spectrum displayed in Fig. 3.10 (lower part) they found half widths of 20 meV for the surface exciton and 80 meV for the bulk part. The result of this fit is shown in the insert. According to Toyozawa's model /1974/ (see Chap. 6) this result indicates that we are dealing with the case of weak exciton - phonon scattering. The derived scattering times are in the order of several 10^{-14} s. It is interesting to contrast this result for Xe with the estimate of 200 meV for the half widths in solid Ne /Saile and Koch, 1978/. The latter value is indicative for the strong scattering case and localized exciton states. The localized nature of the $n = 1$ excitons in solid Ne is also in accord with the current view of exciton dynamics obtained from luminescence experiments.

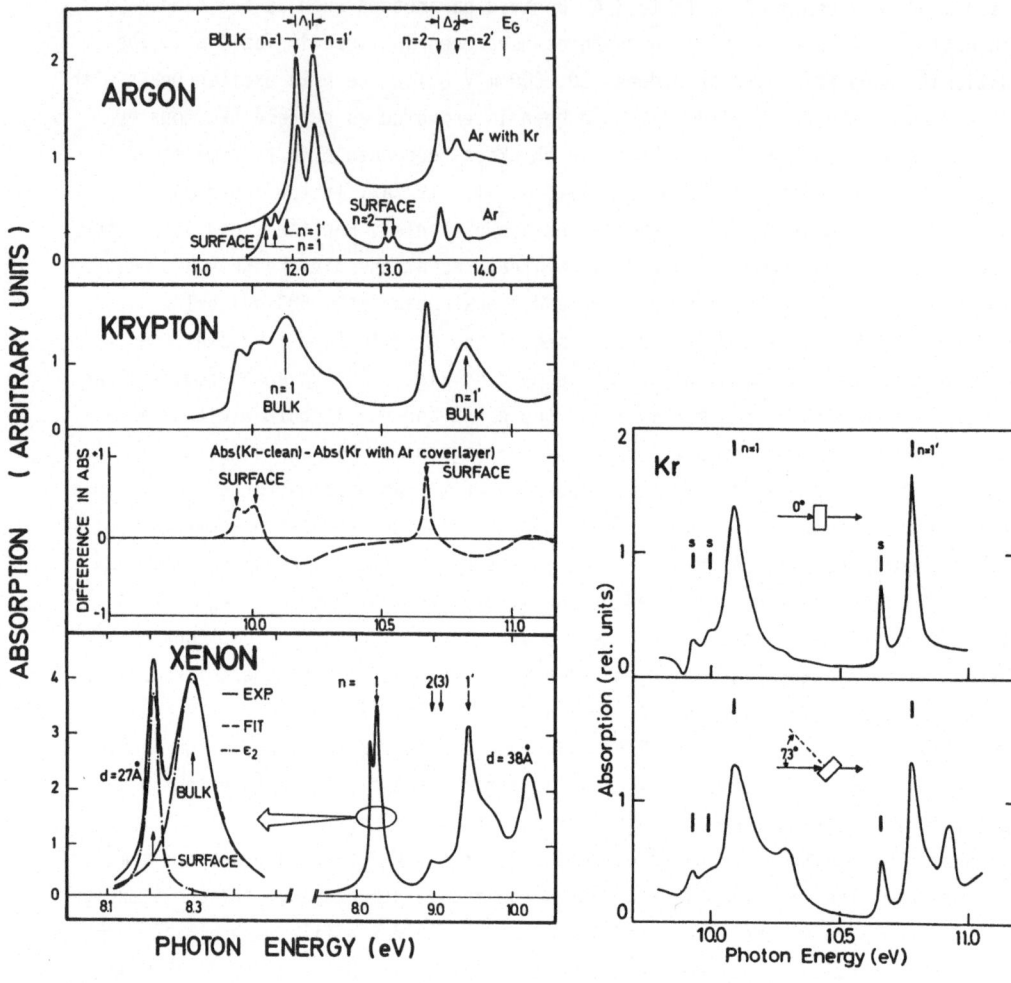

Fig. 3.10   Fig. 3.11

Fig. 3.10. High resolution absorption spectra for solid Ar, Kr and Xe in the range of the valence excitons. Volume and surface excitons are observed for all three samples. For Ar and Kr the results of surface coverage experiments are also shown. For Xe the experimentally determined spectrum in the range of the n = 1 surface and volume exciton is displayed on an expanded scale together with a lineshape analysis using two Lorentzians (after Saile /1976/)

Fig. 3.11. Absorption spectrum for thin Kr films (d = 30 Å). The n = 1 and n = 1' transverse TO volume excitons as well as the surface excitons (s) can be seen in the spectrum for normal incidence in the upper panel. At non-normal incidence (*lower panel*) two new features associated with the n = 1 and n = 1' longitudinal (LO) bulk excitons appear (from Saile et al. /1980/)

We have already mentioned at several points the occurrence of longitudinal excitons in RGS. Bulk excitons in cubic crystals are split by the long-range dipole-
dipole interaction into longitudinal (LO) (with $\underline{E}$ parallel to $\underline{k}$, where $\underline{E}$ is the
electric field vector and $\underline{k}$ the momentum of the exciton) and transverse (TO)
($\underline{E}$ perpendicular to $\underline{k}$) excitons. Observation of longitudinal bulk excitons is usually restricted to electron energy loss spectroscopy where they show up as maxima in
$\text{Im}\{-1/\varepsilon\}$ (see, for example, /Raether, 1965,1974,1980/). In optical ($k \approx 0$) normal
incidence transmission experiments the longitudinal excitons normally do not couple
to the incident transverse electromagnetic field. However, in reflection geometry
and for rough surfaces (see, e.g., /Filinski, 1972/) optical excitation of longitudinal modes becomes possible. The summary of LO-TO splittings in Table 3.5 has been
obtained from the widths of the quasi-metallic reflection bands in reflectance
spectra from thick films and from the analysis of optical transmission data from
thin rare-gas films. From these two sources a consistent set of LO-TO splitting
emerges. In the case of Ne the TO-LO splitting has been theoretically calculated
(Table 3.5). The result is in favourable agreement with the experiment. In principle one could also obtain the splitting from the difference in the peak positions
in ε_2 spectra (optical data) as compared to the peak positions in electron energy
loss spectra. Inspection of Table IV, V, VI and VII in Sonntag's review /1977/,
where the available experimental data have been compiled, shows that the scatter in
the experimental data is still too large for that purpose.

Table 3.5. Longitudinal transverse splittings for bulk excitons in RGS. All values
are in eV

n = 1	Ne		Ar		Kr		Xe	
	3/2	1/2	3/2	1/2	3/2	1/2	3/2	1/2
width of the reflectance band	-	≈0.3 [a]	0.06[b]	0.32[b]	~0.2 [b]	≈0.17[b]	0.2 [b]	0.15[b]
opt. Δ_{LO-TO}	-	0.27[a]	-	0.26[c]	0.12[c]	0.09[c]	0.06[c]	-
cal. Δ_{LO-TO}	0.04 [d] 0.236[e]	0.232[d] 0.252[e]						

[a] Saile and Koch /1979/
[b] see the review by Sonntag /1977/
[c] Saile et al. /1981/
[d] Perrot and Bassani /1976/
[e] Chandrasekharan and Boursey /1978/

This overall description of a part of the extra structures in the bulk exciton bands observed under certain experimental conditions is quite satisfactory in view of the long debates about these additional bands. To further substantiate this point Saile et al. /1980/ have studied the transmission of thin Kr films under non-normal incidence (Fig. 3.11). In the upper part the spectrum for normal incidence shows the n = 1 and n = 1' TO-bulk excitons as well as the surface excitons denoted by s. For the n = 1 exciton there is some additional feature barely discernible at around 10.3 eV. Upon tilting the sample by 73° this feature gains intensity and develops into a separate peak and a new maximum associated with the n = 1' peak appears. These two maxima are assigned to the LO-bulk excitons. For the reststrahlen bands in the infrared it has been shown by Berreman /1967/ that this effect can be derived from a dielectric theory by solving Maxwell's equations with appropriate boundary conditions for slight surface roughness. It is worthwhile to point out here that this effect is independent of other types of anomalies that may arise from differences between dielectric properties in the bulk of the material and those near the surface. Such differences have been proposed by Andreoni et al./1978/ in order to explain the dip in the reflection stop band between $h\nu_{TO}$ and $h\nu_{LO}$ in terms of spatial dispersion and a dead layer for bulk excitons at the surface.

3.2.3 Core Excitons

An interpretation of the fine structure at the onset of transitions from inner shells (Fig. 3.12) is possible in terms of the joint density of states where the valence-band density of states has to be replaced by dispersionless core states. This approach has been described by Rössler /1976/. Thus the optical function is merely a superposition of the conduction band density of states with the spin-orbit splitting and degeneracy of the core levels taken into account (e.g. /Rössler 1971; Kunz and Mickish, 1973/). In principle, a deconvolution of the optical absorption spectra into the weighted contribution from spin-orbit split core levels should yield the density of conduction-band states. It turned out, however, that this simple one-electron approach applied to the available data does not give meaningful results /Saile et al., 1977/.

We mention a few other theoretical calculations which recently took up the problem. Altarelli et al. /1975/ calculated the core excitons at the onset of the 2p soft X-ray threshold for solid Ar. The binding energies and relative transition amplitudes for the lowest allowed exciton states were computed by formulating the problem in terms of Wannier functions of the conduction bands and solving the resulting integral equation in the one-site approximation already applied successfully to the fundamental exciton spectrum of Ar /Andreoni et al., 1975/. The results obtained allow one to locate the onset of interband transitions at an energy of a few eV's above previous theoretical determinations.

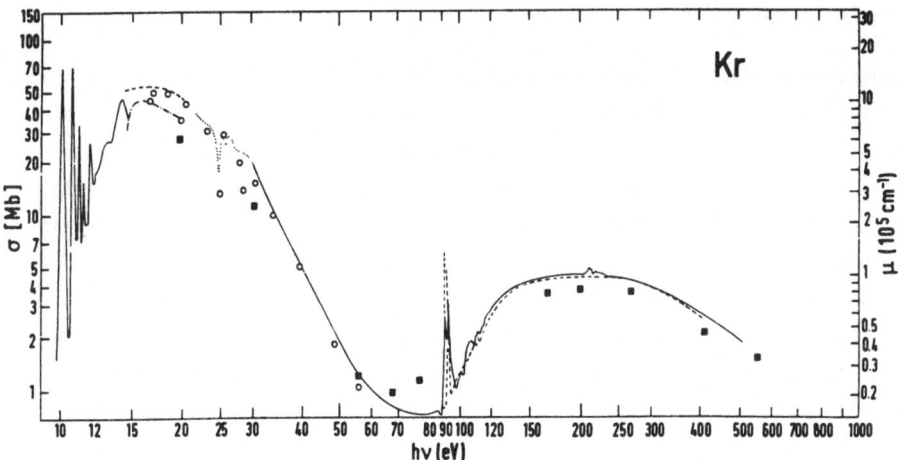

Fig. 3.12. Absorption coefficient of solid (———) and gaseous (– – –) Kr determined by reflectance /Haensel et al., 1970c; Skibowski, 1971/ and absorption measurements /Haensel et al., 1969b/. (·····) Samson /1963/; (∘ ∘ ∘) Rustgi et al.
/1964/; (■) Lukirskii et al. /1964/; (–·–·–) Cook and Metzger /1965/; (⊢----⊣)
between 15 and 20 eV Huffmann et al. /1963/ (from Sonntag /1977/)

Therefore, the sharp fine structure at the onset of absorption, previously interpreted in terms of conduction-band density of states, was attributed to discrete
excitonic transitions. This interpretation had already been strongly suggested by
the close analogy of the optical absorption with the atomic absorption in this
range /Haensel et al., 1969a/. Using the same approach, Grosso et al. /1978/ re-
examined the core exciton spectra from 3d levels of Kr and 4d levels of Xe, both
in the isolated atoms and in the crystal and found support of the earlier interpretation of these resonances as due to 1s, 2p and 3p-like excitons.

Finally, we mention the work by Resca et al. /1978/ extending the theory for
valence excitons to the calculation of core excitons at the onset of Kr 3d transitions, and Xe 4d transitions, and for solid Ne and Ar at the 2s and 2p and 3s onset, respectively. The paper by Baroni et al. /1979/ deals with the Ar 3p core excitons within the envelope function formalism (see also /Baroni et al., 1980,1981/).
For a complete summary of experimental data for core spectra, we refer the reader
to the review by Sonntag /1977/.

3.3 Rare-Gas Alloys

In this section we deal with impurity excited states in the low concentration limit. We discuss also the few experimental results available for concentrated rare-gas alloys. In the discussion, we confine ourselves to rare-gas guest atoms. Thus we are concerned with deep impurity states whose binding energies are comparable to the band gap of the host crystal. Most of the information about the energy levels has been obtained by absorption spectroscopy, but we shall use information also obtained from photoemission and luminescence experiments. We shall find this information instrumental to the following discussion of the electron affinity (V_0) scale as well as for an understanding of the dynamical processes discussed in later chapters of this book.

3.3.1 Deep Dilute Impurities in Rare-Gas Solids

Rare-gas guest atoms in a rare-gas matrix are the best-known examples of deep impurity states in insulators. For a schematic description of the energy levels we refer to Fig. 1.2. These systems were first studied in absorption experiments by Baldini and Knox /1963/ (Xe in Ar) and by Baldini /1965/ (Ar, Kr and Xe in Ne, Kr and Xe in Ar, and Xe in Kr). From the more extensive later investigations we mention the work by Gedanken et al. /1973/ (Xe in Ne, Ar and Kr and C_6H_6, C_2H_4 and CH_3I in Ne, Ar, Kr and Xe), by Pudewill et al. /1976/ (Ar, Kr and Xe in Ne), Hahn et al. /1980,1982/ (Ar, Kr, Xe in Ne and Kr, Xe in Ar) using synchrotron radiation. For dilute rare-gas alloys the pertinent experimental data have been summarized and discussed in detail by Jortner /1974/ and Jortner et al. /1983/.

One of the most informative results for the system Xe, Kr and Ar in a Ne matrix is illustrated in Fig. 3.13 /Pudewill et al., 1976/. Long impurity Wannier series containing up to five lines are observed. In each case two series can be assigned, split by spin-orbit interaction and converging to the bottom of the conduction band of the host matrix. For n = 2 the energies of the series members are almost perfectly described by a Wannier formula (3.3) where E_n, E_G and B are now replaced by the corresponding values for the impurity:

$$E_n^i = E_G^i - \frac{B^i}{n^2} \; .$$

(3.11)

The electron binding energy is given by

$$B^2 = \frac{m_e}{\varepsilon_0^2} \frac{e^4}{2\hbar^2} \; .$$

(3.12)

The excellent fit of the excitation energies to this formula is demonstrated graphically in Fig. 3.14 where E_n^i is plotted versus $1/n^2$. Thus, from the convergence of these series, the gap energies E_G^i of the impurity states as well as the spin-

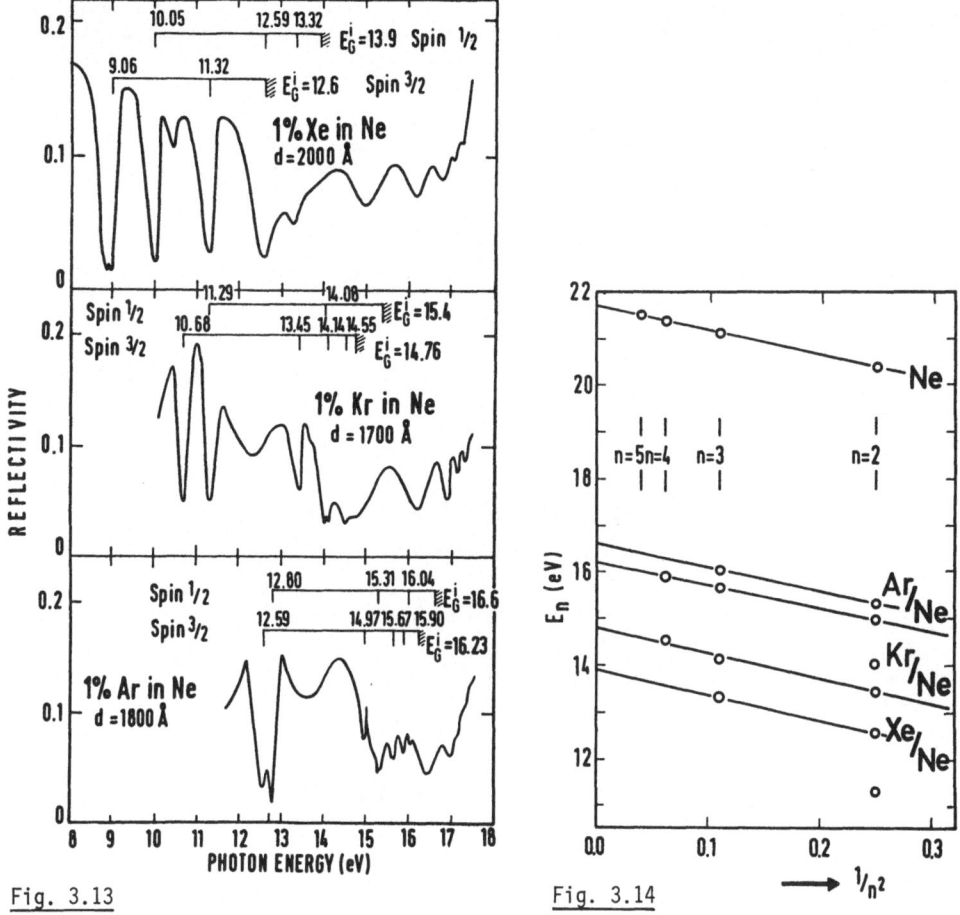

Fig. 3.13. Reflectance spectra of Ne films doped with 1% Xe, d = 2000 Å (*upper panel*); 1% Kr, d = 1700 Å (*middle panel*); and 1% Ar, d = 1800 Å (*lower panel*) (d: film thickness) evaporated onto a gold substrate. The spin-orbit split impurity-exciton series are indicated on top of each spectrum (from Pudewill et al. /1976/)

Fig. 3.14. Excitation energies of excitons in pure Ne (E_n) and of the impurity states of Ar, Kr and Xe in a Ne matrix (E_n^i) plotted versus $1/n^2$ according to the Wannier model (from Pudewill et al. /1976/)

orbit splitting of the guest atoms in the Ne matrix can be derived with an accuracy of 0.2 eV. A summary of the energy levels, together with their assignments, is presented in Table 3.6. This table also includes results for the heavier rare-gas matrices Ar, Kr and Xe and recent values deduced from luminescence excitation spectra /Hahn and Schwentner, 1980/.

We note that in the system Xe in Ne the observation of two series including also higher members enabled Pudewill et al. /1976/ to give a more satisfactory assign-

Table 3.6. Exciton energies of Xe, Kr and Ar guest atoms in Ne, Ar and Kr matrices from absorption (a,b,c) and luminescence excitation spectra (d). The halfwidth is given in brackets

matrix

guest	exciton	Ne a)	Ne b)	Ne c)	Ne d)	Ar a)	Ar b)	Kr a)	Kr b)
Xe	n = 1	9.08 (0.135)	9.12	9.06	9.10	9.22 (0.095)	9.2	9.01 (0.07)	9.0
	n = 1'	10.04 (0.095)	10.08	10.05	10.06	10.53 (0.085)	10.6		
		10.40	10.51		10.35				
			10.69		10.95				
	n = 2	11.28 (0.115)		11.32	11.31	9.97 (0.075)	10.0	9.76 (0.065)	9.8
					11.53				
					11.75				
					11.84				
	n = 3				12.00	10.25			
	n = 4				12.19	10.36			
	n = 5				12.43				
	n = 2'			12.59	12.60	10.80	10.8		
					12.83				
	n = 3'			13.32	13.34				
Kr	n = 1	10.62 (0.15)		10.68	10.60	10.79 (0.11)			
					10.85				
	n = 1'	11.22 (0.14)		11.29	11.22	11.36 (0.10)			
					11.54				
					13.12				
	n = 2			13.45	13.32				
	n = 2'			14.06	13.94				
	n = 3			14.14	14.05				
					14.18				
	n = 4			14.55	14.48				
Ar	n = 1	12.5 (0.2)		12.59	12.48				
	n = 1'	12.7 (0.2)		12.80	(12.74)				
	n = 2			14.97	14.84				
					15.00				
					15.17				
	n = 2'			15.31	15.25				
	n = 3			15.67	15.10				
	n = 4			15.90	15.75				
	n = 3'			16.04	15.90				

a) Baldini /1965/
b) Gedanken, Raz and Jortner /1973/
c) Pudewill, Himpsel, Saile, Schwentner, Skibowski and Koch /1976/
d) Hahn and Schwentner /1980/

ment for the impurity exciton series, with the n = 1 (3/2) exciton at 9.06 eV, the
n = 1 (1/2) exciton at 10.05 eV and the n = 2 excitons at 11.32 eV (3/2) and
12.59 eV (1/2) respectively, thus avoiding some problems with the earlier interpretations by Baldini /1965/ and Raz et al. /1970a/; Pantos et al. /1977/ and Hahn et
al. /1980/ have assigned further structures in these spectra to nd-final states.
However, some of the spectral features (see Table 3.6) remain unexplained.

As shown in Fig. 3.14 the term values $T_n^i = E_G^i - E_n^i$ coincide with those for pure
Ne. Deviations are observed only for the n = 1 excitations where the electron and
hole are fairly localized. It follows that the binding energies B^i determined from
excitons n ⩾ 2 excited at the guest atoms reflect the properties of pure Ne and are
independent of the individual impurity. This observation very strongly supports the
interpretation of the impurity excited states for n ⩾ 2 in terms of a Wannier model.
In Table 3.1 we reproduce the values for E_G and E_G^i and the binding energy B, respectively, the spin-orbit splitting of the impurity levels ΔE_{SO}^i and the central cell
correction for the n = 1 states.

Within the effective mass approximation the exciton binding energy B^i (3.3 - 3.5)
depends only on the dielectric constant ε and the effective mass μ. For excitons
with radii larger than the nearest neighbour distance (n ⩾ 2) and at the low concentrations employed, ε of pure Ne may be used for the calculation of B^i. In this
case electrons excited at the guest atom move in the conduction bands of the host,
leading to a common value of m_e, whereas the hole will be bound to the guest atom
($m_h \cong \infty$). Pudewill et al. /1976/ obtained an upper limit of B^i = 5.8 eV taking typical values for m_e from band-structure calculations /Kunz and Mickish, 1973/ and
a B value of B = 5.24 eV for pure Ne from their experiment with $1/m_h$ = 0. In order
to determine B^i from the experiment one can use n = 2, n = 3 and n = 4 of the $\Gamma(3/2)$
exciton series of Ar in Ne and Kr in Ne. One obtains B^i = 5.27 eV for Ar and B^i =
5.32 eV for Kr (Table 3.1). The experimental accuracy for the energy positions is
approximately 0.05 eV at 20 eV and results in an uncertainty of 0.5 eV for the B^i
values. Thus, the predictions of the Wannier model for the binding energies of
n ⩾ 2 excitons of these deep impurity states in solid Ne are in good agreement with
the experiment.

We turn now to the energetics of the n = 1 intermediate-type impurity excitations.
The central cell correction Δ determined from the experiments appear in Table 3.1.
The calculated relationship between Δ and B shows a monotonic increase of Δ with
the binding energy /Hermanson and Phillips, 1966/. In fact, Gedanken et al. /1972,
1973/ were able to describe their experimental results by a linear relationship
$\Delta = \alpha + \beta B^i$, where the constants α and β are determined by the impurity. It is interesting to note that this linear relationship apparently holds only for a range of
small B values up to roughly B = 2.5 eV. The Δ value for Xe in Ne (see Table 3.1)
is considerably too low to fit into the linear dependence. Finally, we note that,

as for the pure materials, the quantum defect theory has also been applied for the impurity exciton states yielding a nice fit to the experimental data /Resca and Resta, 1979/.

Next we have to consider the line width of the intermediate and Wannier type atomic impurity excitations. Compared to the line width of the free gas atom these apparent absorption bands are considerably broadened (see Table 3.6). To illustrate one broadening mechanism we reproduce in Fig. 3.15 the change of the line shape with concentration in a "modified absorption" experiment for the n = 1 (3/2) and (1/2) states of an Xe impurity in Ne /Pudewill et al., 1976/. With higher concen-

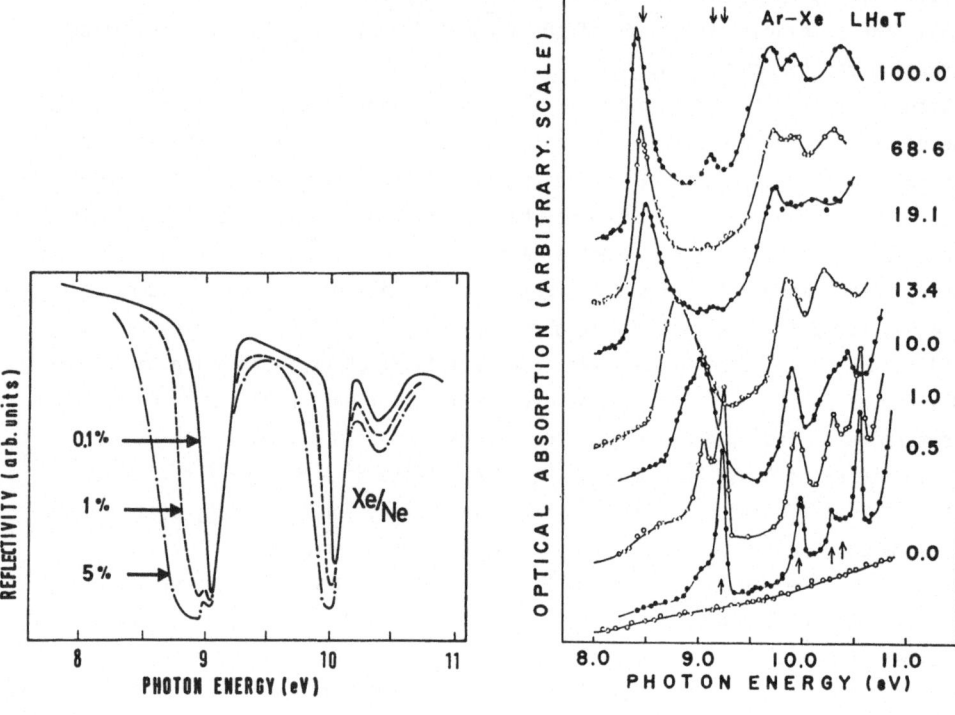

Fig. 3.15   Fig. 3.16

Fig. 3.15. Reflectance of Ne films doped with ≅ 0.1% (———), ≅ 1% (– – –) and ≅ 5% (—·—·) Xe, respectively evaporated onto a gold substrate in the range of the n = 1 Xe impurity excitations (from Pudewill et al. /1976a/)

Fig. 3.16. Optical absorption spectra of thin solid films of Ar-Xe mixtures deposited onto LiF plates cooled at 4.2 K. The number at the right side of each curve indicates the Xe concentration in mol%. The ordinate of each curve is shifted for the convenience of comparison. Upward arrows ↑ indicate the position of the localized Wannier exciton bands due to the Xe impurity /Baldini, 1965/ and downward arrows ↓ mark the intrinsic Wannier exciton bands of the undoped Xe solid /Haensel et al., 1970/ (from Nagasawa et al. /1972/)

tration of the guest atoms the line profile becomes broader, and a splitting of the band into two distinct subbands is observed for the (3/2) n = 1 band for Xe in Ne at a concentration of 1% (Fig. 3.15). This observation strongly suggests that in most of the impurity spectra broadening of the absorption bands due to formation of clusters of the impurity atoms starts already at fairly low impurity concentration around 1%.

As the work by Nanba et al. /1974/ on Xe in Ar matrices has shown, the study of the intensity variation of different bands with concentration in the low concentration regime is helpful for distinguishing between monomer and dimer absorption bands. For Xe in Ar Baldini /1965/ had observed a small shoulder on the low energy side of the n = 1 Xe impurity absorption band (see also the low concentration curve in Fig. 3.16 which is similar to the extra band appearing in the Xe/Ne mixture discussed above /Pudewill et al., 1976/. The origin of this band was not clear until Nanba et al. /1974/ investigated in detail the intensity variation of this band at about 9.03 eV. It has a quadratic dependence on the Xe concentration over the range 0 - 2% molar fraction of Xe, if one assumes that the n = 1 Xe impurity state has a linear intensity variation. Thus, it was concluded that the band at 9.03 eV is due to Xe dimers in the Ar matrix. We shall return to these experiments in the next section.

For a discussion of the lineshapes, the following line-broadening mechanisms, which are discussed in more detail in Chap. 6, should be considered:

1) Inhomogeneous broadening due to distribution of different trapping sites. This broadening effect results in a Gaussian lineshape. It appears that in well annealed samples of the molecular solids considered here this effect is small.

2) Electron-phonon interaction which originates from the displacements of the medium normal modes and from changes in the vibrational frequencies between the ground and the excited state. This problem can be handled by multiphonon theory. For weak electron-phonon coupling a zero phonon line will be exhibited, followed by multiphonon transitions on the high energy side, while for strong electron-phonon interaction a Gaussian lineshape will result. Experimentally, the zero phonon line has never been observed for rare-gas impurities in rare-gas matrices. For a molecular impurity, for example N_2 in Ne, Gürtler and Koch /1980/ were able to resolve a detailed fine structure in valence transition bands which they could assign to a zero phonon line, librational modes of the N_2 molecule and a selective coupling (depending on the symmetry of the excited state) to phonon modes of the Ne lattice.

3) Electronic relaxation between a high Wannier state to lower n states. This electron-phonon relaxation process results in a Lorentzian contribution of the lineshape.

For the lowest extravalence atomic impurity excitation only mechanism (2) is important, while for the higher Wannier states mechanisms (2) and (3) prevail. Provided that the electronic-nuclear coupling is strong we expect a Gaussian lineshape for the n = 1 excitation /Kubo and Toyozawa, 1955; Markham, 1959/.

3.3.2 Concentrated Rare-Gas Alloys

A challenging problem in the theoretical study of mixed molecular crystals involves the understanding of the band structure and the optical properties of heavily doped molecular crystals. Two limiting cases for the different types of mixed crystal systems may be distinguished /Onodera and Toyozawa, 1968; Hoshen and Jortner, 1972/: (i) in the case of the persistence-type mixed crystals, the energies of the electronic states of the two constituents remain almost unchanged against the change in composition. Thus, although the mixed crystal itself is homogeneous, the electronic properties corresponding to the two constituent substances persist. (ii) The second class of mixed crystals is usually called the amalgamation type. In this case, only one set of electronic states is observed for the mixed molecular crystal which is characteristic of the new material where the two substances are completely amalgamated.

Examples for the persistence type of mixed crystals are provided by excitons in halogen substituted alkali halides, e.g., KBr - KCl and for the amalgamation type by excitons in alkali-substituted alkali halides. The phenomenological parameter, which determines to which type a given mixed crystal belongs, is the ratio of the difference in the energies of two neighbouring valence bands of the two individual components to the width of these bands /Onodera and Toyozawa, 1968/. When this ratio is large two energy bands persist. In the opposite limit, when the band width is very large, the two bands merge into a single band characteristic for the amalgamation-type systems.

Only a few systems of concentrated RGS have been investigated so far; Nagasawa et al. /1972/ have measured the optical absorption of Ar-Xe solid solutions in the photon energy range 8 eV up to 11 eV, i.e., the range of Xe 5p excitations, whereas Haensel et al. /1973/ investigated the optical properties of rare-gas mixtures in the photon energy range 40 - 260 eV, i.e., in the region of Ne 2s, Ar 2p. Kr 3d and Xe 4d transitions. In addition to these optical investigations, Nürnberger et al. /1977/ have studied the band formation in Xe-Ar alloys by photoelectron spectroscopy.

In Fig. 3.16 we reproduce the absorption spectrum for Xe in Ar in the concentration range of 0 - 100 mole per cent (mol%) Xe /Nagasawa et al., 1972/. For low impurity concentrations the hydrogen-like $\Gamma(3/2)$ Xe impurity series with terms n = 1 up to n = 4 dominates the spectrum. Additionally, a shoulder at about 9.03 eV is observed which grows into a distinct separate peak in the 1 mol% spectrum, which has been ascribed to Xe-dimer absorption (see previous section). A gradual change of the spectrum starts at concentrations above 1 mol% Xe. The maxima of the Wannier series are broadened and a new broad maximum appears on the low energy side of the n = 1 peak. When the concentration of Xe increases to about 15 mol%, the spectrum undergoes severe changes and becomes similar to the spectrum of undoped solid Xe. The positions of the resolved structures connect smoothly to those of pure Xe. This observation suggests that the impurity Wannier exciton changes gradually from a localized excitation to a band-type excitation at higher concentrations up to solid Xe. Further, this observation is consistent with the many optical studies of valence excitons for alkali halide solid solutions. The concentration dependence of impurity absorption spectra was discussed theoretically by Cho and Toyozawa /1969/.

The study of the absorption spectra of solid solutions of rare gases in the extreme ultraviolet by Haensel et al. /1973/ was undertaken to check experimentally the interpretation given for the core level spectra of pure RGS. Selected results for Xe in Kr and Xe in Ar in the range of the Xe 4d excitation appear in Figs. 3.17,18. In most cases a continuous shift of the energetic position of the absorption peaks of the pure RGS upon addition of another rare gas or of N_2 was observed, thus indicating a dilute mixing of both components over the whole range of concentrations.

As mentioned in Sect. 3.2.3, the core excited spectra of pure RGS show excitonic as well as density-of-state effects. For instance, the features B and C in Figs. 3.17,18 have been interpreted as core excited Frenkel-type excitons. By comparing the experimental spectra with the calculated curves for the joint density of states /Rössler, 1971/, Haensel et al. /1973/ concluded that these Frenkel excitons in Kr 3d and Xe 4d are above the interband transition threshold. This conclusion is supported by the fact that the oscillator strength of these excitons in Kr and Xe is smaller than that of the corresponding gas lines owing to interactions with the underlying continuum. The concept underlying the analysis of core-excited optical spectra in terms of the conduction-band density of states has been described in Sect. 3.2.3. In their discussion of the effect of density of states on the spectra of rare-gas mixtures, Haensel et al. /1973/ used the known density of states in the pure crystals. For Xe-Kr as an example (Fig. 3.17), we note that the structures D, E, F, G and H and their spin-orbit partners shift with changing composition but do not split. They may be classified as almagamation-type structures /Onodera and Toyozawa, 1968/. Thus the pure Xe spectrum in the lowest part of Fig. 3.17 should change more and more and become finally the pure Kr spectrum in the upper curve of Fig. 3.17, a spectrum which is to be understood in terms of the density of states of the Kr conduction bands,weighted, of course, by the spin-orbit splitting of the Xe 4d initial states. A rigorous discussion of theoretical and experimental results is aggravated by the absence of density-of-states calculations for the mixed crystals. However, a comparison of the maximum shifts of the observed features with calculated maximum shifts of the five lowest conduction bands of Kr /Haensel et al., 1973/ supports the interpretation of the pure RGS spectra as due to density-of-states structures.

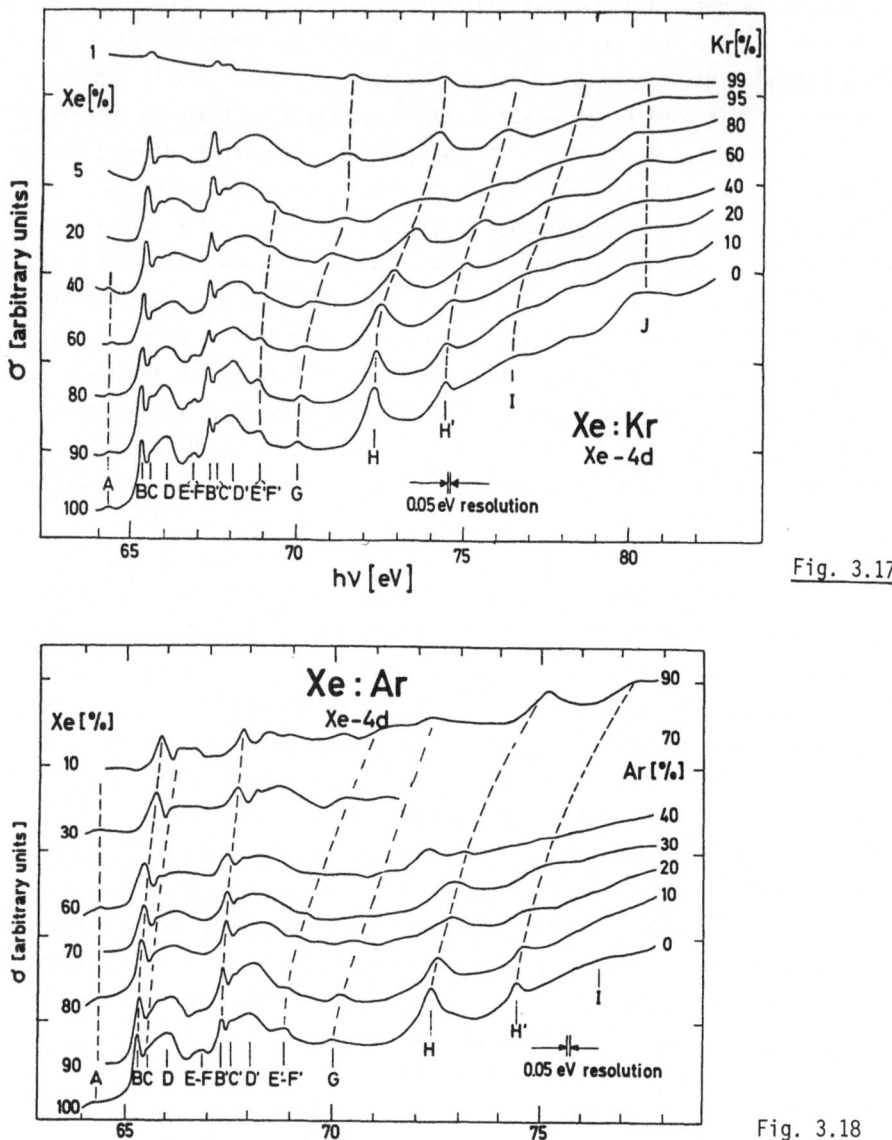

Fig. 3.17. Fine structure of the Xe 4d absorption in the energy range 64 - 90 eV
for different Xe : Kr mixtures ranging from 100 at.% to 1 at.% Xe. The corresponding
peaks are connected by (– – –) (from Haensel et al. /1973/)

Fig. 3.18. Fine structure of the Xe 4d absorption in the energy range 64 - 80 eV
for different Xe : Ar mixtures ranging from 100 - 10 at.% Xe. The corresponding peaks
are connected by (– – –) (from Haensel et al. /1973/)

The Xe-Ar mixtures (Fig. 3.18) as well as Kr-Ar mixtures also exhibit a widening of the spectra with increasing admixtures of the lighter rare gas. In contrast to this the conduction band density of states does not follow this trend when going from Xe or Kr to Ar. For example, the first and second density-of-states maxima as calculated by Rössler /1971/ are at a lower energy in Ar than in Xe or Kr, and the concept which led to a sound interpretation of the Xe-Kr spectra did not work here with the same success. Thus, we conclude again that the original hopes cherished for core excited spectra being a tool to map out conduction band density of states did not materialize. On the other hand, these results stress once more the importance of matrix element and relaxation effects in optical spectra which are beyond the simple one-electron picture employed here.

Further insight into the band formation in rare-gas alloy systems can be obtained from photoelectron spectroscopy. In Figs. 3.19,20,21 the results of such a study on Xe-Ar alloys are shown /Nürnberger et al., 1977/. In Fig. 3.19 photoelectron energy distribution curves for concentrations ranging from 0 - 100% are displayed and are measured by excitation with synchrotron radiation at $h\nu$ = 13.8 eV, 16.5 eV and 18.0 eV. With increasing Xe concentrations the gradual formation of Xe valence bands starting from the atomic Xe $5p_{1/2}$ and Xe $5p_{3/2}$ states is observed. Similarly, with Ar the 3p states are broadened with increasing Ar concentration. The width of the

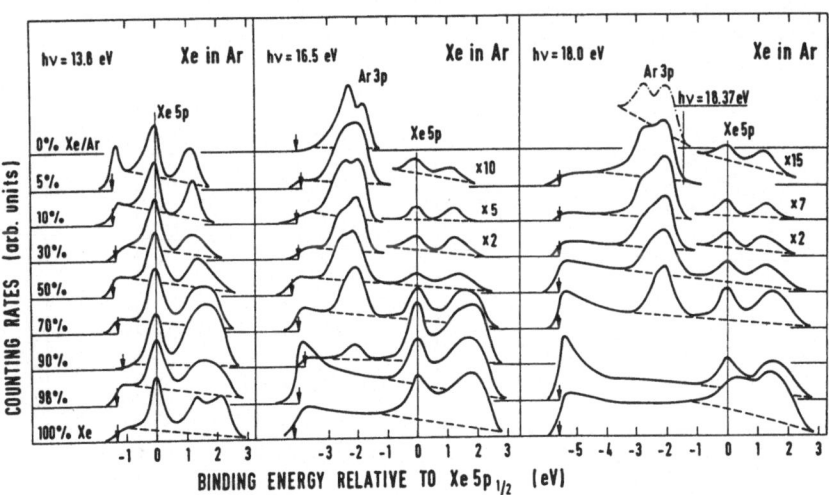

Fig. 3.19. Photoelectron energy distribution curves for Xe/Ar alloys for three different photon energies with the Xe concentration as a parameter. All spectra are normalized to the same maximum counting rate. The estimated background is indicated by (— — —). The energy scale for all EDC's has been fixed at the Xe $5p_{1/2}$ maximum. The arrows indicate the onset of photoemission at E_{kin} = 0. (— · — ·) for pure Ar ($h\nu$ = 18.37 eV) was taken from Schwentner et al. /1975/ (from Nürnberger et al. /1977/)

energy bands is comparable to their separation. Nevertheless, the energy bands corresponding to the two individual constituents are well separated and persist through all concentrations. It was noted by Nürnberger et al. /1977/ that rather high concentrations of Xe or Ar are necessary in order to reach the fully developed Xe or Ar valence bands, respectively.

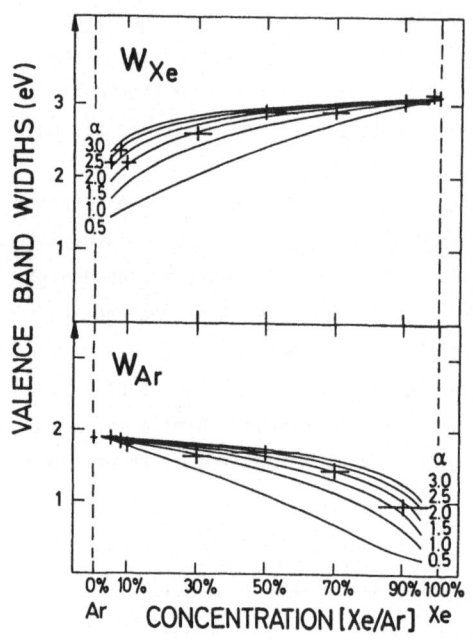

Fig. 3.20. Comparison of the measured total valence band widths in eV for Xe (W_{Xe}) and Ar (W_{Ar}) as a function of concentration including estimated errors (crosses) with calculated band widths. The free parameter α is used in the following expression: $pp\pi(c) = pp\pi(0) \cdot e^{-(d/\alpha)d_0}$ and an analogous expression for $pp\sigma$, with (c) and (0) denoting the band structure parameters at a given concentration and for the pure material. d_0 is the Xe nearest neighbour distance in pure Xe (from Nürnberger et al. /1977/)

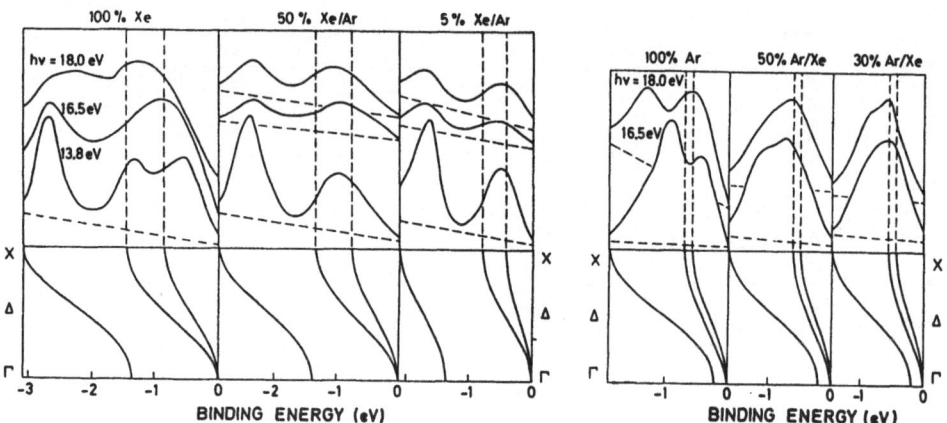

Fig. 3.21. Comparison of the experimentally determined photoelectron energy distribution curves for Xe at $h\nu$ = 13.8, 16.5, 18.0 eV (*left part*) and for Ar at $h\nu$ = 16.5 and 18.0 eV (*right part*) for three different concentrations with the calculated valence bands between the Γ and X point using α = 2 for the calculation (from Nürnberger et al. /1977/)

The concentration dependence of the band width has been compared to concentration dependent band-structure calculations. The results are displayed in Figs. 3.20,21. For simplicity, Nürnberger et al. /1977/ used band-structure calculations in the tight binding scheme /Slater and Koster, 1954/. The parameters ppσ, ppπ and the spin-orbit splitting ΔE_{SO} needed for the calculation have been determined from the photoelectron energy distribution curves of pure Xe by Rössler /1977/ to be ppσ = +0.33, ppπ = -0.038 and ΔE_{SO} = 1.37 eV. The change of ppσ and ppπ with concentration due to the reduced overlap integral of the wave function has been represented by an exponential decrease with increasing mean separation of the Xe atoms using the nearest neighbour distance in solid Xe d_0 multiplied by α as a reference length. The mean separation has been correlated with the concentration. In Fig. 3.20 a comparison of the measured valence band width with calculated band widths is shown using α as a free parameter. For the calculated band width the energy difference between Γ and X has been used because this is the largest width of the bands in the tight-binding approach given the above parameters. For both the Xe and Ar valence states a value of α = 2 agrees satisfactorily with the experimentally determined concentration dependence (Fig. 3.20).

In Fig. 3.21 the valence bands along the direction from the centre of the Brillouin zone Γ to the X point at the boundary are compared with the measured EDC's for Xe (left part) and Ar (right part) at three concentrations using α = 2 for the calculation. The common feature in Fig. 3.21 for the three-photon energies is the appearance of two maxima which correlate quite well with the maxima of the density of states of the lower valence band and of the centre of the density of states of the split upper valence bands. The maximum of the density of states is expected to lie at the flat region near the X point. The relative shift of the maxima with reduced Xe concentration is well reproduced by the position of the upper and lower bands at X. Changes in the EDC's with photon energy are attributed to structures in the final states. According to Fig. 3.21 such changes are only significant for pure Xe.

This observation can be taken as a hint that the conduction band states are disturbed by a relatively small concentration of guest atoms. Similar tendencies were found for the Ar valence bands but the smaller spin-orbit splitting yields a less clear-cut case for the separation of the different contributions.

Finally, we mention the calculation by Parinello et al. /1977/ who have attributed the considerable difference between the calculated band width for Xe and the width of the valence bands observed in photoelectron spectra to different relaxation energies of the relaxed holes for the Xe $5p_{3/2}$ and Xe $5p_{1/2}$ valence states.

On the basis of a two-band Koster-Slater model the following qualitative predictions for doped rare-gas samples have been made by Parinello et al. /1977/:
(i) For heavy rare-gas atoms (Xe) in light rare-gas matrices (Ar) no bound state is expected but at higher concentrations of the heavy atoms (Xe) the 3/2 state

should shift to higher energies, as observed in the experiment by Nürnberger et al. /1977/. (ii) On the other hand, a strong bound state is expected for the case of light impurities (Ar) in a matrix of heavy atoms (Xe). Therefore at small Ar concentrations in Xe a large apparent spin-orbit splitting is predicted. However, taking a width for the Ar states comparable with that of the Xe states in Ar, a much larger splitting than the real spin-orbit splitting of 0.2 eV does not fit the experimental results.

In summary, there seems to be an inconsistency insofar as absorption experiments favour an amalgamation-type classification of rare-gas alloys, whereas EDC measurements suggest a persistent-type model. Both results can, however, be reconciled if we recall that the photoemission experiments probe essentially the initial state density of states, which is persistent (Fig. 3.19). The absorption bands on the other hand are determined by both the initial and final states where the latter may introduce amalgamation-type character.

3.4 Molecular Wannier Impurity States in Rare-Gas Solids

The observation of intermediate and Wannier-type atomic impurity states in RGS and in fluids implies that analogous molecular excitations should be amenable to experimental observation. There has been a lively controversy as to whether molecular Rydberg states are amenable to observation in RGS /Robin et al., 1968; Katz et al., 1969/. Extensive experimental work /Gedanken et al., 1973a-e; Jortner, 1974/, some of which is presented in Figs. 3.22,23, has established that intermediate-type and higher Wannier-type molecular excitations are observed in RGS. The following features of molecular Rydberg states in RGS should be emphasized:

1) The first molecular extravalence excitation corresponds to an intermediate case. The tight-binding description in terms of the first molecular Rydberg state subjected to large non-orthogonality correlations implies large blue spectral shifts of the lowest molecular Rydberg state in the matrix /Jortner, 1974/. Furthermore, extensive line broadening is exhibited originating from phonon broadening and also from some inhomogeneous broadening effects /Jortner, 1974/.

2) Higher Rydberg states, corresponding to an $n = 2$ Wannier state of an electron bound to a positive ion in an RGS, were observed /Katz et al., 1969; Gedanken et al., 1973a-e/.

3) Vibrational structure for all n molecular intermediate and Wannier-type excitations was observed /Katz et al., 1969; Gedanken, 1973a-e/. This spectroscopic feature of molecular extravalence excitations in a dense medium corresponds to vibrational excitations of the positive ion core.

4) Overlap of the molecular Wannier-type excitations with other broad structure-less transitions, which correspond to dissociative states, complicate the observation of molecular extravalence excitations in condensed phases.

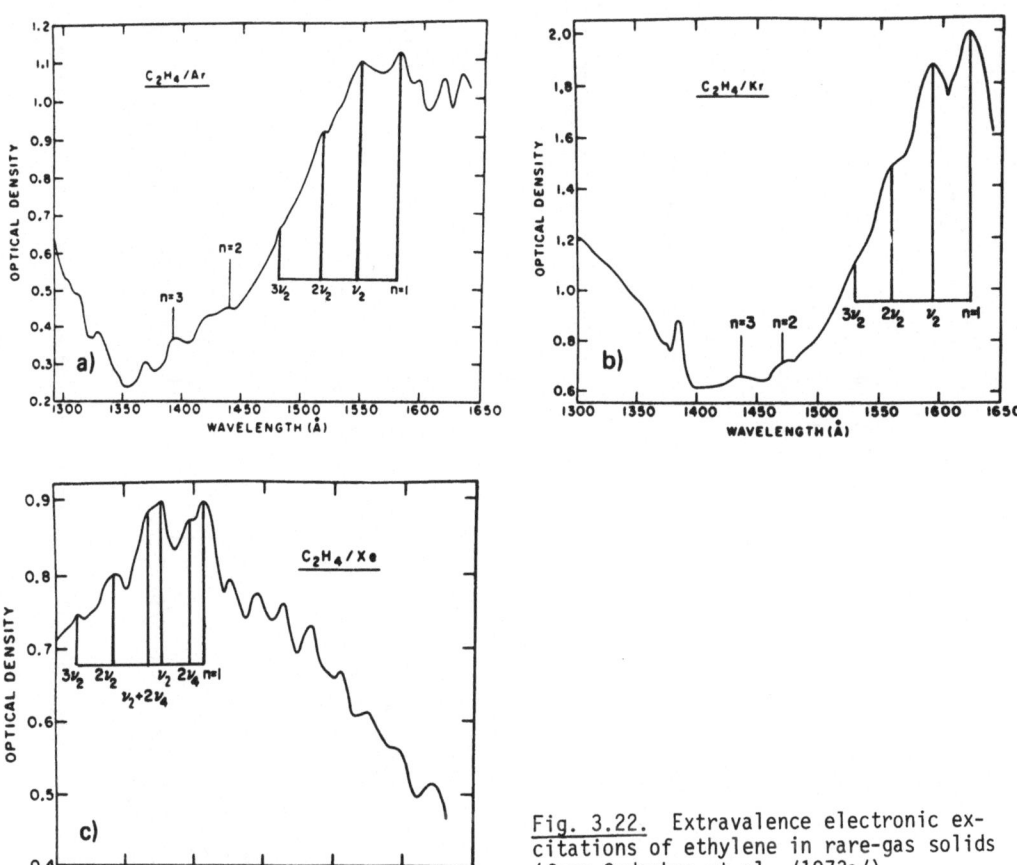

Fig. 3.22. Extravalence electronic excitations of ethylene in rare-gas solids (from Gedanken et al. /1973a/)

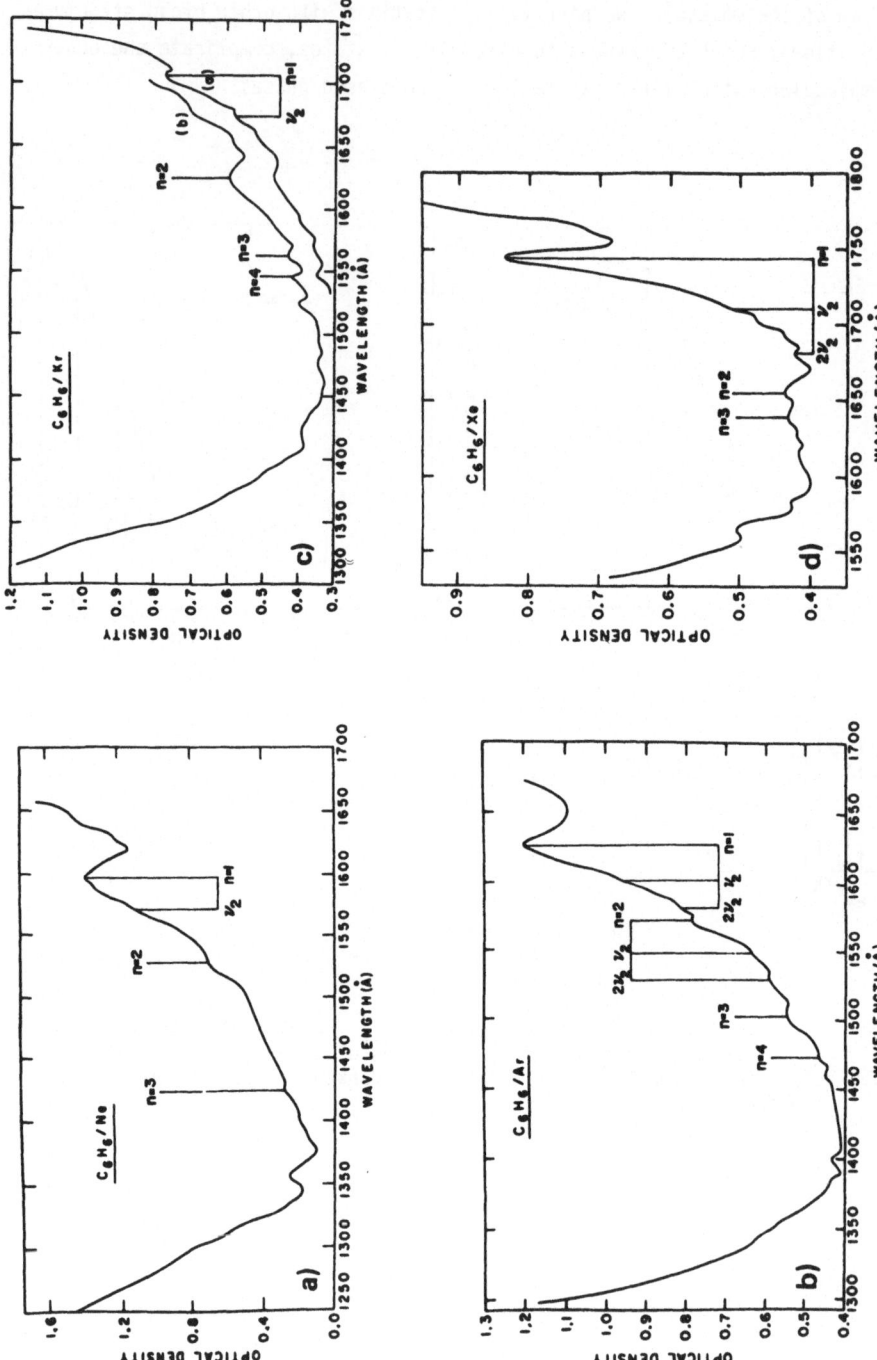

Fig. 3.23. Extravalence electronic excitations of benzene in rare-gas solids (from Gedanken et al. /1973a/)

4. Electronic Excitations in Rare-Gas Liquids

The central features of bound one-electron excitations in pure RGS and in dilute rare-gas alloys are characterized by Frenkel-type exciton states, Wannier-type excitons, or impurity states converging to the bottom of the conduction band (see previous chapter). As we have seen, the theoretical treatment of such bound excited states involves essentially the translational symmetry of the crystal lattice (e.g. /Knox, 1963; Phillips, 1966; Rice and Jortner, 1967/). These solid state concepts break down when excitons and impurity states with extended electron orbits are considered in the liquid phase. The following general questions arise in relation to bound excited electronic states of simple dense fluids:

i) Do exciton states "exist" in liquids, how should such states be described, and what physical information can be gathered from the optical spectrum? More specifically, can tightly bound excited states of a liquid be described in terms of a dispersion relation appropriate for the disordered system? Such Frenkel type states have a unique parentage in the states of the constituents and will thus be definitely experimentally observed. Can Wannier-Mott-type exciton and impurity states be experimentally observed in a liquid? Furthermore, if Wannier states exist in a liquid, it is interesting to inquire what is the relation between these liquid electron-hole pair states and the conduction-band states in the disordered system.

ii) How do modifications of interatomic interactions affect the spectrum of these excited states? Studies of the density and temperature dependence of optical spectra from pure and doped liquids may shed some light on this question.

It was suggested by Rice and Jortner /1966/ and by Rice, Nicolis and Jortner /1968/ that large radius Wannier-type exciton states are amenable to experimental observation in a dense fluid where electron-medium interaction is sufficiently weak. Extensive theoretical studies of Frenkel-type tightly bound exciton states in liquids were performed by Rice and coworkers /Popielawski and Rice, 1967; Nicolis and Rice, 1967; Fischer and Rice, 1968/.

Pure rare-gas liquids and impurities in rare-gas liquids provide suitable candidates for the study of excitons in liquids and are of interest in relation to the

Table 4.1. Absorption-, reflection- and photoconductivity experiments on pure rare-gas liquids and atomic and molecular impurity states in rare-gas liquids. For the spectroscopic results see text

System	Method	Spectral range	Technical remarks	Reference
pure liq. He	reflectance	12.4 eV – 23.9 eV	windowless reflectance cell, He bath	Surko et al. /1969/
pure liq. Xe	reflectance	7.8 eV – 11.0 eV	LiF window, 165 K	Beaglehole /1965/
	reflectance	7.9 eV – 9.6 eV	LiF window	Asaf and Steinberger /1971/
	reflectance	7.7 eV – 10.3 eV	LiF window, temperature dependence of exciton line positions	Steinberger and Asaf /1973/
	photoconductivity	6.5 eV – 10.0 eV	LiF window, variation of electric field and light intensity	Roberts and Wilson /1973/
	photoconductivity	8.0 eV – 10.3 eV	MgF_2 window, interlaced combs of gold electrodes	Asaf and Steinberger /1974/
	reflectance	7.5 eV – 10.8 eV	MgF_2 window, density dependence	Laporte and Steinberger /1977/
	electron injection	–	metal substrate, V_0 analysis	Reininger et al. /1982/
	photoconductivity	8.0 eV – 11.8 eV	MgF_2	Reininger et al. /1983/
pure liq. Ar	reflectance	11.3 eV – 12.3 eV	LiF window, density dependence	Bernstorff et al. /1983/
Xe in liq. Ar / in liq. Kr	absorption	8.4 eV – 10.6 eV	LiF window	Raz and Jortner /1970a,b/
Xe in liq. Ar	absorption	8.2 eV – 9.1 eV	MgF_2 window, Ar density variation 0.1– 1.4 gcm^{-3}	Messing et al. /1972e/
Xe in liq. Ar / in liq. Ne / in liq. He	absorption density dependence high pressure	7.7 eV – 10.5 eV	MgF_2 window, T = 24 – 300 K pressures up to 400 atm.	Messing et al. /1977a,f/
in liq. Ar	photoconductivity	10.4 eV – 11.8 eV	LiF window, density dependence	Reininger et al. /1984/
CH$_3$I in liq. Kr	absorption	6.1 eV – 8.3 eV	LiF window	Gedanken et al. /1973a/
CH$_3$I in liq. Ar / H$_2$CO in liq. Ar	absorption	6.1 eV – 7.3 eV	MgF_2 window, Ar density variation 0.1– 1.4 gcm^{-3}, T = 90 – 300 K, medium effects on vibrational structure	Messing et al. /1977b/
CH$_3$I in liq. Ar / in liq. Kr / CS$_2$ in liq. Ar / H$_2$CO in liq. Ar	absorption	6.1 eV – 10.3 eV	MgF_2 window, medium density variation, perturbation of molecular extravalence excitations	Messing et al. /1977d/

above-mentioned general problems as well as for the investigation of ionization processes in simple liquids (see, for example, /Schmidt, 1974; Tauchert, 1975; Jortner and Gaathon, 1977/). For liquid Ar and Kr electron mobility data /Schnyders, Rice and Meyer, 1966; Miller, Howe and Spear, 1968/ indicate that the electron-atom interaction is sufficiently weak and that the transport properties of an excess electron can quite well be described in terms of disorder scattering of a nearly free electron. Although transport properties do not give conclusive evidence about the details of the density of conduction-band states in a disordered system /Edwards, 1965/, it appears that a reasonable working hypothesis is that the conduction-band states in liquid rare gases can be taken as free electron-like, or nearly so.

Furthermore, in rare-gas liquids, the weak electron-atom coupling /Lekner, 1967; Cohen and Lekner, 1967; Springett, Jortner and Cohen, 1968/ ensures that the line widths of the Wannier states will not be excessive.

Quantitative experimental information concerning Wannier exciton states in liquid rare gases is still meagre and, so far, only a few groups have tackled this problem. A summary of the optical absorption and reflection experiments performed up to date on both the pure liquids and impurities in rare-gas liquids appears in Table 4.1. The theoretical and experimental information concerning atomic- and molecular-type states in liquid He, Kr, Ar and Ne, and especially in liquid He, will be considered in Chap. 6 where we shall discuss results from transient absorption and luminescence experiments. In this chapter, we focus attention on excitons in pure liquid Xe, Xe impurity states in liquid Kr, Ar and He and, in particular, on the density dependence of these states studied by optical absorption and reflection measurements. Further, we discuss one example for molecular impurity states in liquid rare gases. Finally, the electron affinities derived from the experimental data for solid and liquid rare gases are compared with calculated V_0 values.

4.1 Excitons in Pure Liquid Rare Gases

Beaglehole /1965/ reported the first optical experiment on rare-gas liquids in the VUV. Reflection measurements on Xe in a LiF-windowed cell were performed in the region 7.8 - 11.0 eV covering the range of the Wannier-type exciton lines observed for solid Xe. The main n = 1 $(6s_{3/2})$ exciton line at around 8.4 eV in solid Xe was found to shift to lower energies and to broaden as the temperature was raised becoming quite markedly wider in the liquid. More important was the observation of a weak and unresolved band around 9.0 eV in the liquid, which was taken as evidence for the fact that Wannier states n $\geq$ 2 do exist in the liquid phase, but, as expected in a much broadened form.

Optical spectra of pure liquid Xe have subsequently been studied in detail by Steinberger and coworkers (see Table 4.1). In a careful study by Steinberger and

Asaf /1973/ spectra of solid and liquid Xe covering the range 7.75 eV to 10.3 eV
have been obtained at temperatures between 6 K and 160 K. Throughout the temperature range, the peak positions of the n = 1 (3/2) and n = 1 (1/2) excitons, as well
as the n = 2 and n = 3 (3/2) Wannier excitons, were found to shift to lower energies with increasing temperature. Steinberger and Asaf /1973/ inferred from the
linear shift of the peak positions as a function of density that these shifts are
caused mainly by the change of energy band edges with thermal expansion. In particular, the experimental points for the liquid fit these linear functions very well.

The influence of the density on the optical spectra was further studied for pure
liquid Xe in the density range from near the critical point to the triple point by
Laporte and Steinberger /1977/. The experiment covered the spectral range from 7.5
to 10.8 eV, that is the atomic transitions in Xe which are observed at wavelengths
above the MgF_2 cutoff: 8.44 eV (1469 Å), 9.57 eV (1295 Å) and 10.4 eV (1192 Å). For
each atomic line, two bands have been observed, i.e., one broadened and somewhat
red-shifted band corresponding to the atomic transition, and another distinct band
corresponding in position to that of the exciton observed in the triple point liquid. These greatly broadened excitonic bands were observed in the liquid near the
triple point at 8.12 eV (1526 Å), 9.35 eV (1326 Å) and 10.34 eV (1199 Å) where the
values are those of the reflectivity maxima. As an example, we present in Fig. 4.1
the evolution of the 8.33 eV excitonic band with increasing density. Fig. 4.1 represents the reflectivity spectra of an Xe/MgF_2 interface at several densities near
the critical point as well as near the triple point. The gradual rise of the excitonic band at 8.33 eV at the expense of the perturbed atomic band at 8.12 eV is evident. Laporte and Steinberger /1977/ performed also a dispersion analysis for the
two $^3P_1$ bands. They assumed Lorentzians for ε_1 and ε_2 in order to obtain via a fit
to the experimental reflectivity data the optical constants ε_1, ε_2, n and k as a
function of photon energy (see the solid curve in Fig. 4.1). They interpreted their
results as evidence for exciton transitions appearing in momentary clusters of the
fluid.

Recently Bernstorff et al. /1983/ have extended the spectral range for optical
studies on liquids up to ≈ 12.3 eV by using a cooled LiF window. Thus they were
able to investigate the first two excitons in Ar separated by the spin-orbit splitting. The method of data analysis from the measured reflectance spectra was very
similar to the one described above for liquid Xe. Again the atomic lines were found
to be broadened with increasing density without being shifted. The excitons appear
at $\rho \approx 2.5 \cdot 10^{21}$ cm^{-3} (n' = 1) and $\rho \approx 6.3 \cdot 10^{21}$ cm^{-3} (n = 1), respectively. Their
energies (maxima in ε_2) show a monotonic shift with increasing density (Fig. 4.2)
and the intensity ratio changes continuously with density from 1:3 (atom) to roughly
1:1 (solid). As shown in Fig. 4.2 the spin-orbit splitting stays constant over the
whole density range.

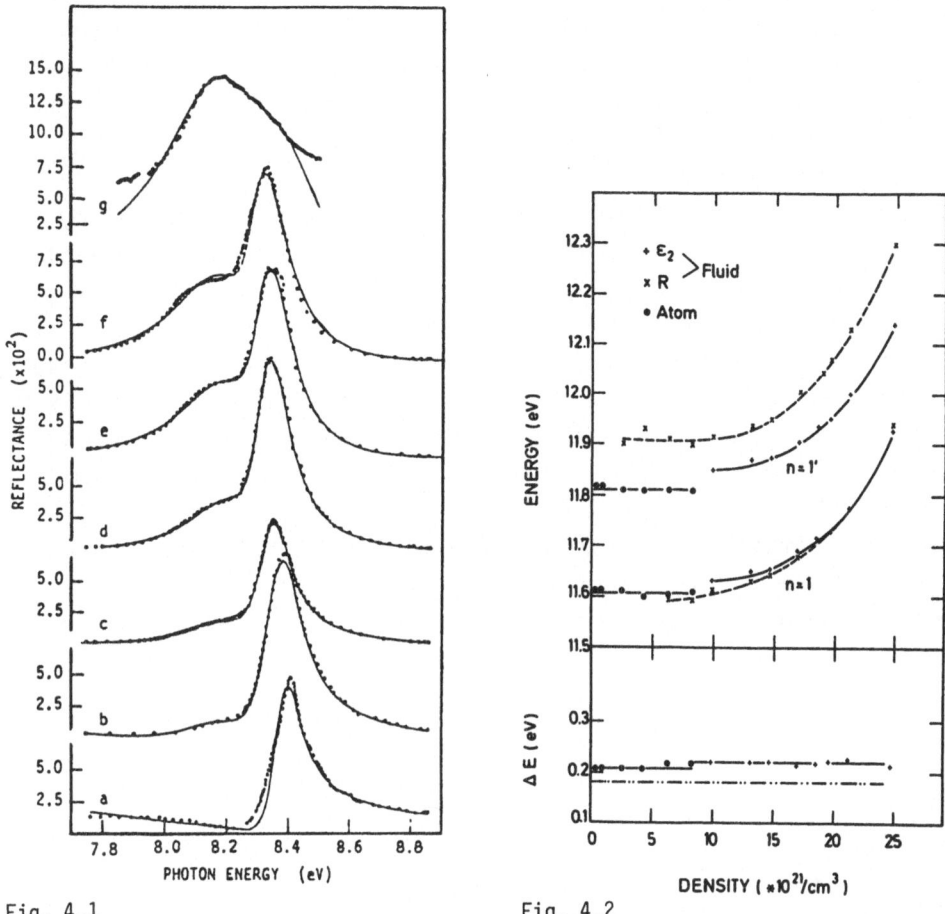

Fig. 4.1   Fig. 4.2

Fig. 4.1.   The reflection of fluid Xe in the range of the $6s_{3/2}$ excitation at various densities. The experimental values are given by the points. The full curves represent the results of a dispersion analysis. The curves a to g correspond to densities ranging from 182 to 500 amagat. The corresponding densities (amagat) pressures (bar) and temperatures (K) are: (a) 182, 61.8, 292.7;   (b) 241, 65.7, 293.9; (c) 283, 61.8, 287.8;   (d) 306, 61.8, 283.4;   (e) 338, 61.8, 274.4;   (f) 350, 83.4, 275.0, and (g) 500, 0.8, 163.0 (from Laporte and Steinberger /1977/)

Fig. 4.2.   Density dependence of the exciton energies in Ar. (Upper part) excitation energies for the n = 1 and n = 1' excitons as measured in reflection experiments (R). The derived peak-positions in the imaginary part of the dielectric function (ε_2) are compared with the atomic excitation energies $5p^6 \to 5p^56s$, $5p^56s^1$. The highest density corresponds to solid Ar. (Lower part) energy separation of the two excitons in the ε_2 spectra and of the two atomic lines. For comparison the spin-orbit splitting of the $5p^6$ ground state in the atom as well as in the solid is given and denoted by (—··—··) (after Saile et al. /1984/)

The only other piece of information concerning optical spectra of excitons in pure rare-gas liquids pertains to liquid He. Here the free exciton states are less well known. They can be derived from the one available reflection spectrum taken by Surko et al. /1969/. It shows two maxima, a strong one at 21.4 eV (580 Å) corresponding to the strongly allowed transition to the $2p^1P_1$ state (at 584 Å in the gas) and a very weak one at 20.8 eV (597 Å) corresponding to the forbidden $2s^1S_0$ transition (at 602 Å in the gas).

Photoconductivity in liquid Xe has been studied by Roberts and Wilson /1973/ and Asaf and Steinberger /1974/. The latter authors found in photoconductivity excitation spectra a threshold of 9.20 eV and a dip at 9.45 eV very near the peak of the competing $n = 1$ $\Gamma(1/2)$ exciton transition. From these results a band gap $E_G = 9.22 \pm 0.01$ eV was determined for liquid Xe. This value is very near the estimate (9.24 eV) based on the density change upon melting. Combining the value for E_G with the measured excitation energy of the $n = 2$ $\Gamma(3/2)$ exciton yielded an exciton binding energy $B = 1.08$ eV and an effective exciton reduced mass $\mu = 0.27$ (in units of electron mass).

In a recent investigation for Xe /Reininger et al., 1982,1983a,1984/ it has been possible to follow the onset of photoconductivity E_{pc} from the dense gas phase through the liquid phase to the solid. A smooth variation with density was observed (Fig. 4.3).

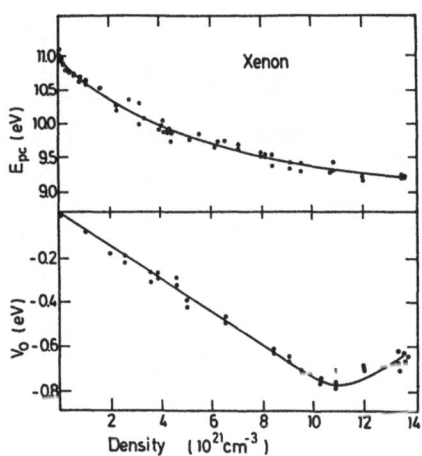

Fig. 4.3. Density dependence of the photoconductivity threshold (*upper part*) and of the energy V_0 of an electron at the bottom of the conduction band with respect to the vacuum level (*lower part*) (after Reininger et al. /1983a,d/)

4.2 Xe-Impurity States in Liquid Rare Gases

The identification of "high" n = 2 Wannier-type exciton states in liquid rare gases
rests on a comparison (perhaps somewhat over-simplified) of the absorption bands
with the optical properties of the corresponding solid. In order to establish the
nature of Wannier states in dense simple fluids, it is of interest to obtain a continuous description of the electronic excitations of a guest atom or molecule in a
host fluid where the fluid density is varied over a broad density range. In this
context, three types of electronic transitions of the guest atom should be considered:

1) *Intravalence excitations*. Studies of spectral perturbations of such transitions
by foreign gases at low and moderate densities reveal (see, for example, /Robin et
al., 1956; Granier, 1969/): (i) Small red or blue spectral shifts and line broadening, (ii) the appearance of "red" or "blue" satellites due to the formation of a
van der Waals diatomic molecule, as well as, (iii) symmetry breaking effects, i.e.
the appearance of symmetry forbidden transitions. At high fluid densities features
(i) and (iii) are expected to prevail while the spectral satellites and resonance
line will merge into a single broad band.

2) *Low extravalence excitations*. These transitions are accompanied by a change
$n_a = 1$ in the principal atomic quantum number n_a. The features of such transitions
(e.g., the $5p^6 \rightarrow 5p^5 6s$ transition in Xe) of rare-gas atoms are: (i) Spectral shifts
to the red at low density and to the blue at high density and line broadening /Rupin
et al., 1967; Messing et al., 1977a-e/, (ii) the appearance of "blue" spectral satellites, originating from continuum-continuum and bound-continuum transitions between the diatomic adiabatic potential surfaces of the guest-host pair /Granier et
al., 1967; Castex, 1974/, and (iii) induction of symmetry-forbidden transitions. At
higher densities of the host fluid large blue shifts are exhibited, the satellites
are smeared out by overlapping the resonance line, while the symmetry breaking effect (iii) is also operative. Apart from quantitative differences regarding the
large magnitude of spectral blue shifts, the gross features of the lowest extravalence transitions of rare gases are qualitatively similar to those of intravalence excitations.

3) *High extravalence excitations*. Such transitions are characterized by $n_a \geqslant 2$.
Only very little is known concerning spectral perturbations of such transitions at
moderate and át high densities of the host fluid and questions arise, such as:
What are the spectral shifts of higher atomic Rydberg states at low and at moderate
densities of the host fluid? Or, can one define a critical density of the host fluid
where these high extravalence excitations disappear being replaced by the Wannier-
type states of the dense medium? Or, alternatively, do they converge to the n = 2
Wannier states?

In an attempt to understand these problems, Messing et al. /1977a/ have conducted
an extensive experimental study of the absorption spectrum in the energy range 8.3 eV
to 10.8 eV of Xe guest atoms perturbed by Ar, Ne and He over a density range of the
host from the atomic limit to the liquid density. The VUV absorption spectra of Xe
with an initial impurity concentration of 0.5 to 1 ppm in Ar, Ne and He have been
studied using the absorption cell as described in Fig. 2.1. We shall discuss here,
as an example, the results obtained for the Xe/Ar system.

In Figs. 4.4 , 5 the experimental results for the absorption spectra of Xe in liquid Ar are shown at low and at intermediate densities (ρ = 0 - 0.6 gcm^{-3}) (Fig. 4.4)

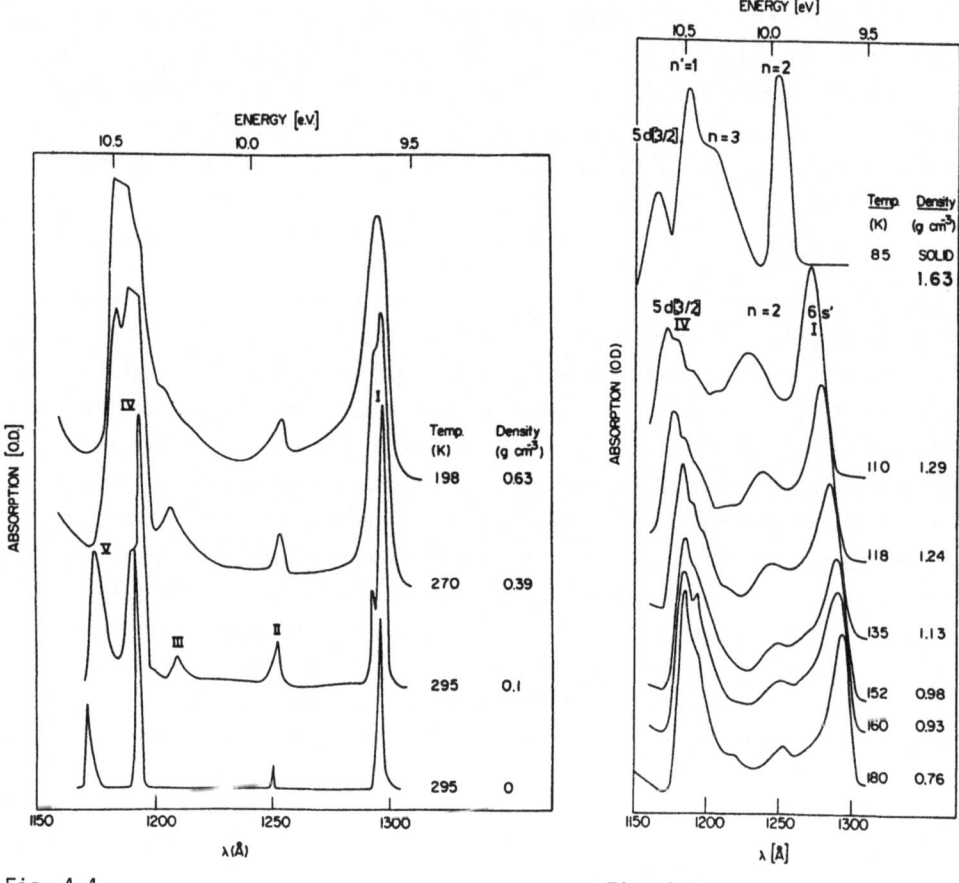

Fig. 4.4

Fig. 4.5

Fig. 4.4.   The absorption spectra of Xe in fluid Ar in the low and intermediate Ar
density range. The absorption curves are horizontally shifted, all being displayed
on the same optical density (OD) scale (from Messing et al. /1977a/)

Fig. 4.5.   The absorption spectra of Xe in fluid Ar at high Ar density and in solid
Ar. The Xe concentration is roughly 1 ppm. The absorption curves are horizontally
shifted, all being displayed on the same optical density (OD) scale (from Messing
et al. /1977a/)

and at high Ar densities (Fig. 4.5), respectively. The following electronic excitations of the guest Xe atom were observed in the spectral range from 9.5 eV (1300 Å)
to 10.8 eV (1150 Å) in order of increasing energy. The transition energies are given
in the atomic limit using Moore's notation /Moore, 1958/:

Transition (I) to 6s'[1/2]J = $1(5p^5\{^2P_{1/2}\}6s)$ at 9.57 eV (1296 Å).
Transition (II) to 5d[1/2]J = $1(5p^5\{^2P_{3/2}\}5d)$ at 9.92 eV (1250 Å).
Transition (III) to 5d[5/2]J = $3(5p^5\{^2P_{3/2}\}5d)$ at 10.25 eV (1210 Å), which are
   symmetry-forbidden in the atomic limit /Castex, 1974/.
Transition (IV) to 5d[3/2]J = $1(5p^5\{^2P_{3/2}\}5d)$ at 10.40 eV (1192 Å).
Transition (V) to 7s[3/2]J = $1(5p^5\{^2P_{3/2}\}7s)$ at 10.60 eV (1170 Å).

Transitions (I) - (IV) correspond to the low extravalence excitations with $\Delta n_a = 1$,
while transition (V) originates from the second member $\Delta n_a = 2$ of a Rydberg series
which starts with the first resonance line 6s[3/2]J = $1(5p^5\{^2P_{3/2}\}6s)$ at 8.44 eV
(1470 Å). The following medium effects have been observed at low and moderate densities (Fig. 4.4).

1) *Satellite bands.* In the density range 0.1 - 0.4 gcm^{-3} a single blue satellite
band accompanies each of the transitions (I) and (IV). The intensity of these blue
satellites increases with increasing density until at Ar density ≈ 0.6 gcm^{-3} they
merge together with the corresponding resonance line into a single band.

2) *Symmetry breaking effects.* The transition (III) to a J = 3 state is symmetry
forbidden in the atomic limit, as is evident from Fig. 4.4, being induced by the
host fluid. No other symmetry-forbidden, pressure-induced transitions were observed
in the spectral region 9.5 eV to 10.8 eV with the possible exception of the broad
structureless absorption exhibited around 9.6 eV to 9.8 eV which may originate from
pressure induced transitions to states of the $5p^56p$ configuration /Castex, 1974/.

3) *Medium-induced shifts of the lowest $\Delta n_a = 1$ extravalence excitations.* These
shifts are summarized in Fig. 4.6. The Xe resonance lines 6s[3/2] (not shown in Fig.
4.6) and 6s'[1/2] (Transition I) are weakly red-shifted at low densities, exhibiting
a large blue-shift at higher densities. A similar behaviour is observed for transition (IV). These features have been explained in terms of a strongly repulsive
Xe-Ar interaction potential in the excited state /Messing et al., 1977a/. In contrast, transition (II) exhibits only a small red-shift throughout the low and the
intermediate density region.

4) *Medium-induced shifts of the higher $\Delta n_a = 2$ extravalence excitation.* The spectral shift of transition (V) to the 7s[3/2] state drastically differs from those revealed by the 6s[3/2]and 6s'[1/2] transitions. While the latter low extravalence
excitations exhibit predominantly a large spectral blue shift (Figs. 4.4 , 6), the

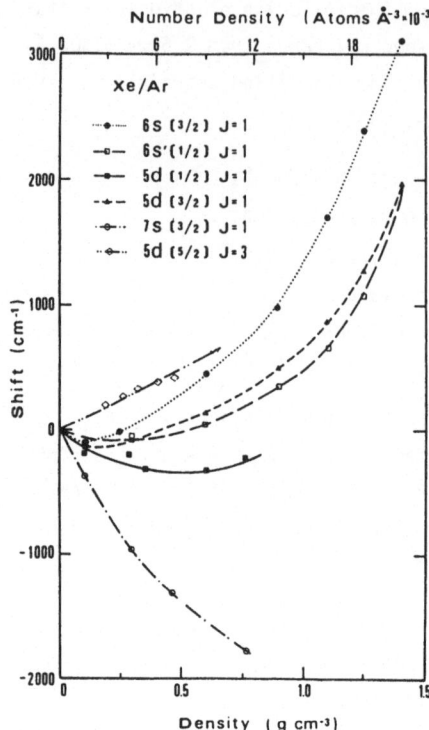

Fig. 4.6. The density dependence of the energies of the maxima of the absorption bands relative to the unperturbed atomic transition energy for the absorption lines of Xe in fluid Ar observed in the spectral range 8.5 eV ($\cong$ 1460 Å) to 10.8 eV (1150 Å); see Figs. 4.4, 5. The spectral lines are designated by the conventional atomic notation (from Messing et al. /1977a/)

former "high" extravalence transition reveals an appreciable spectral red shift, even at high densities, overlapping the blue-shifted band (IV) at a density of $\approx$ 0.4 gcm^{-3} and appearing as a low energy shoulder of band (IV) at a density of $\approx$ 0.6 gcm^{-3}.

In the high density range (Fig. 4.5) the first transition at $\approx$ 9.7 eV (1265 Å to 1290 Å) was assigned by Messing et al. /1977a/ to an excitation of the 6s'[1/2] state (transition I) and the third main transition at $\approx$ 10.5 eV (1170 Å to 1190 Å) to the 5d[3/2] state (transition IV). These transitions exhibit large blue shifts (Fig. 4.6) at high densities.

Medium perturbations of the lowest Δn_a = 1 extravalence excitations in the medium and high density range are manifested by blue spectral shifts and by appreciable line broadening. These spectral perturbations of atomic-type excitations can be accounted for in terms of the statistical theory /Morgenau, 1951/ as modified to account for correlation effects between the positions of the perturbating atoms /Saxton and Deutsch, 1974; Messing et al., 1977e,f/. The line-shape, L(ω), at the frequency ω is given by

$$L(\omega) = \frac{1}{2\pi} \int_{-\infty}^{\infty} dt \exp\left[it(\omega - \omega_0)\right] f(t) \tag{4.1}$$

where ω_0 is the excitation energy of the free guest atom, while the generating function, $f(t)$, can be expressed in terms of an exponential density expansion /Messing et al., 1977e,f/

$$f(t) = \exp[A_1(t) + A_2(t) + \dots] \tag{4.2}$$

with the leading terms in the exponential being

$$A_1(t) = \int d^3R_1 \; F_1(R_1) \; U(R_1)$$

$$A_2(t) = \frac{1}{2} \int d^3R_1 \; d^3R_2 \; F_2(R_1,R_2) \; U(R_1) \; U(R_2) \tag{4.3}$$

where

$$U(R_j) = \exp[-i\Delta V(R_j)t] - 1 \tag{4.4}$$

being expressed in terms of the difference of the guest-host pair potentials, $\Delta V(R_j)$, between the excited and the ground states.

The functions

$$F_1(R_1) = \rho g_{12}(R_1)$$

$$F_2(R_1,R_2) = \rho^2 [g_{11}(|R_1 - R_2|) - 1] \; g_{12}(R_1) \; g_{12}(R_2) \tag{4.5}$$

at the density ρ are the Ursell distribution functions, being determined by the guest-host radial distribution function $g_{12}(R_1)$, and by the guest-guest radial distribution function $g_{11}(R)$. This semiclassical theory of line-shapes /Messing et al., 1977d,e/ provides a proper account for the marked temperature dependence of the spectral shifts and line broadening of the absorption spectra of Xe in Ar. This theory also provides reliable estimates of the first and second moments of the absorption bands. A reasonable account of the entire line profile has been provided (Fig. 4.7) using the exponential density expansion of the generating function up to second order. The statistical theory can also be utilized as well for the calculation of spectroscopic parameters and line profiles of emission spectra of "atomic" impurities.

The most interesting feature of the spectra of Xe in Ar (Fig. 4.5) is the absorption band at 9.9 eV to 10.2 eV. This second band centred at 10.12 eV (1225 Å) in the high density fluid was assigned by Messing et al. /1977a/ to an n = 2 Wannier state, because it is close in energy to the n = 2 Wannier impurity state in solid Xe/Ar alloys which peaks at 9.95 eV (1246 Å) /Baldini, 1965/. As Messing et al. /1977a/ pointed out, this is only a qualitative argument in favour of the n = 2 Wannier exciton assignments and they therefore also considered alternative assignments. They pointed out that the absorption intensity around 10.1 eV in the Xe/Ar

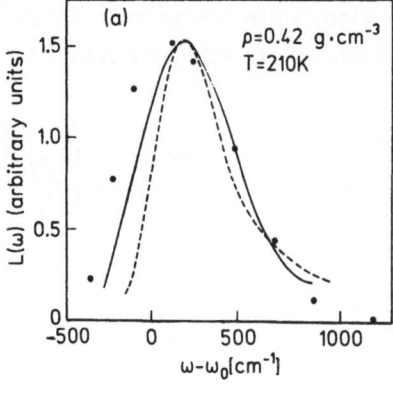

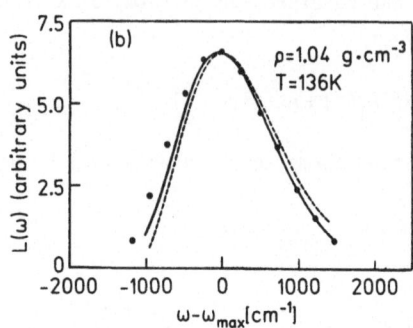

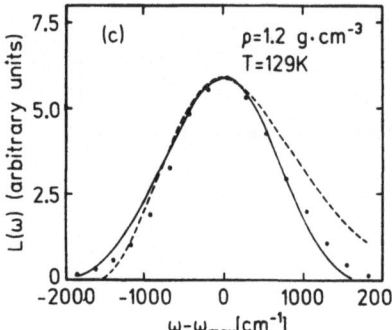

Fig. 4.7a-c. Experimental (points) and calculated lineshape (————,— — —) for the
$^1S_0 \rightarrow {}^3P_1$ transition of Xe in liquid Ar. The
panels a, b, and c correspond to low, medium,
and high density with $\hbar\omega_{max}$ of 68045 cm^{-1},
69450 cm^{-1}, and 70100 cm^{-1} respectively. The
calculated curves with generating functions
$f(t) = \exp[A_1(t)]$ (— — —) and $f(t) =
\exp[A_1(t) + A_2(t)]$ (————) incorporate two
body and three body correlation effects
(from Messing et al. /1977e/)

system at high ($\geqslant 7 \cdot 10^{-3}$ atoms Å^{-3}) densities is anomalously enhanced, indicating
the appearance of a new absorption band at these densities. This behaviour is unique
to the Xe/Ar system and was not observed for Xe/Ne and Xe/He in the same range of
atomic densities. It was taken as a strong support for the n = 2 exciton assignment. The assignments are summarized in the correlation diagram in Fig. 4.8.

The assignment of the n = 2[3/2] Wannier impurity state of Xe in liquid Ar at
10.12 eV enables one to draw an energy correlation diagram which relates the high
extravalence atomic excitations and the "high" Wannier states (Fig. 4.8). With the
n = 2 exciton, the ionization energy and polarization energies for the electrons
and Xe ions (hole states) for Xe in liquid Ar have been estimated (see Table 4.2)
/Raz and Jortner, 1970a,b; Messing et al., 1977a/. Furthermore, Messing and Jortner
/1977c/ have calculated the density dependence of the polarization energy P_+ of Xe$^+$
in Ar using Lekner's screening function /Lekner, 1967/ and, alternatively, a simple
Born charging formula.

The spectroscopic investigation described above has been complimented by photoconductivity measurements for Xe in Ar over a large density range /Reininger et al.,
1984/. In this way the density dependence of the band gap for the Xe impurity E_G^i
and the hole polarization energy P_+^i has been obtained. The experimental P_+^i values

Table 4.2. Experimental V_0 scale for solid and liquid rare gases obtained from optical and photoemission experiments (see Table 3.1) compared to the calculation of V_0 based on the Springett, Jortner, Cohen (SJC) /1968/ theory. A superscript i denotes values for the impurity states. $V_0 = E_G - E_{TH} = U_p + T$. U_p: background polarization potential; T: kinetic energy term

System		Reference	Experiment V_0 (eV)	Theory a) b)		
				T (eV)	U_p (eV)	V_0 (eV)
liquid ^{4}He	4.2 K	c)	+1.05	1.66	−0.38	+1.28
solid Ne	4 K	f)	+1.3	2.44	−1.85	+0.60
liquid Ne	25 K	d)	+0.67 ± 0.05	1.91	−1.46	+0.45
Ar in solid Ne		f)	+1.1			
Kr in solid Ne		f)	+1.1			
Xe in solid Ne		f)	+1.0			
solid Ar	6 K	f)	+0.4	3.46	−3.54	−0.08
liquid Ar	87 K	d)e)	−0.2 ± 0.02	2.53	−2.73	−0.20
Kr in solid Ar		f)	+0.3			
Xe in solid Ar		f)	+0.3			
solid Kr		f)	−0.3	3.84	−4.05	−0.21
liquid Kr	123 K	g)d)	−0.45 ± 0.05	2.73	−3.11	−0.38
Xe in solid Kr		f)	−0.2			
solid Xe			−0.4	4.12	−4.64	−0.52
liquid Xe	165 K	g)d)	−0.61 ± 0.05	2.85	−3.46	−0.61

a) Springett, Cohen and Jortner /1968/
b) Jortner and Gaathon /1977/
c) Sommer /1964/
d) Tauchert /1975/ and Tauchert and Schmidt /1975/
e) Leckner et al. /1972/
f) see Table 3.1
g) Reininger et al. /1982/

have been confronted with theoretical calculations /Messing and Jortner, 1977c/. Concerning the E_G^i values, an agreement with the results of Messing et al. /1977a/ for the liquid phase has been found whereas for the solid the value of E_G^i = 10.98 ± 0.04 eV is larger than previous results (Table 3.1). This raises some doubts concerning the previous assignment of the n' = 1 and n = 2 excitons.

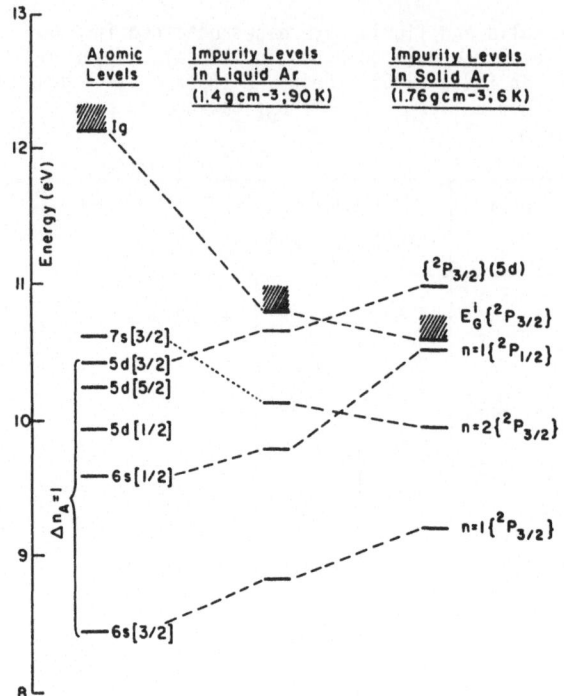

4.3 Molecular Impurity States in Liquid Rare Gases

VUV spectroscopy of molecular impurity states in liquid rare gases was initiated by the study of Miron et al. /1972/ on the spectra of ethylene in liquid Ar and Kr. In this section we shall discuss only one of the few selective examples (Table 4.1) for a molecular impurity in liquid rare gases, namely the perturbation of extra-valence excitations of methyl iodide in liquid Ar and Kr, as reported by Messing et al. /1977d/, since the Rydberg excitations of CH_3I are similar to the lowest Xe atomic states. The molecular impurity states can roughly be classified into three categories analogous to the atomic impurity states:

A) *Valence transitions.*

B) *Molecular Rydberg states.* These correspond in the low pressure limit to the lowest Rydberg states of the molecule, which are characterized by a change $\Delta n \geqslant 1$ of the principal quantum number.

C) *Wannier states.* Wannier-type molecular impurity states were identified for CH_3I, C_2H_4 and C_6H_6 in solid rare gases /Katz et al., 1969; Gedanken et al., 1972a,b, 1973a-d; Jortner, 1974/ and evidence for these highly excited states in a liquid was obtained for CH_3I in liquid Kr by Gedanken et al. /1973e/.

The detailed experiments of Messing et al. /1977d/ on molecular impurities in liquid Ar and Kr in the spectral range from 6.2 eV (2000 Å) to 10.3 eV (1200 Å) were performed over a broad density range of the host liquid from the low pressure limit ("isolated molecule") up to the liquid density. The absorption spectra of CH_3I in Ar in the low energy range are displayed in Fig. 4.9. CH_3I is a suitable and interesting molecular system to study because of two reasons. First, the lowest Rydberg transitions in the gas phase correspond to the Xe atomic $^1S_0 \rightarrow {}^3P_1$ and $^1S_0 \rightarrow {}^1P_1$ transitions. They originate from the iodine $np\pi$ (5p) non bonding orbital /Price, 1936; Herzberg, 1966; Boschi and Salahub, 1972; Tsai and Baer, 1974/. Most of the absorption bands were assigned to three Rydberg series of different symmetry converging to the same ionization limit. Each of the Rydberg series is further split into two components because of the strong spin-orbit coupling within the positive hole when the CH_3I^+ ion core is left in either $^2E_{3/2}$ or $^2E_{1/2}$ state. Only the

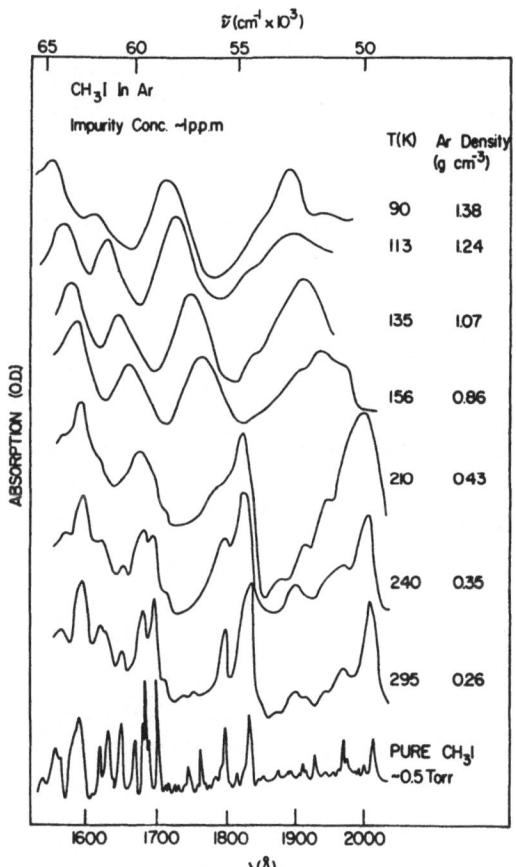

Fig. 4.9. Absorption spectra of CH_3I in liquid Ar in the low energy range 5.9 eV (≅ 2100 Å) to 8.0 eV (≅ 1550 Å). The absorption curves are horizontally shifted, all being displayed on the same optical density (OD) scale (from Messing et al. /1977b/)

lowest broad and structureless absorption band located at 4.8 eV is assigned to a valence $n p \pi \rightarrow \sigma^*$ transition /Robin, 1974/. Second, in view of the low gas phase ionization potentials of this molecule, I_{gas}^1 = 9.49 eV, I_{gas}^2 = 10.11 eV /Potts et al., 1970/, the Wannier series in the matrix or liquid should be located at moderately low energies amenable to experimental observation in a liquid.

The results of the spectroscopic study of CH_3I in liquid Ar and Kr by Messing et al. /1977a/ can be summarized as follows:

1) Medium-induced spectral shifts can provide a diagnostic tool for the distinction between Rydberg-type and valence excitations (see also the discussion of this method by Robin /1974/). The magnitude of the blue spectral shifts decreases in the following order: (a) The largest blue spectral shifts are exhibited for the lowest n = 1 Rydberg transitions of an s or p excited-state configuration, (b) the blue shift for an n = 1 type configuration is slightly lower, (c) mixed type valence-Rydberg transitions exhibit even a smaller blue shift, and (d) the lowest spectral shift is revealed by a pure valence transition.

2) Medium effects on large radius n > 1 Rydberg transitions can be followed only at moderate densities where these high extravalence molecular excitations can be unambiguously identified. In general, these high excitations in fluid Ar exhibit a red shift which increases with increasing density, being larger for higher term values. This trend implies that in this fluid where V_0 < 0 the impurity ionization potential $E_g(\rho)$ decreases with increasing density ρ.

3) The impurity absorption spectrum at high fluid density is characterized by a small number of broad bands, some of them quite distinct and intense, some of them are diffuse and some appear as shoulders or humps. Only those impurity states originating from the lowest Rydberg transitions (n = 1) of different symmetries are amenable to a reliable assignment as they develop continuously from the "isolated" molecule spectrum (see Fig. 4.9). This definite molecular parentage justifies the classification of the latter transitions as intermediate states. Some of these intermediate states are accompanied by a vibrational structure persisting up to the highest density (e.g., the n = 1 impurity states of CH_3I in Ar (Fig. 4.9) at 6.16 eV, 2012 Å). The relation of these lowest extravalence impurity excitations to the molecular states leads to the conclusion that Liehr splitting of the n = 1 states (the splitting of states belonging to s , p , d excited-state configurations) persists up to the liquid density, in contrast to the conventional interpretation of the analogous solid matrix spectra.

4) Several impurity absorption bands at high fluid density were interpreted in terms of large radius n ⩾ 2 impurity states. In particular, experimental evidence was obtained for the appearance of an n = 2 impurity state in the spectrum of CH_3I in Ar. In the spectra of CS_2 and H_2CO there is a tendency of the molecular n = 2

Rydberg transition originating from different series to converge into a single band in the high density fluid. The significance of this observation is that the Liehr splitting for n = 2 states gradually disappears, being vanishingly small in the liquid, as is appropriate for a hydrogenic series of impurity states in a dense medium.

In an accompanying study, Messing et al. /1977b/ have also studied the medium effects on the vibrational structure of the low extravalence (Rydberg) excitations of CH_3I and of H_2CO in liquid Ar. The statistical theory was utilized for model calculations of the lineshape of the medium-perturbed molecular transitions. It was concluded from these studies that independent asymmetric broadening of individual vibronic components accounts for the modification of the apparent intensity distribution with increasing density of the host fluid.

4.4 Electron Affinities for Solid, Liquid and Fluid Rare Gases

The combination of spectroscopic, photoemission and photoconductivity data results in a reliable V_0 (= $-E_A$) scale for solid rare gases. In Table 4.2 we compare these V_0 values obtained for pure and doped RGS with the experimental data for liquid rare gases determined by Tauchert /1975/, Tauchert et al. /1977/ and Reininger et al. /1982 , 1983b,c/. Negative V_0 values mean that energy is released by bringing an electron from the vacuum into the solid.

Interest in V_0 values arises from the interrelationship between the electronic level structure and electron transport in solids and liquids. According to the simple SJC theory /Springett et al., 1968/, the magnitude of V_0 determines whether an extended or a localized state of the electron in the liquid is energetically the most favourable one. It is apparent from Table 4.2 that in all cases V_0 in the solid is higher than in the corresponding liquid, indicating that in the former case the contribution of short-range repulsive interactions is somewhat higher. In liquid He the electron medium interaction is repulsive and the localized bubble state is energetically favoured. The notion of the repulsive e-He interaction is in agreement with the V_0 value /Sommer, 1964/. The positive V_0 values for solid and liquid Ne predict, according to the SJC model, localized electron states to be stable which is corroborated by the experimentally observed low electron mobilities. High electron mobilities were observed experimentally in liquid Ar, Kr and Xe /Miller et al., 1968/ and the reported V_0 value for liquid Ar /Leckner et al., 1967/ favours the extended electron state. These observations are consistent with the negative V_0 values determined by Tauchert et al. /1977/ for these materials.

It is of interest to confront the experimental V_0 data with the prediction of the simple SJC theory /Springett et al., 1968/. The energy of the conduction-band

minimum for the excess electron consists of two contributions, the background polarization potential U_P and the kinetic energy term T arising from short-range repulsion from nearest neighbours, so that $V_0 = T + U_P$. The SJC model provides a simple recipe for the estimate of U_P including the screened polarization field of the
surrounding atoms and for the evaluation of T by the Wigner-Seitz approximation.
Table 4.2 summarizes the results of such calculations, where the scattering length,
a, of the Hartree-Fock atomic field for He and Ne was taken from theoretical scattering data /Springett et al., 1968/, while for Ar, Kr and Xe the value of a was
adjusted to fit the liquid V_0 value. These semiquantative theoretical estimates of
V_0 are fraught with difficulties as Table 4.2 emphasizes how much smaller V_0 is
than the T and the U_P terms. These theoretical estimates clearly indicate that the
increase of T in the solid is somewhat higher than the corresponding decrease in
U_P; however, the SJC model underestimates the relative increase of the short-range
repulsive interactions in the solid, as is apparent from the result for Ar. Thus,
this semiquantitative approach of the SJC model should not be considered as an ulti-
mate theory and further theoretical work in this field is required.

The density dependence of V_0 of rare gases over a broad density range of the
fluid is of considerable interest because of two reasons. Firstly, the interrogation of V_0 from the dense gas through the liquid phase will provide information on
the formation of a conduction band in a fluid. Secondly, the density dependence of
the mobility for excess electrons is related to the V_0 values /Basak and Cohen,
1979/. Theoretical estimates of the density dependence of V_0 in Ar were performed
by Jahnke et al. /1972/, who found V_0 to be practically independent of density over
the range from $\rho = 0.28$ gcm^{-3} to $\rho = 1.43$ gcm^{-3} with an average value of -0.5 eV.
The SJC theory /Springett et al., 1968/ predicts the occurrence of a minimum in the
density dependence of V_0. The density dependence of V_0 for liquid Ar has been discussed by Messing and Jortner /1977c/. They reported on calculations of the adiabatic electrostatic polarization energy P_+ of Xe^+ in fluid Ar over a density range
from 0.1 gcm^{-3} to 1.4 gcm^{-3}. P_+ is connected to V_0 by the impurity ionization energy
E_{TH}^i of an atom or molecule in a dense fluid and the gas phase impurity ionization
energy I_{gas} via $E_{TH}^i = I_{gas} + P_+ + V_0$ (see, for example, /Raz and Jortner, 1969/). The
density dependence of P_+ can quite well be approximated (within ~ 10%) by the Born
charging energy with the effective ionic radius being identified with the effective
hard-core diameter for the solute-solvent separation. As an upper limit for V_0 in
liquid Ar, Messing and Jortner /1977c/ estimated a value of -0.2 eV based on their
calculations of P_+ and spectroscopic data. This value is in good agreement with the
experimental value of $V_0 = -0.22 \pm 0.02$ eV at $\rho = 1.4$ gcm^{-3} determined by Tauchert
and Schmidt /1975/. More interestingly, V_0 was found by Messing and Jortner /1977c/
to be practically constant over a wide density range, which is in accord with the
theoretical calculations of Jahnke et al. /1972/. The weak density dependence of
V_0 in liquid Ar over a broad density range in the fluid is in contrast to the pro-

nounced increase of V_0 from -0.2 eV in liquid Ar to +0.3 eV in solid Ar (Table 4.2).
The density, dependence of the electron affinity V_0 in Xe was recently measured
directly /Reininger et al., 1982/ by electron injection from a metal substrate
(Fig. 4.10). Again a smooth variation with density and with a minimum near the mobility maximum was found.

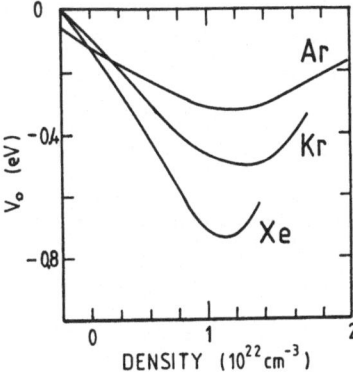

Fig. 4.10. Experimental electron affinity V_0
for liquid Ar, Kr and Xe versus density (after
Reininger et al. /1982 , 1983c,d/)

Using a theory by Basak and Cohen /1979/, it is possible to determine the general trends of the density dependence of the mobility from V_0 data. A comparison
of the measured mobility /Huang and Freeman, 1978 , 1981/ with the Basak-Cohen theory /1979/ gives a qualitative agreement concerning the steep decrease of the mobility near the critical point /Reininger et al., 1983c/. The experimental density
dependence of V_0 for fluid Kr and Xe (Fig. 4.10) was found to be in overall good
agreement with the SJC theory /Reininger et al., 1982/.

4.5 Ionization of Clusters

The photo-electron spectra of Xe clusters reported by Dehmer and Dehmer /1978a,b/,
to which we have already alluded in Chap. 3, show qualitatively how the ionization
potential changes with increasing cluster size. The threshold energy for photoemission of solid Xe is 9.7 to 9.8 eV, as compared to the adiabatic ionization potential of Xe_2 of 11.127 eV. The highest pressure spectrum in Fig. 3.4 yields a
threshold energy of $\approx$ 10.5 eV which is already considerably below the ionization
of the dimer, but still higher than the photo-emission threshold of solid Xe. A
systematic experimental study of the ionization potentials of Ar_n (n = 2 - 6) dimers
/Dehmer and Pratt, 1982/ has provided quantitative information concerning the role
of hole trapping in clusters. The ionization potential of Ar_2 is lower by about

1.13 eV than that of Ar, which is close to the binding energy of Ar_2^+ (1.25 eV). The local hole binding effects are reflected in the gradual reduction of the ionization of small Ar_n (n = 2,3) clusters (Fig. 4.11). This hole trapping mechanism bears a close analogy to excimer formation in gaseous liquid and solid rare gases, and will be discussed in Chap. 6. For larger Ar_n (n = 4 - 6) clusters, additional small polarization effects contribute to the cluster stabilization energy. While the electronic level structure of rare-gas clusters will be explored, it is important to bear in mind that these clusters can exhibit both the characteristics of a rigid "solid-state"-type structure at low temperatures and the features of a non-rigid "liquid"-type structure at higher temperatures /Briant and Burton, 1975; Klaberer and Etlers, 1977/, with the melting temperature increasing with the increase of the cluster size /Natanson et al., 1983/. The study of ionization processes and electronic excitations in clusters under controlled conditions will provide significant information on the level structure of species which are in the transition range between the gas and condensed phases.

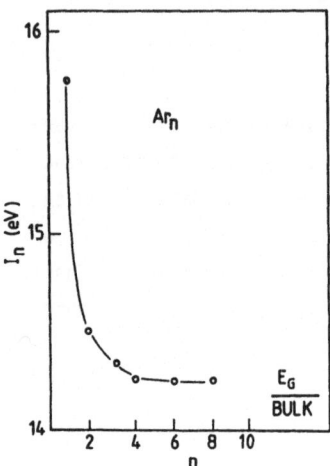

Fig. 4.11. Ionization energy I_n of Ar clusters versus number n of Ar atoms in the cluster. The bulk band gap E_G is shown for comparison (after Dehmer and Pratt /1982/)

5. Metal Rare-Gas Mixtures

Matrix isolation techniques (e.g. /Pimentel, 1958,1978; McCarty and Robinson, 1959; Meyer, 1971/) utilize rare-gas matrices (RGM's) as inert host solids for the trapping of a variety of metastable atoms and molecular species. During the last few years extensive studies were reported for binary mixtures of metals (M) and RGSs, explored over the entire composition range, with the metal atomic fraction (X) being varied in the range $X = 0 - 1.0$. These systems are of interest, because of several reasons:

1) Matrix isolation at moderate metal concentrations ($X = 0.01 - 0.05$) results in the production of interesting weakly bound diatomic M_2 or MM' homonuclear and hetero-nuclear molecules (e.g. /Ozin, 1977,1983; Miller et al., 1977; Leutloff and Kolb, 1979; Danor et al., 1979; Kolb, 1981; Kolb and Forstmann, 1981/).

2) Matrix isolation at higher metal concentrations ($X \cong 0.02 - 0.1$) results in the stabilization of clusters of metal atoms M_n with $n = 3 - 5$ trapped in solid rare gases (e.g. /Schulze et al., 1978; Ozin and Huber, 1978; Welker, 1978; Welker and Martin, 1979; Danor et al., 1979; Kolb, 1981; Kolb and Forstmann, 1981; Ozin, 1983/) which are of considerable fundamental interest in the fields of nucleation, chemisorption and catalysis.

These two points pertain to interesting but still rather conventional extensions of the traditional matrix isolation techniques. Even more interesting phenomena are exhibited at high metal concentrations:

3) At concentrations $X > 0.1$ metal - RGS mixtures provide model systems for dis-ordered amorphous semiconductors /Mott and Davis, 1971/.

4) A well dispersed metal RGS mixture at even higher X can be considered as an expanded metal, where the rare-gas atoms serve as "spacers" which keep the metal atoms apart. With increasing X the two-component mixture will undergo a metal-nonmetal transition (MNMT) /Mott, 1974; Friedmann and Tunstall, 1978/ from a zero conductivity system to a finite conductivity system. Metal-RGS systems are proto-types for the experimental and theoretical studies of the nature of the MNMT in-duced by compositional changes in a disordered material.

5) At high metal concentrations above the metal composition which marks the MNMT the transport and optical properties of metal-RGS mixtures provide information on the electronic structure and transport in disordered metals /Mott, 1974; Cate et al., 1970; Even and Jortner, 1972; Cohen and Jortner, 1974/.

From the foregoing discussion it is apparent that studies of metal-RGS mixtures can yield direct information regarding the electronic states, transport properties and MNMTs in disordered materials, a field of research which is of considerable theoretical and experimental interest /Mott and Davis, 1971; Mott, 1974/. In metal-RGS mixtures both structural disorder and compositional disorder can prevail. The characteristic structural features of the density-of-states function of the perfect lattice will be reduced and disappear in disordered systems and, in particular, the van Hove singularities will be smeared out. Gaps in the density of states of the perfect crystal may be filled up with a finite density of states, and sharp band edges are smeared out to form band tails /Mott, 1967; Cohen et al., 1969/. However, the concept of density of states is still useful and applicable for disordered materials. The density of states in such disordered systems is characterized by a mixed distribution of localized states and of extended states. The localized electronic states, specified by an envelope wave function, which decays exponentially at large distances, are associated with finite atom clusters. On the other hand, extended electronic states, which are characterized by a plane wave-like envelope wave function, extend to infinity, so that for a binary disordered material these states are associated with an infinite cluster. It has been argued that there is no coexistence of localized and extended states at the same energy /Mott, 1967; Cohen et al., 1969/. Therefore, there exist a finite number of well defined energies which segregate between localized and extended states. Since at zero temperature the localized states do not contribute to d.c. electronic transport, these energies separating localized and non-localized states are referred to as mobility edges. The transport and optical properties of disordered materials stem from the basic differences between extended and localized states.

The nature of the electronic states of disordered materials is determined not only by the general characteristics of disorder but is dominated by the microscopic structure of these solids. In RGS mixtures at low X localized states corresponding to single atoms, diatomic molecules and small clusters will be exhibited. The cardinal question is whether the distribution of the metal atoms is statistical or whether preferential clustering of the metal atoms occurs. Obviously the experimental preparatory conditions of the sample have a profound influence on its microscopic structure which is affected by the nature of the substrate, deposition temperature, deposition rate, the kinetic energy of the metal atoms, surface diffusion, etc (see, for example, /Kolb, 1981/). Two general classes of disordered two-component materials can be distinguished (Fig. 5.1) in this context:

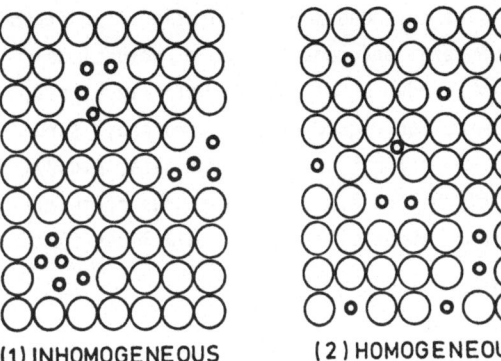

(1) INHOMOGENEOUS   (2) HOMOGENEOUS

Fig. 5.1. Scheme for inhomogeneous and homogeneous disordered two-component materials

a) *Microscopically inhomogeneous metal-RGS mixtures*. Here clustering is preferential. At low metal concentrations it can happen that a single metal atom cannot be matrix-isolated as efficient surface diffusion prefers the production of M_2 diatomics, as is the case for Be /Miller et al., 1977/. At higher metal concentrations the material consists of metallic clusters, i.e., microclusters, embedded in the insulator. At even higher metal concentrations the inhomogeneous metal-RGS mixture becomes metallic; its properties are expected to be analogous to those of granular metals /Abeles et al., 1975/.

b) *Microscopically homogeneous metal-RGS mixtures*. In such randomly substituted material only statistical clustering of metal atoms exists. When increasing the metal concentration, statistical clusters of increasing size will be formed, while at even higher X the homogeneous disordered material will become metallic.

The information currently available regarding the microscopic structure of metal-RGS is meagre. A preliminary study using electron diffraction techniques was re-

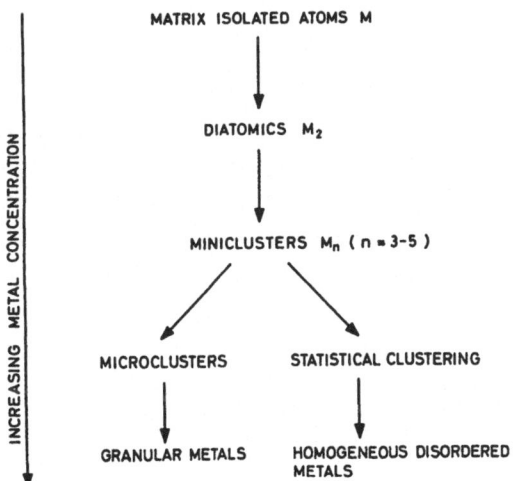

Fig. 5.2. Flow diagram for the different structural features of metal-RGS mixtures developing with increasing metal concentration

ported by Quinn and Wright /1976/. Analysis of extended X-ray absorption fine structure (EXAFS) is another promising method. Several programs for the structural determination of matrix systems are under way /Purdum et al., 1982; Malzfeld et al.,
1983/. Presently most of the available information regarding the microscopic structure is still indirect. It is derived from the optical and transport data. Figure
5.2 presents a scheme for the diverse structural and electronic features of metal-
RGS mixtures over the entire composition range which will provide guidelines for
the subsequent discussion of these systems.

5.1 Single Metal Atoms

The pioneering studies of the electronic spectra of single metal atoms in RGS were
conducted for Na and Hg by McCarty and Robinson /1959/. Since then more than forty
metal atom systems have been studied (see, for example, the reviews by /Meyer, 1971;
Gruen, 1976; Meyer, 1978/). A summary of the individual atomic systems studied by
optical spectroscopy /McCarty and Robinson, 1959; Weymann and Pipkin, 1964; Meyer,
1965; Kupferman and Pipkin, 1968; Andrews and Pimentel, 1967; Belyaeva et al.,
1969/ and by ESR techniques /Coufal et al., 1974a,b,c; van Zee et al., 1981 , 1982/
is presented in Fig. 5.3. In view of the extensive reviews available, we shall restrict ourselves to a brief selective survey of the energetics and microscopic structure of the single atom M centres in RGS which are pertinent for the subsequent discussion.

Considering absorption spectroscopy two general features should be noted: First,
in all cases the optical absorption is exhibited in the proximity of the lowest
electronic excitations of the free atom. The proper theoretical treatment for these
excitations should rest on the tight-binding picture considering small matrix perturbations of the free-atomic levels. This approach is similar to the current fash-
ionable description of the lowest electronic excitations in pure RGS and rare-gas
alloys (Chap. 3). Twenty years ago, an absorption of Na in Ar at 3300 Å was reported
/McCarty and Robinson, 1959/ which was assigned to a 3s → 4p atomic transition. It
may be viewed as an n = 2 Wannier state. Such Wannier impurity states will be of
interest for determining the ionization energy of the M centre and in connection
with nonradiative processes between the high energy bound states of this centre.

Next, we would like to emphasize that while the general spectroscopic assignment
of the observed optical spectra of the M centre is straightforward, the interpretation of the details of the structure of these absorption spectra are not yet understood. The optical absorption spectra of the M centre reveal broad lines which exhibit fine structure. A characteristic example is shown in Fig. 5.4 for the spectra
of Na in RGS /Balling et al., 1978/. Those spectra of alkali atoms in RGS reveal

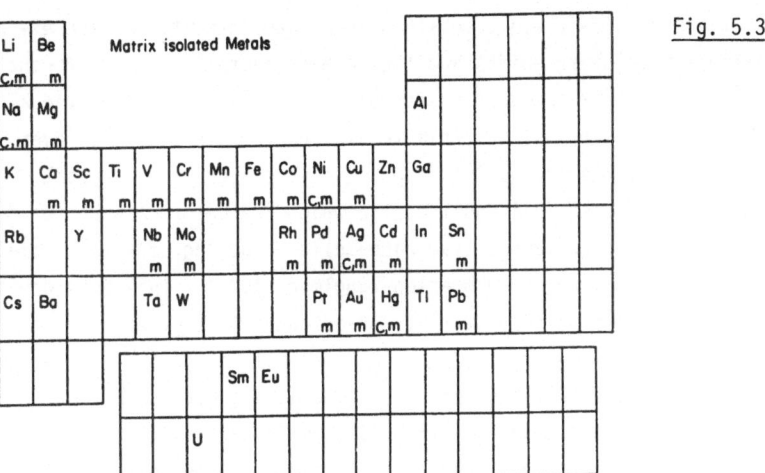

Fig. 5.3

Fig. 5.3. Summary of UV absorption spectra of matrix isolated metals. m or c means that also diatomic molecules or clusters have been studied (after Meyer, /1979/)

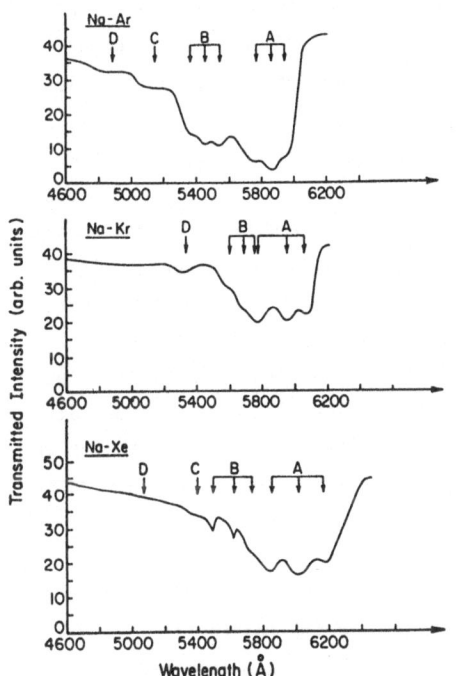

Fig. 5.4. Normalized scans of the absorption by Na atoms trapped in Ar, Kr and Xe at 10 K. The arrows indicate the position of absorption peaks determined by laser excitation (from Balling et al. /1978/)

two triplets labelled A and B and one or two additional bands /Meyer, 1978/. Qualitatively similar effects of broadening and splitting are exhibited in many M centres. The following effects have been considered for an interpretation: (i) Line broadening of individual spectral features, i.e., a single component of an alkali triplet may originate either from intrinsic phonon broadening or from inhomogeneous broadening due to statistical Gaussian distribution of trapping sites. (ii) Site

splitting; different trapping sites, e.g., substitutional and interstitial are expected to exhibit different spectral shifts and thus are characterized by different
excitation energies. (iii) Matrix spectral shifts exhibited within each individual
trapping site. These can be conventionally attributed to short-range repulsive overlap interactions and to long-range attractive dispersive coupling. (iv) Crystal
field splitting effects within each individual centre which are determined by the
local symmetry /Weymann and Pipkin, 1964/. (v) Jahn-Teller coupling effects originating from the coupling of a degenerate vibration /Englman, 1972/. (vi) Spin-orbit
coupling effects which may be modified by overlap non-orthogonality effects /Nagel
and Sonntag, 1978/.

The effects of inhomogeneous broadening and of site splitting have to be sorted
out before the intrinsic electronic effects (iii) - (vi), which are all of comparable
magnitude as well as the effect of phonon broadening within a single well-defined
centre, can subsequently be considered. The techniques of laser-induced fluorescence
line narrowing /Powell, 1978/ can be utilized to discriminate between the effects
of inhomogeneous and homogeneous broadening. Balling et al. /1978/ have applied these
methods to the M centres of alkali atoms in RGS obtaining definite identification
and characterization of individual trapping sites.

Absorption spectra for ns-np transitions in the case of alkali, Cu, Ag and Au
atoms provide an appropriate starting point for a discussion of the optical properties. The appearance of two triplets (A and B in Fig. 5.4) and of some other bands
(D, C) indicates that at least three different trapping sites are occupied. The
trapping site associated with the red triplet A is less stable and can be removed
by annealing /Nagel and Sonntag, 1978; Welker and Martin, 1979; Ozin and Huber,
1979; Hormes and Karrasch, 1982/. Ozin and Huber /1979/ converted the stable blue-
type sites of Na centres to red-type sites by illumination with visible light at
570 nm wavelength and achieved a reverse conversion by 598 nm light. Excitation in
the absorption bands of each triplet yields identical fluorescence spectra, but the
fluorescence spectra from both triplets are distinctly different /Balling et al.,
1970/. Hence a safe band assignment of the triplets is possible.

Ossicini and Forstmann /1981/ derived matrix shifts of the involved ns and np
states from a pseudo-potential calculation. By comparing the theoretical results
for different site geometries with the experimentally observed matrix shifts of the
centre of gravity of the triplet bands A and B, respectively, they found that the
stable trapping site B consists of four vacancies for Na and Kr atoms in rare-gas
matrices except for a substitutional site for Na in Xe. The metastable red triplet
A was assigned to three vacancies. This analysis establishes considerable matrix
dependent shifts of the ground state and excited-state electronic energy. The pseudo-
potential for an Ag atom in a substitutional site of a Kr matrix and the resulting
matrix shifts are illustrated in Fig. 5.5a,b.

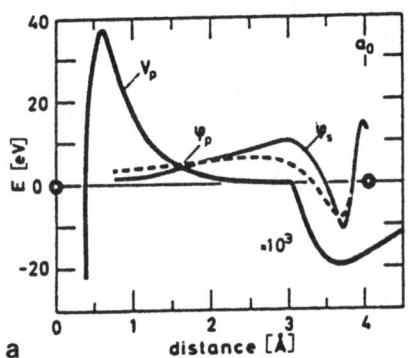 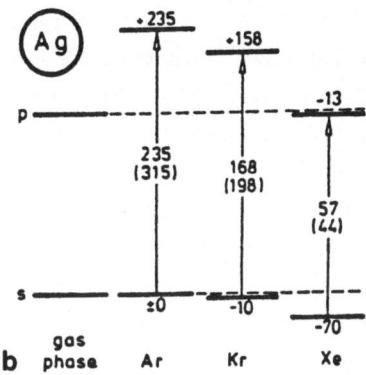

Fig. 5.5a,b. (a) Local approximation Vp to the pseudo-potential of a Kr atom (note the scale change in the attractive region). For comparison the 5s and 5p wave functions of Ag at a substitutional distance in Kr are also shown (after Forstmann and Ossicini /1980/). (b) Absolute shifts in meV for the 5s and 5p levels of Ag isolated in Ar, Kr and Xe, with respect to gas phase, as calculated by the pseudo-potential model. The calculated transition energies are compared with experimental values (in parentheses) (after Forstmann and Ossicini /1980/)

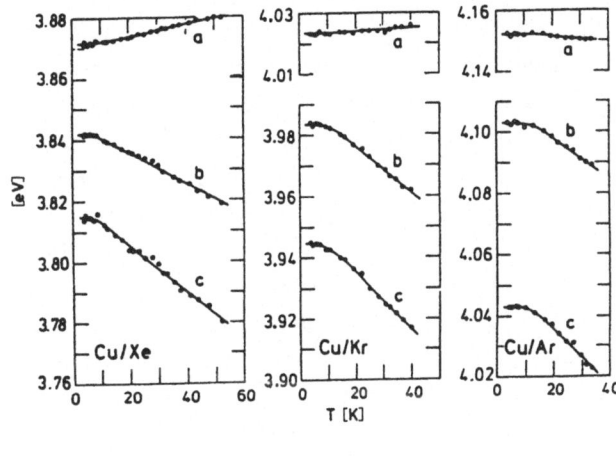

Fig. 5.6. Temperature dependence of the peak positions in the absorption spectra of Cu in Ar, Kr, and Xe (after Forstmann et al. /1977/)

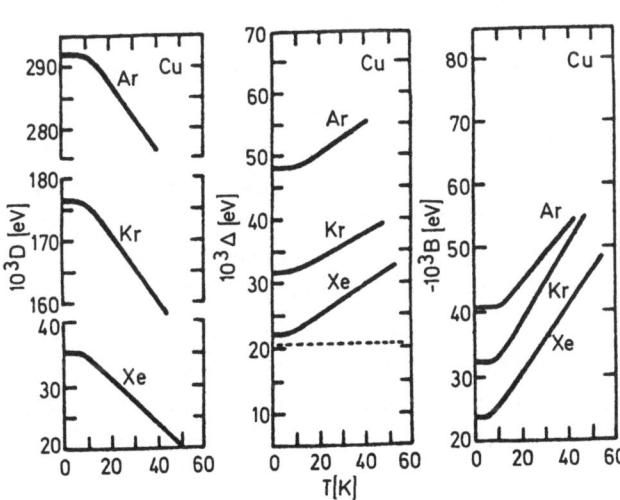

Fig. 5.7. Temperature dependence of the crystal field parameters D, Δ, and B for Cu in Ar, Kr, and Xe. (– – –) spin-orbit splitting Δ for the free Cu atom (after Forstmann et al. /1977/)

The magnitude of the splitting of the ns-np transition into a triplet has been discussed in a crystal field model /Ammeter and Schlosnagle, 1973; Gruen et al., 1974; Forstmann et al., 1977/. This model contains the relative level shift D, the spin-orbit splitting Δ and the crystal field splitting B as parameters. Absorption spectra of Cu in Ar, Kr and Xe show an increase of the energy of components a and a decrease for b and c with temperature (Fig. 5.6). The analysis yields a uniform variation of the three parameters D, Δ and B with temperature in the three matrices /Forstmann et al., 1977/ (Fig. 5.7). Qualitative estimates of the modification of the site geometry have been derived from this analysis /Kolb and Forstmann, 1981/.

Finally, the lineshape of the individual absorption bands has to be considered. A detailed analysis of the moments of the non degenerate $S_{1/2} - P_{1/2}$ transition of Au centres in Ar, Kr and Xe matrices /Weinert et al., 1982/ yields a satisfying description of the temperature dependence of the half width (second moment) by a configuration coordinate model with one effective lattice mode. But a fit of the temperature dependence of the first and the third moments requires at least a second mode, coupling quadratically to the electronic transition and still some discrepancies remain. Even in the simple case of ns-np transitions basic questions like the validity of the crystal field or the Jahn-Teller model are open and demand additional experimental information, such as magnetic circular dischroism (MCD) spectra /Miller et al., 1981; Amstrong et al., 1981; Weinert et al., 1982/.

To illustrate the difficulties involved in the interpretation of more complex transitions we restrict ourselves to Ni and Cu atoms. The distribution of peaks in the absorption spectra of Ni in Ne matrix is completely different from that of Ni in Ar, Kr and Xe matrices /Breithaupt et al., 1983/. The ground state of Ni in Ne matrix has the same configuration $3d^8 4s^2$ as in the gas phase. The matrix-guest interaction in Ar, Kr and Xe matrices stabilizes the $3d^9 4s^1$ state as ground state, which lies in the gas phase 200 cm^{-1} above the $3d^8 4s^2$ state. The $3d^9 4s^1$ ground state of Ni atoms in Xe matrix has also been identified in photo-electron spectra /Jacobi et al., 1980/. Absorption spectra of Cu in rare-gas matrices contain also higher lying transitions besides the 4s-4p transition near 4 eV, which has been mentioned above /Hormes et al., 1983/. A group of bands between 5 and 7 eV has been assigned to fine structure components of the 3d-4p transitions. Bands at 6.357 eV in Xe and at 8.3 and 8.5 eV in Kr have been tentatively ascribed to 4s-5p and 3d-5p Rydberg transitions, respectively. The difficult interpretation of the complicated 3d-4p multiplet (Fig. 5.8 bottom) has been backed by MCD spectra (Fig. 5.8 top). The results of an involved model calculation are shown in Fig. 5.8 as solid lines /Hormes et al., 1983/.

By fluorescence spectroscopy of the M centre /Belyaeva et al., 1973; Balling et al., 1978; Leutloff and Kolb, 1979/ excited-state potential surfaces, relaxation processes and photochemically induced diffusion processes /Ozin and Huber, 1978; Ozin, 1978/ have been studied. We consider first the relevant energetics. While the attractive interaction between a ground-state guest atom and the host is of the weak

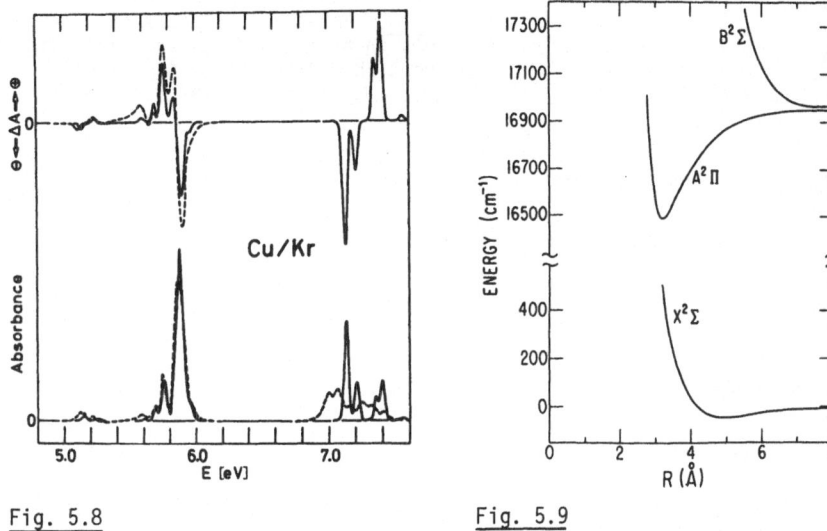

Fig. 5.8   Fig. 5.9

Fig. 5.8. Comparison of the theoretical (——) and experimental (– – –) absorption and MCD spectra of Cu atoms in Kr (from Hormes et al. /1983/)

Fig. 5.9. Ground and lowest excited electronic states of NaAr as calculated by Baylis (from Smalley et al. /1977/)

van der Waals type, the pair interaction between the rare gas and the excited metal atom can lead in many cases to quite a stable molecule M*R which can be called a heteronuclear excimer. For example, the dissociation energy of the $Na(^2P_{3/2,1/2})$ Ar diatomic molecule is 400 cm^{-1} /Smalley et al., 1977/ (Fig. 5.9). Trapping of an M* atom to form a heteronuclear excimer bears a close analogy to the formation of homonuclear excimers via excitation trapping in pure RGSs (Chap. 6). Provided the excimer is stable and that vibrational relaxation to form the M* rare-gas diatomic molecule is fast on the time scale of the radiative decay of M*, one expects the emission spectra of many M centres to be characterized by broad structureless excimer emission bands. This expectation is borne out by the emission spectra of Na in Ar, Kr and Xe /Balling et al., 1978/ reproduced in Fig. 5.10, and of Ag in Kr and Xe /Leutloff and Kolb, 1979/ which indeed exhibit broad emission bands. In addition, large Stokes shifts (3000 cm^{-1} for Na, 8000 cm^{-1} for Ag) are observed which are characteristic for excimer emission. This Stokes shift is illustrated in a comparison between absorption and emission spectra (Fig. 5.11).

The formation of M* rare-gas excimers requires that from the energetic point of view this molecule is stable, while the dynamic point of view requires that vibrational relaxation is effective. A violation of the energetic condition is encountered for Au centres in Kr and in Xe which are characterized by a very narrow emission line /Leutloff and Kolb, 1979/. It was assigned to an atomic-type emission from a d^9s^2 configuration which does not form an excimer.

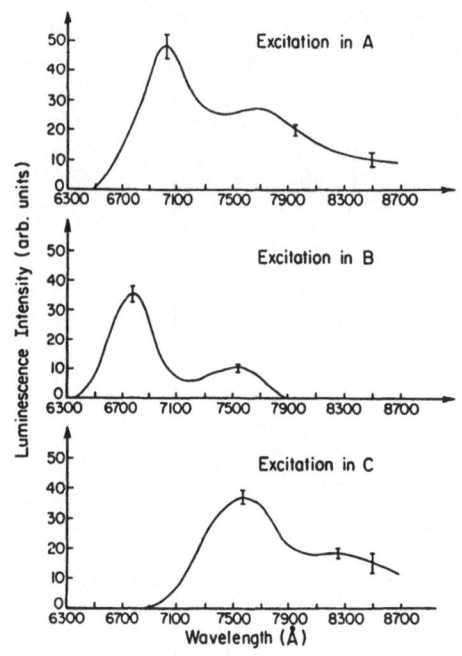

Fig. 5.10. Normalized luminescence of Na atoms in Ar at 10 K for excitation in different absorption bands A, B and C (from Balling et al. /1978/)

Fig. 5.11. Absorption and fluorescence spectra of Ag atoms isolated in Kr. The arrow marks the excitation energy for the emission spectrum (from Kolb /1981/)

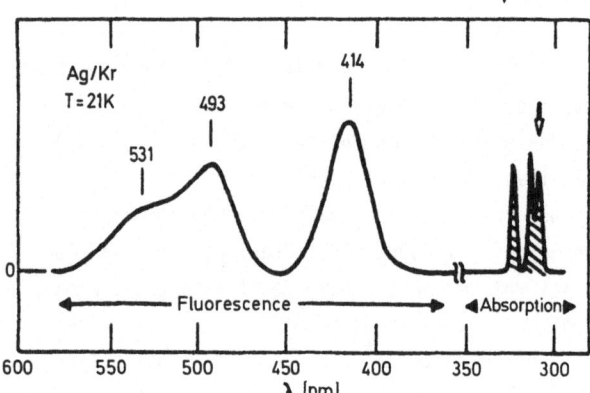

In systems where the energetic and dynamic conditions for excimer formation are satisfied, the strongly repulsive ground state of the M^* rare-gas excimer contributes substantially to the red Stokes shift. Thus, as a consequence of the excimer emission, the translational energy of the M-rare-gas pair is locally dissipated. Such an energy dissipation process results in local heating of the matrix which will enhance the diffusion of the metal atom. The photo-induced bulk diffusion of atoms in RGS /Ozin and Huber, 1978; Welker and Martin, 1979/ originates from this local heating effect. It is worthwhile to point out that the same considerations apply to pure RGS where excimer radiative decay to the repulsive ground state (Chap. 6) can result in appreciable local heating and possibly to photo-induced bulk self-diffusion in RGS.

Spectroscopic studies provide an important diagnostic tool to probe the microscopic structure of the material. From the concentration dependence of absorption bands information can be derived on whether the material is microscopically homogeneous or whether preferential clustering occurs. In Fig. 5.12 /Danor et al., 1979/ the composition dependence of the absorption intensity of a single Hg atom in solid Xe is shown for the 4.85 eV peak, which corresponds to the $^1S_0 \rightarrow {}^3P_1$ transition. The

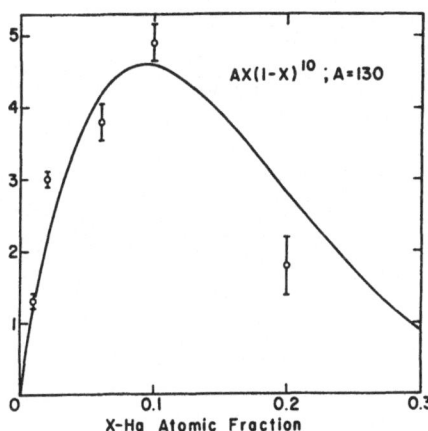

Fig. 5.12. Composition dependence of the height of the single Hg atom absorption. The experimental points represent the peak height above the background absorption. (——) given by (5.1) with z = 10 (from Danor et al. /1979/)

concentration (X) dependence can be accounted for in terms of the probability $P_1(X)$ of finding a single He atom

$$P_1(X) = X(1 - X)^Z \tag{5.1}$$

where z is the coordination number. The Hg/Xe experimental data of Fig. 5.12 /Danor et al., 1979/ can be described by this relation with z = 10 - 12, as is appropriate for an amorphous rare-gas solid. These experimental data demonstrate that for this system up to X = 0.20 the distribution of the M centres is statistical.

5.2 Diatomic Metal Molecules

The study of homonuclear and heteronuclear diatomic metal molecules in RGS is expected to provide a wealth of new spectroscopic information complementary to that available for the spectra of diatomic molecules in the gas phase /Herzberg, 1966/. The interpretation of the spectra of such diatomic molecules in RGS rests on two types of criteria. First, the vibrational structure (e.g. /Miller et al., 1977/) is used and matrix bands are compared to the spectra in the gas phase (e.g. /Danor et al., 1979/). Second, one can use a compositional criterion. For materials where the distribution of the metal atoms is statistical the intensity of the absorption band corresponding to a homonuclear dimer is proportional to X^2 at moderately low values of X. Figures 5.13,14,15 show the concentration dependence of the absorption spectra of Li and Na in Ar /Welker and Martin, 1979/ and of Hg in Xe /Danor et al., 1979/ which at moderate (X = 0.02 - 0.10) metal concentration exhibit M_2 absorption

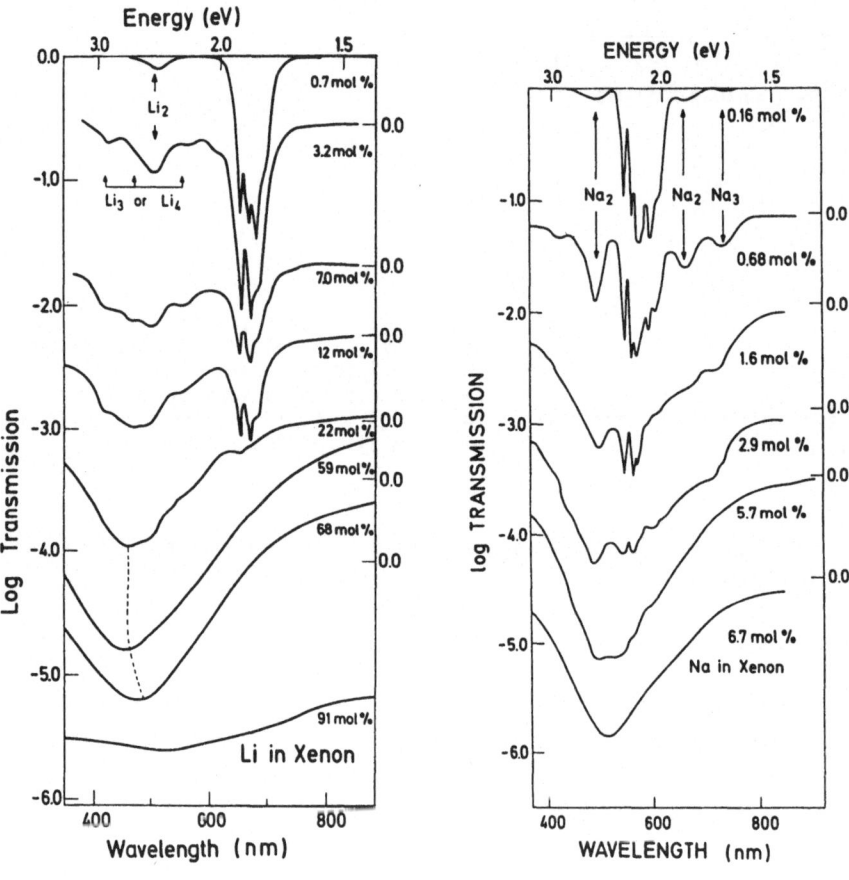

Fig. 5.13   Fig. 5.14

Fig. 5.13. Optical transmission of Li in various stages of aggregation. The in-dicated concentration of lithium vapour is mixed with Xe and the mixture condensed onto a 5 K substrate. As the concentration of lithium increases, the degree of ag-gregation increases (from Welker and Martin /1979/)

Fig. 5.14. Optical transmission of Na in various stages of aggregation. The indi-cated concentration of sodium vapour is mixed with Xe and the mixture condensed onto a 5 K substrate (from Welker and Martin /1979/)

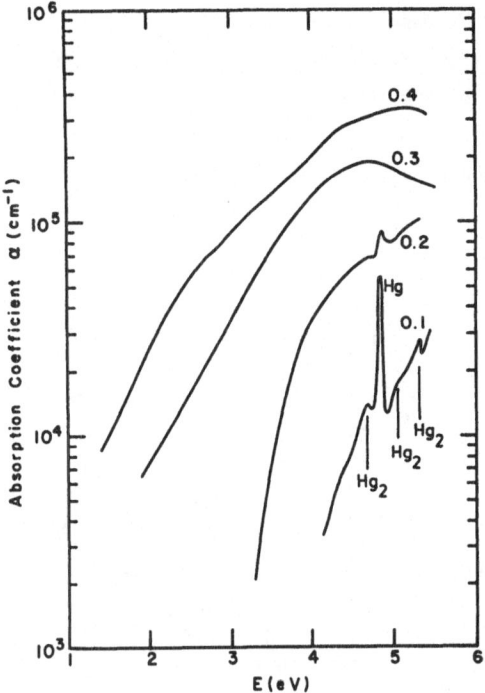

Fig. 5.15

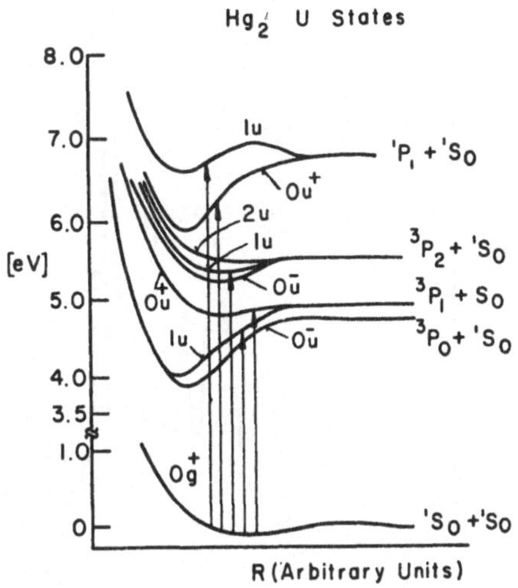

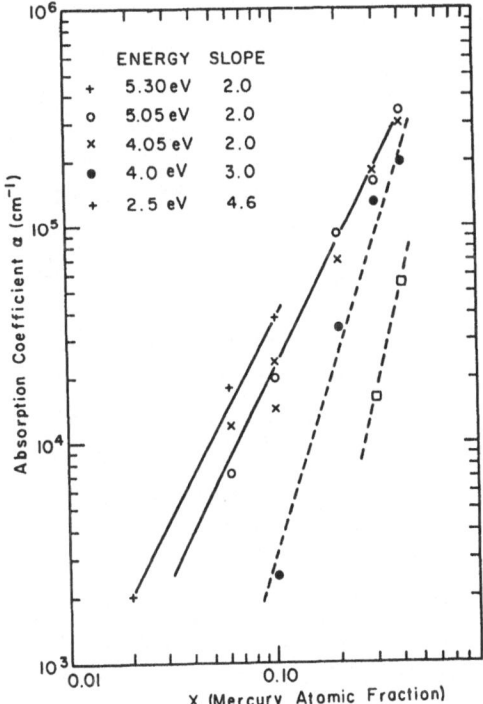

Fig. 5.16

Fig. 5.15. Absorption spectra of Hg/Xe samples at 6 K for X = 0.1 - 0.4 (from Danor et al. /1979/)

Fig. 5.16. Potential curves for the lowest u-type states of Hg_2 from Stock et al. /1978/. Allowed transitions are marked by arrows (from Danor et al. /1979/)

Fig. 5.17. Dependence of the absorption coefficient of Hg/Xe samples on the Hg concentration at several energies (see Fig. 5.15) (from Danor et al. /1979/)

bands. An unambiguous identification of the M_2 bands is crucial. In Fig. 5.16 an example for the spectroscopic criterion is shown for Hg in Xe /Danor et al., 1959/ where the transitions in the range 4.7 – 5.7 eV are identified. The application of the compositional criterion for the identification is demonstrated in Fig. 5.17 for Hg /Danor et al., 1979/ and in Fig. 5.18a,b for Ag in Kr /Schulze et al., 1978/.

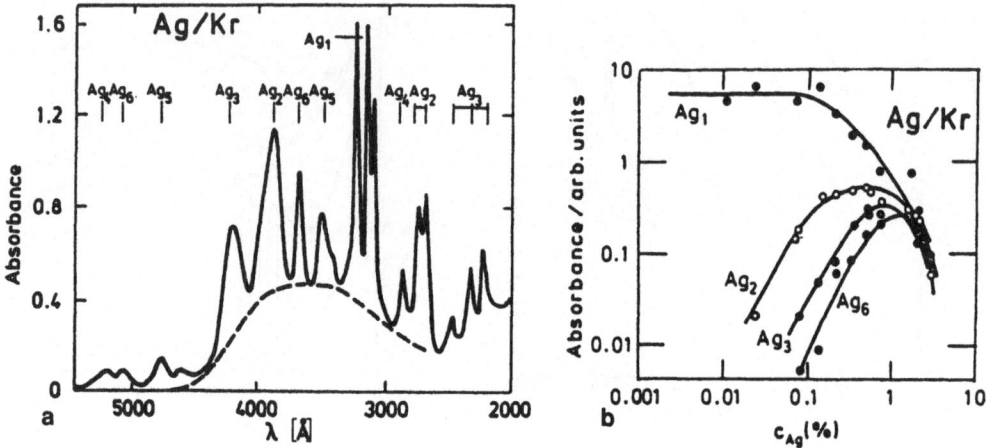

Fig. 5.18a,b. (a) Absorption spectrum of Ag clusters in a Kr matrix at c_{Ag} = 2%. (b) Relative yield of Ag_n in Kr as a function of concentration derived from the intensities of the absorption bands (from Kolb /1981/)

The absorption bands of M_2 molecules in RGS fall into two distinct classes: (1) unstructured bands as is the case for the spectra of alkali dimers (Fig. 5.13, 14) and of Hg_2 (Fig. 5.15), and (2) structured bands which exhibit well defined vibrational structure, as in the situation for matrix isolated alkaline earth metal diatomic molecules (e.g. /Miller et al., 1981/). A typical example is reproduced in Fig. 5.19 for Ca_2. Characteristic vibrational frequencies are 474 cm^{-1} for Be_2, 190 cm^{-1} for Mg_2, 111 cm^{-1} for Ca_2 and 66 cm^{-1} for Sr_2, all being recorded in Ar /Miller et al., 1981/. The nature of the matrix exerts small spectral shifts and small modifications of the vibrational frequency. These general features of the absorption spectra can be rationalized in terms of different potential energy curves for the ground and excited state (Fig. 5.20). When the ground state is characterized by well defined minima due to valence forces, as is the case for alkali diatomic molecules, the excited state can be either strongly bound (Case a) or weakly bound or even repulsive (Case b). The unstructured spectra of the alkali molecules correspond to Case (b). For many cases the ground-state potential curves are shallow and only stabilized by weak van der Waals forces, as is the situation for alkali

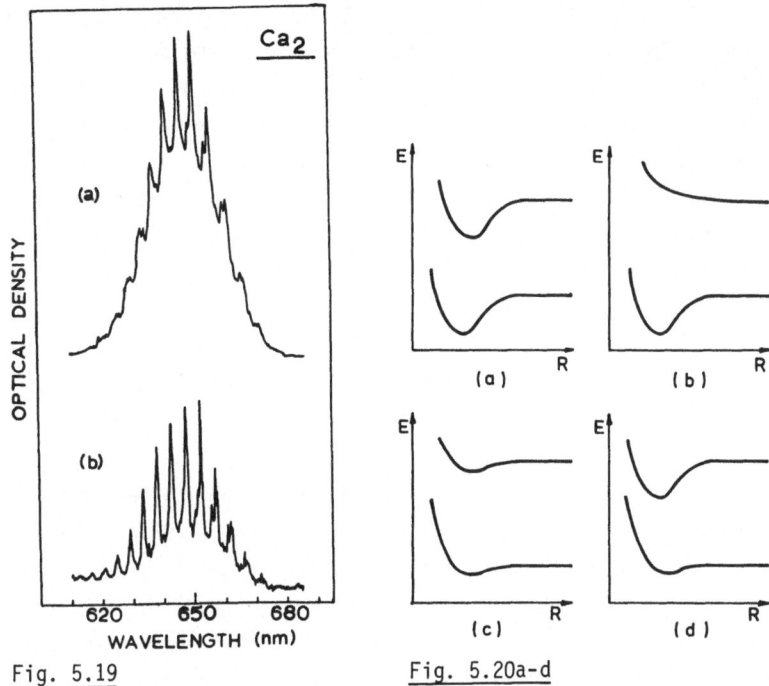

Fig. 5.19

Fig. 5.20a-d

Fig. 5.19. Absorption spectra of Ca$_2$ in an argon matrix before (a) and after (b) annealing (from Miller et al. /1977/)

Fig. 5.20a-d. Scheme of combinations (a , b , c , d) of weakly and strongly bound potential curves for the ground state and the excited state (see text)

earth atoms and for Hg. Again, the excited state can be either shallow (Case c) or deep (Case d). For Case (c) a well defined vibrational structure will be exhibited provided that the minima of the two potential curves are not too far displaced, as is the case for the alkaline earth diatomics. On the other hand, when in Case (c) the minima of the two curves are well displaced (e.g., for some transitions of Hg$_2$ (Fig. 5.15)), an unstructured peak is observed. For Case (d) only an unstructured peak is expected.

5.3 Miniclusters

The search for model systems, constituting the missing link between bulk materials and isolated atoms has focused interest on the physical properties of miniclusters. Nowadays, gas phase miniclusters can be produced by supersonic nozzle beam techniques (see e.g. /Wöste, 1982; Sattler, 1983/ and their properties are studied by many groups. Due to the still fairly low abundances, an enrichment and stabilization

in inert matrices is often favourable for a detailed study of their physical properties. Molecular aggregates M_n (n = 3 - 5) containing a few metal atoms were identified by optical spectroscopy in RGS /Schulze et al., 1978; Danor et al., 1979;
Welker and Martin, 1979/. The identification of these species on the basis of the
order of their appearance /Ozin and Huber, 1968/ with increasing X is not reliable
/Welker and Martin, 1979/. The compositional criterion for the assignment of absorption in a certain energy range /Danor et al., 1979/ to specific M_n clusters is
more reliable. The application of this compositional criterion to the identification of Ag_{2-6} and of Hg_3 and Hg_{4-5} is presented in Figs. 5.18 and 17, respectively.
The list of well identified clusters in RGS includes Ag_3, Ag_4, Ag_5 and Ag_6 /Schulze
et al., 1978/, Na_3 /Welker and Martin, 1979/, Li_{3-4} /Welker and Martin, 1979/, Hg_3
and Hg_{4-5} /Danor et al., 1979/. In all cases broad bands devoid of any vibrational
structure are observed.

Ab initio and semiemperical calculations of energy levels and optical excitation
energies of such clusters /Welker and Martin, 1979; Paccioni et al., 1983/ provide
general correlations for the rationalization of the experimental data. For a qualitative discussion of the general features of such molecular clusters we would like
to point out that matrix isolation is expected to give a variety of isomers for any
given M_n molecular cluster with a fixed value of n > 2. Consider for example the
M_3 molecule. It is apparent that the ground state energy, the ionization potential
and the excitation energies of the triangular and of the linear M_3 species are different. Since energetically metastable molecules can be trapped in the RGS, we expect a distribution of excitation energies due to different isomers for each size
n of the minicluster. Consequently, one cannot just inquire what the trends are of
the excitation energy of miniclusters M_n with increasing size of the metallic molecule without specifying the structure of the clusters. For example, for chains M_n
of M atoms, the lowest excitation energy decreases monotonically with increasing n.
On the other hand, the lowest excitation energy for a two dimensional structure of
a given M_n molecule is in general higher than that for the corresponding linear
chain of the same n. These considerations stress the importance of the coordination
number in determining the electronic excitations for these molecular miniclusters.
With increase of the cluster size the average coordination number is crucial in
determining the "transition" from molecular properties to solid-state characteristics
which can correspond to those of metallic microclusters or to statistical clustering.

5.4 Microclusters and Statistical Clustering

When the concentration is increased beyond a few ppm, clustering of the metal atoms is expected. In the following discussion microscopically inhomogeneous and homogeneous systems have to be considered separately.

In *microscopically inhomogeneous* metal RGS (non statistical clustering) the material consists of islands of metal atoms embedded in the insulator. For the "transition" from molecular properties of a minicluster (see below) to metallic properties of the islands, two conditions have to be satisfied /Kubo, 1962/: Firstly, the size of the metallic regions has to be sufficiently large so that the level spacing of the highly-filled states is small relative to the thermal energy. The particles are locally metallic. Secondly, the correlation length for the metallic regions has to be large relative to the mean free path (or the coherence length) of the conduction electrons in such metallic regions. This situation is analogous to granular mixtures of small metallic particles in insulators /Abeles, 1975/. As long as the metallic particles are mutually isolated, that is when the concentration is below the percolation threshold /Shante and Kirkpatrick, 1971/, then the material is an insulator.

A striking characteristic of the optical properties of small metallic particles dispersed in an insulator is the appearance of Mie resonances /Maxwell-Garnett, 1906; Cohen et al., 1973; Gittelman and Abeles, 1977; Webman et al., 1977/. At low values of the volume fraction of the metal, a Mie resonance is expected at the energy E_{MI} which satisfies the relationship

$$\varepsilon_1^M(E_{MI}) = - 2\varepsilon_1^0(E_{MI}) \tag{5.2}$$

where ε_1^M and ε_1^0 correspond to the real part of the dielectric functions for the metal and for the insulator, respectively. At higher metal concentrations the Maxwell-Garnett theory /1906/ or the effective medium theory /Webman et al., 1977/ can be applied. The Mie resonance exhibits only a weak concentration dependence below the percolation threshold. Thus it indicates selective clustering of the metal atoms. An example has been discussed by Welker and Martin /1979/ for Ag in Ar (Fig. 5.21). The observed broad Mie resonance bears a close analogy to Mie resonances of granular mixtures of silver particles in Al_2O_3 /Cohen et al., 1973; Abeles et al., 1975/.

In *microscopically homogeneous* metal RGS the metal atoms are statistically clustered. A diagnostic criterion for microscopically homogeneous materials is the lack of Mie resonances in the optical spectra. The average size of the metal clusters in homogeneous materials can be estimated from the percolation theory /Shante and Kirkpatrick, 1971/. The average cluster size is $S = (1 - p / p^*)^{-(j+1)}$ where $j = 11/16$ for a three-dimensional system, p is the fractional site occupation probability and

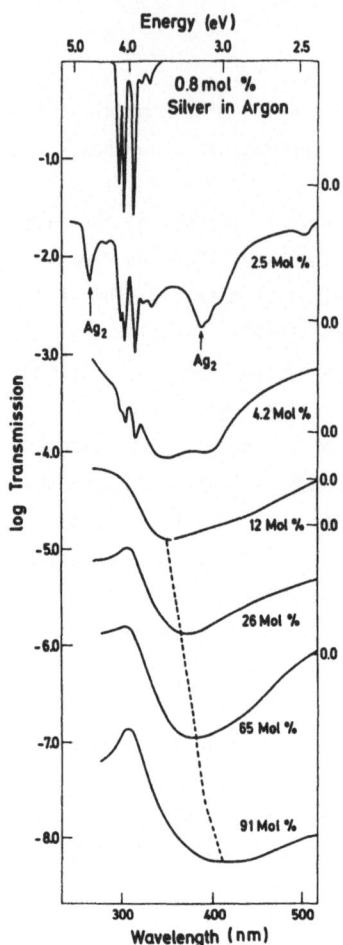

Fig. 5.21. Optical transmission of silver in various stages of aggregation. The indicated concentration of silver vapour is mixed with argon and the mixture condensed onto a 5 K substrate (from /Welker and Martin, 1979/)

p^* is the percolation threshold. For an amorphous system containing two types of atoms of different size, e.g., a metal and rare gas, it was proposed /Danor et al., 1979; Cheshnovsky, 1979/ that one can replace (p/p^*) by (C/C^*) where C is the atomic volume fraction of M while $C^* = 0.15$ corresponds to the continuous percolation threshold /Zallen and Scher, 1971; Webman et al., 1976/. For the Hg/Xe system, where a Mie resonance is not observed /Cheshnovsky et al., 1979/, Danor et al. /1979/ estimated that S = 3 - 4 at X ≅ 0.4 while S diverges at the percolation threshold at X ≅ 0.7. Thus, statistical clustering results in rather small cluster sizes up to reasonably high (X ≅ 0.4) metal concentration.

Danor et al. /1979/ have assigned the low energy absorption tails of Hg statistical clusters in Xe to charge-transfer excitations within a single cluster. This assignment is of considerable interest as these charge transfer states are analogous to Wannier-type states in a large cluster, which will converge to the bottom of the conduction band. When the size of the Hg_n cluster will increase further in the concentration range $X > 0.4$ the excitation energy will be reduced by the gradual increase of the coordination number increasing the charge-exchange stabilization energy. Large three-dimensional Hg_n clusters will be characterized by low energy absorption tails which will then be classified as valence-band to conduction-band transitions in the amorphous Hg/Xe material. The charge-transfer states in small Hg_n miniclusters are the molecular precursor to the interband optical transition in amorphous semiconductors consisting of a highly doped metal - RGS /Danor et al., 1979/. Thus, microscopically homogeneous metal RGS systems bridge the gap between the molecular description of localized clusters and the band structure model applied to amorphous semiconductors.

5.5 Metal-Nonmetal Transitions in Metal Rare-Gas Mixtures

Metal-RGS mixtures undergo a metal-nonmetal transition (MNMT) /Mott, 1949,1966,1967, 1972,1974/ with increase of the metal concentration. At low metal concentrations the material is an insulator while at high metal concentrations the metal-RGS is an amorphous disordered metal. The metallic state is characterized by a high finite d.c. electrical conductivity in the zero temperature limit. The variation of the electrical transport properties of typical metal-RGS mixtures while changing the metal concentration is demonstrated in Figs. 5.22 , 23. This transition from zero conductivity at low X to metallic conductivity at high X is referred to as the MNMT /Mott, 1974/. A complementary and equivalent description is that metal-RGS mixtures are "expanded metals" which, by decreasing X, can be continuously expanded from normal metallic density to very low metal densities and which undergo a MNMT. The list of such "expanded metals" that have been experimentally studied involves Na /Cate et al., 1970; McNeal and Goldman, 1977/, Cs, Rb /Phelps et al., 1975; Phelps and Flynn, 1976/, Cu /Endo et al., 1973; Hunderi and Rydberg, 1975; Rydberg and Hunderi, 1977/, Ga /Rydberg and Hunderi, 1977/, Pb /Eatah et al., 1975; Hilder and Cusack, 1977; Quinn and Wright, 1977/, Co /Quinn and Wright, 1977/, In /Hilder and Cusack, 1977/, Fe /Shanfield et al., 1975/ and Hg /Raz et al., 1972; Cheshnovsky et al., 1977,1979; Danor et al., 1979/. Usually the conductivity has been measured.

The variation of the electrical conductivity σ, with the metal atomic fraction X, falls into two categories: (i) The *abrupt conductivity transition* is characterized by a sudden drop of σ at a certain value of X, where $d\sigma/dX$ is discontinuous. The experimental evidence for such an abrupt transition in disordered materials reported for Cu/Ar /Endo et al., 1973/ and for Pb/Ar (Fig. 5.22) /Eatah et al., 1975/ is not conclusive. It may originiate from spurious crystallization effects /Mott, 1978/. (ii) The *continuous conductivity transition* is characterized by a gradual

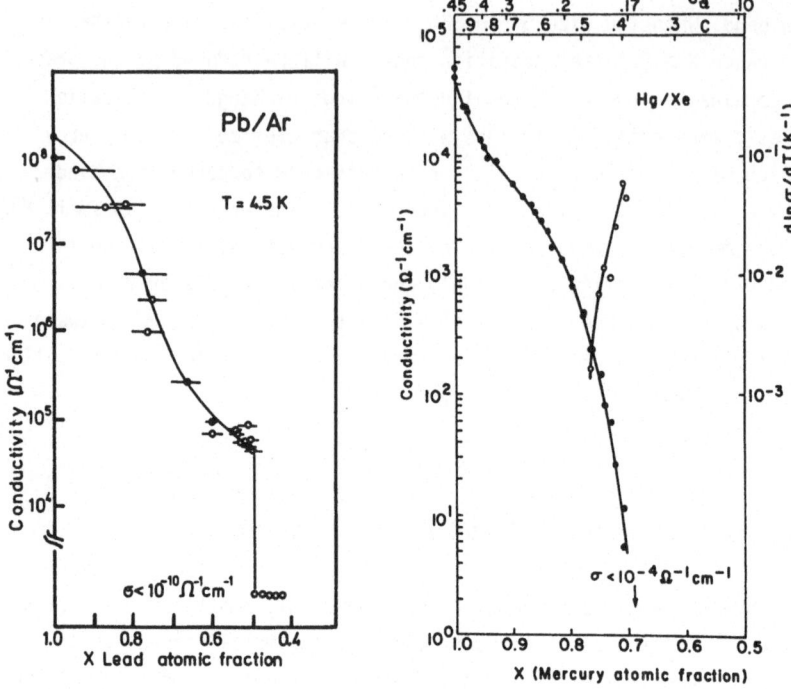

Fig. 5.22   Fig. 5.23

Fig. 5.22. Conductivity versus composition for mixtures of lead and argon deposited at 4.2 K (after /Eatah et al., 1975/)

Fig. 5.23. The composition dependence of the conductivity (●) and of the tempera-ture coefficient of conductivity (○) in Hg/Xe mixtures at 6 K. Three scales of mercury concentrations are presented: X - the atomic fraction of Hg, C - the vol-ume fraction of Hg and C_a - the atomic volume fraction of Hg (from /Cheshnovsky et al., 1977/)

decrease of σ with decreasing X. A typical example is presented in Fig. 5.23. Such a behaviour was observed for a variety of systems, e.g., Na/Ar /McNeal and Goldman, 1977/, Cs/Xe /Phelps et al., 1975/, Rb/Kr /Phelps et al., 1975/, Hg/Xe /Cheshnovsky et al., 1977/ and Fe/Xe /Shanfield et al., 1975/. In general, the conductivity tran-sition was found at X = 0.2 - 0.6 for monovalent metals and at X = 0.4 - 0.8 for poly-valent metals. Figure 5.24 provides an overview of the composition characterizing the conductivity transition in a variety of such materials.

Can the conductivity transition in low temperatures metal-RGS mixtures be iden-tified with the MNMT? Cheshnovsky et al. /1979/ have argued that the conductivity transition in metal-RGS mixtures cannot be identified in general with the MNMT, as contributions from thermally activated hopping and hopping between large radius localized states cannot be disregarded even at low temperatures. For a meaningful interpretation it is necessary to distinguish again between microscopically inhomo-geneous and homogeneous materials.

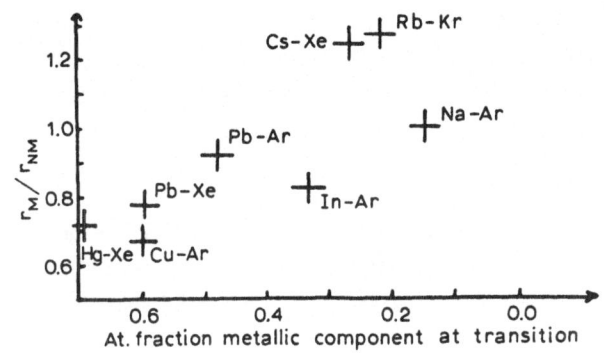

Fig. 5.24. Composition of the conductivity transition for several systems. The ordinate is the metal atom radius / gas atom radius (from /Hilder and Cusack, 1977/)

The MNMT in inhomogeneous materials containing randomly dispersed finite volumes of conducting materials can be adequately accounted for in terms of the classical percolation theory /Shante and Kirkpatrick, 1971; Kirkpatrick, 1972; Cohen and Jortner, 1974; Cohen et al., 1979/. The classical percolation theory is valid, provided that the correlation length for local clustering exceeds the mean free path of the conduction electrons. In the case of mixtures of conducting and insulating particles, there is a critical concentration of metal above which there is a finite probability for the creation of a conducting path through the sample. This concentration depends only on the topology of the system and marks the MNMT in classical systems. The critical concentration occurs at a metallic volume fraction C, denoted C^*, which is referred to as the percolation threshold. For continuous percolation /Zallen and Scher, 1971; Webman et al., 1976/ $C^* = 0.15 \pm 0.02$. The d.c. conductivity in the vicinity of C^* is expected to be /Kirkpatrick, 1971/

$$\sigma = 0 \qquad \text{for } C < C^* \tag{5.3}$$

$$\sigma \propto (C - C^*)^p \quad \text{for } C > C^* \tag{5.4}$$

with $p \cong 1.6$. The applicability of the percolation theory for the MNMT requires that the power law (5.4) is fulfilled. It is plausible on the basis of the optical data (Fig. 5.21) that the electronic structure, the transport and the nature of the MNMT in Ag/Ar mixtures correspond to a granular metal and can be handled in terms of the classical percolation theory.

In this case one should be aware of the disappointing fact that there is no unique mechanism for the MNMT. Notable mechanism proposed to account for the occurrence of the MNMT /Mott, 1974; Friedman and Tunstall, 1978/ are, (a) the MNMT due to band overlap, (b) MNMT due to electron-electron interaction resulting in short-range screening /Mott, 1979/, (c) MNMT due to localization in a disordered material /Anderson, 1958/. In real life, several of these mechanisms combine to induce the

MNMT in a particular disordered material. As mentioned above, all the available reliable data show a continuous conductivity transition for metal RGS. Therefore, it
is important to outline supplementary experimental evidence for the characterization
of the MNMT in metal RGS which are believed, on the basis of admittedly indirect
optical evidence, to be microscopically homogeneous:

1) *Screening of Wannier exciton states of the insulator.* Raz et al. /1972/ reported
that the Wannier excitons of Xe prepared at 40 K in Hg/Xe mixtures (Fig. 5.25) are
smeared out at some high value of $X \cong 0.55 - 0.60$ indicating the onset of short-range

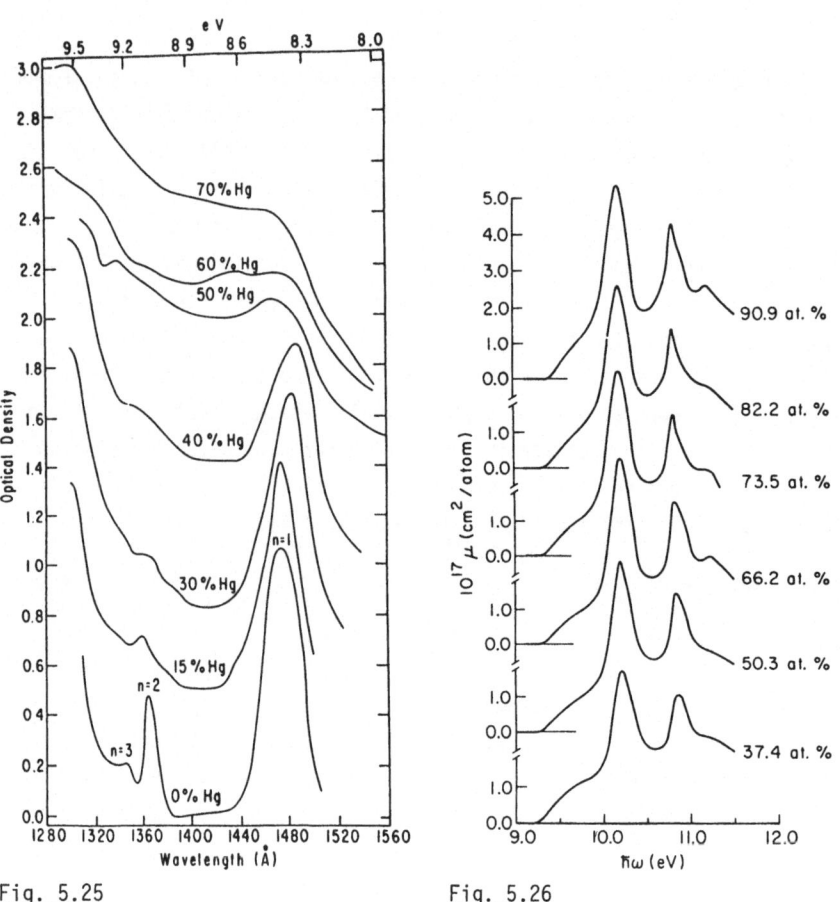

Fig. 5.25   Fig. 5.26

Fig. 5.25. Vacuum UV absorption spectra of Hg/Xe solid films deposited at 40 K and
measured at 10 K. The spectra were practically temperature independent in the range
10 - 40 K. The percentages represent the mole fractions of Hg. The spectra were ver-
tically displaced on the optical density scale; the residual absorption at 1550 Å
in each case was 0.1 - 0.2 optical-density units (from /Raz et al., 1972/)

Fig. 5.26. Optical absorption per Kr atom in thin Rb/Kr films spanning the composition of the metal-insulator transition (from /Phelps and Flynn, 1976/)

screening and thus marking an MNMT. The high deposition temperature raises some serious questions concerning the structure of this material and a quantitative conclusion cannot be drawn. On the other hand, this screening effect was not observed in Rb/Kr and in Cs/Xe /Phelps et al., 1975; Phelps and Flynn, 1976/ where peaceful existence of Wannier excitons of the rare-gas atoms was observed at metal concentrations far above the MNMT (Fig. 5.26). One way out of this contradictory and mutually inconsistent data is to propose that while the Hg/Xe system is microscopically homogeneous, the Rb/Kr and the Cs/Xe samples are microscopically inhomogeneous. However, more work is required to resolve these problems.

2) *Vanishing of the temperature coefficient of the electrical conductivity.* The temperature coefficient for the Hg/Xe system (Fig. 5.23) vanishes at $X_M = 0.79$ ± 0.02 /Cheshnovsky et al., 1977/, which corresponds to a concentration high above the onset $X_M = 0.69 \pm 0.02$ of the conductivity transition in this material. Thus, in the composition range $X = 0.69 - 0.79$ d.c. conductivity by thermally activated hopping is exhibited in Hg/Xe at 6 K /Cheshnovsky et al., 1977,1979/. The onset of temperature independent conductivity at $X_M = 0.79$ marks the MNMT.

3) *Optical data.* The optical data for the Hg/Xe system in the visible and near UV (Fig. 5.27a,b) /Cheshnovsky et al., 1979,1982/ show the features of an amorphous semiconductor, i.e., positive ε and vanishing low-frequency optical conductivity at low ($X < 0.70$) and the characteristics of a metallic system at high ($X > 0.83$) concentrations. In particular, the real part of the dielectric function at the lowest energy measured (Fig. 5.27a) exhibits an abrupt discontinuity at $X_M = 0.80 \pm 0.02$ which marks /Cheshnovsky et al., 1982/ the MNMT, and which is in perfect agreement with the analysis of the temperature coefficient of the conductivity /Cheshnovsky et al., 1979/.

Cheshnovsky et al. /1977,1979/ have proposed that in microscopically homogeneous materials, such as Hg/Xe, the continuous conductivity transition should be interpreted in terms of the Mott-Anderson transition /Mott, 1974/ intermediated by low-temperature thermally-activated hopping between large-radius localized states. A different point of view was advanced by Phelps, Avci and Flynn /1975/, who proposed for the onset of conductivity a description in terms of a classical percolation theory. This proposal is incompatible with the observation of activated conductivity in Hg/Xe and with the optical properties in the vicinity of the conductivity transition /Cheshnovsky et al., 1977,1979/. Recently, Cheshnovsky et al. /1982/ have attempted to provide a uniform picture for the MNMT in microscopically inhomogeneous metal-RGS mixtures. They suggested that with increasing X the topological percolation threshold marks the onset of activated transport corresponding to the conductivity transition, while for $X = X_M$ the electronic states become extended. Obviously, more experimental and theoretical work for the characterization of the MNMT transition associated with a continuous conductivity transition is required.

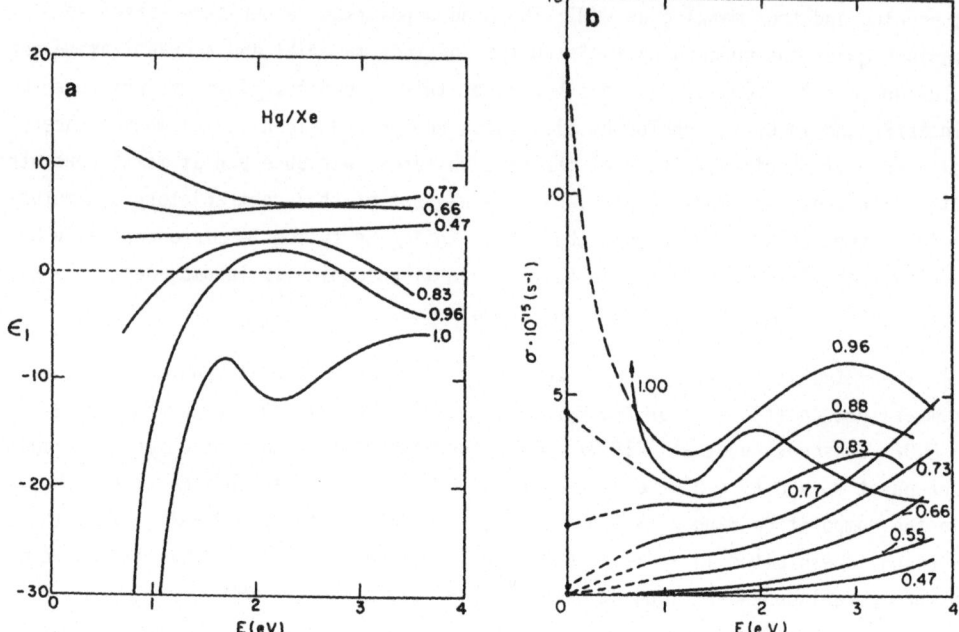

Fig. 5.27a,b. (a) The real part $\varepsilon_1(E)$ of the dielectric function for Hg/Xe mixtures at 6 K for different atomic fractions of mercury (from /Cheshnovsky et al., 1979,1982/). (b) The optical conductivity $\sigma(E)$ of Hg/Xe mixtures at 6 K for different atomic fractions of mercury. $\sigma(E) = \omega\varepsilon_2(E)/4\pi$, where E is the photon energy while $\omega(s^{-1})$ is the angular frequency. (———) Experimental data from transmission measurements. (———) Extrapolation to the d.c. conductivity values which are marked by dots (from /Cheshnovsky et al., 1979,1982/)

6. Excited-State Dynamics

Up to this point we have been concerned with the electronic structure in pure and doped condensed rare gases. We shall now proceed to discuss dynamic processes in electronically excited states of RGS, alloys, and liquid rare gases. The electronic structure provides the essential input data which serves now as the starting point for the discussion of excited-state dynamics. Optical excitation of an electron in an insulator results in two general effects: (A) The system is produced in a non-equilibrium nuclear configuration so that the balance of interatomic (or inter-molecular) interactions is destroyed. (B) The electronically excited state is degenerate with electronic-nuclear excitations (e.g., phonons) of lower electronic configurations including the electronic ground state.

The first effect leads to two important consequences:

A1) *Intrastate Spectroscopic Implications.* Simultaneous excitation of electronic and nuclear states or, more generally, simultaneous excitation of various kinds of elementary excitations, e.g., excitons and phonons in a pure solid or a localized electronic state and phonons in an impurity centre.

A2) *Intrastate Dynamic Implications.* The medium produced in a nonequilibrium con-figuration relaxes to a new nuclear equilibrium configuration. Typical examples involve modulation of exciton motion by phonons, exciton trapping in pure solids, lattice relaxation around an electronically excited impurity state and medium re-laxation around an electronic excitation in a liquid. These occur within a single electronically excited configuration of excitonic or impurity type and are thus referred to as intrastate dynamic processes.

The central implications of effect (B) are:

B) *Interstate Dynamic Implications.* Electronic relaxation processes can occur between different electronic configurations. These processes include diverse phenomena which in the order of decreasing energy are:

a) Slowing down of free electrons by electron-electron and electron-phonon scat-tering (Chap. 7).

b) Autoionization of metastable excitons located above the interband threshold /Philips, 1966; Jortner, 1968/.

c) Thermal ionization of exciton and impurity states /Kubo and Toyozawa, 1955/.

d) Electron-hole recombination /Kubo and Toyozawa, 1955/.

e) Electronic relaxation of exciton and impurity states into lower electronic configurations /Jortner, 1974/.

f) Electronic energy transfer between free and localized exciton and impurity states /Förster, 1948; Dexter, 1953; Knox, 1963/.

All these interstate relaxation processes between two electronic configurations involve the dissipation of electronic energy into phonon energy.

The relaxation processes can be interrogated by time-resolved measurements, quantum-yield studies and for ultrafast (ps or shorter) processes by the uncertainty broadening of spectral lines. Usually, the physical situation is complicated and several elementary relaxation processes occur simultaneously (Fig. 6.1). Obviously,

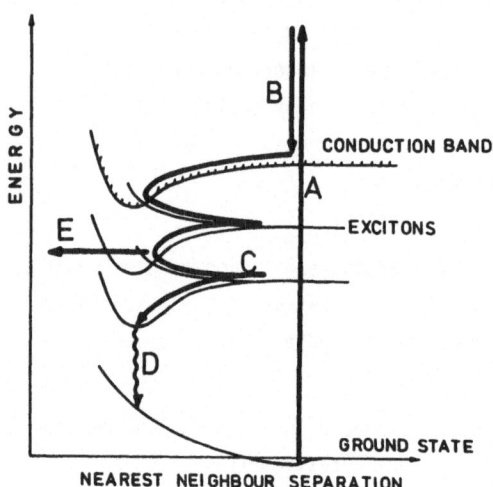

Fig. 6.1. Schematic sketch for the relaxation cascade of a free electron-hole pair in a RGS. (A) excitation; (B) scattering in the conduction bands; (C) radiative and radiationless relaxation in free and localized exciton states; (D) radiative decay to the ground state; (E) energy transfer to guest atoms and boundaries

the basic concepts are interrelated but different for excitons in pure RGS and for impurity states. The problem of exciton-phonon coupling in the pure solid requires in principle a "global" starting point for the description of the electronic states, while intrastate medium relaxation around an impurity centre in a solid or in a liquid will start from a "local" description. It is still an open question whether intrastate relaxation in a liquid requires a "global" or a "local" description. Starting with intrastate relaxation we shall discuss exciton trapping in pure RGS and then proceed to phonon relaxation in impurity centres in solids and in liquids. Subsequently, interstate processes involving electronic relaxation for localized impurity states and energy transfer between either free or localized exciton states in pure crystals and in localized impurity states will be discussed.

6.1 Modulation of Exciton Motion by Phonons

Following Toyozawa /1958,1961,1974/, Kotani and Toyozawa /1979/, we shall consider the implications of exciton-phonon coupling on the nature of exciton states formed by optical excitation in a pure solid. The nth exciton band is described in the tight-binding model by the states $|nk>$, the dispersion curve $E_n(k)$ and bandwidth 2B (Fig. 6.2). Phonon-coupling effects are introduced by lattice distortions from

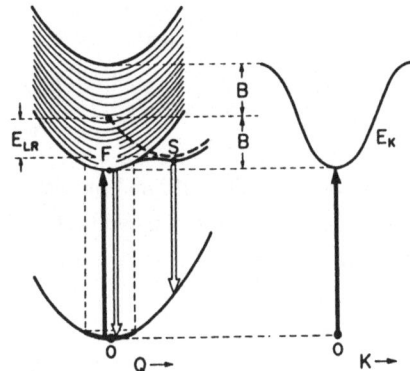

Fig. 6.2. Configuration coordinate models for a localized exciton and an exciton band in a rigid lattice (after /Toyozawa, 1974/)

equilibrium at each lattice site R_j, the coupling between nuclear and electronic motion being $V = \sum_j V_j = \sum_j \gamma Q_j$, with Q_j being the nuclear coordinate at site R_j while γ determines the local nuclear distortion upon excitation so that the lattice nuclear relaxation energy is $E_{LR} = \gamma^2/2$. Exciton scattering is determined by the matrix element

$$V_{kk'} = \frac{1}{N} \sum_j \exp[i(k - k')R_j] \, V_j \quad . \tag{6.1}$$

The thermal average of the coupling $<V_{kk'}>_T = 0$ for a harmonic lattice. As the average value of the coupling vanishes, what is required is the square of the exciton-phonon coupling strength which, by neglecting correlations between different lattice sites, is: $|V_{kk'}|^2 \cong V_j^2 = \gamma^2 Q_j^2$ and its thermal average, D^2, for a harmonic lattice being

$$D^2 \equiv <|V_{kk'}|^2>_T = \begin{cases} 2E_{LR}(\hbar<\omega>/2) & k_B T \ll \hbar<\omega> \\[2mm] 2E_{LR} k_B T & k_B T \gg \hbar<\omega> \end{cases} \tag{6.2}$$

where $<\omega>$ is the characteristic phonon frequency. D represents the statistical dispersion of the energies of the electronic excitation due to nuclear motion. It is a measure of the thermal fluctuations which damp the band motion of the exciton according to the free exciton band in k space. One can now express the exciton-scattering time τ_k and the line width Γ_k in terms of perturbation theory taking an average value for the exciton density of states

$$\frac{2\Gamma_k}{\hbar} = \frac{1}{\tau_k} \cong \frac{\pi D^2}{\hbar B} \quad . \tag{6.3}$$

The parameters D and B reflect the local and the band nature of the exciton, respectively. One can distinguish two limiting situations: (1) strong scattering $D \gg B$, when the rate of excitation transfer is slow relative to thermal fluctuations. The modulation of the exciton motion is so extreme that the exciton is essentially localized. (2) Weak scattering $D \ll B$. The exciton stays on a given lattice site only a small fraction of the time being required to switch on the nuclear fluctuation, so that the effect of nuclear motion is smeared out.

The local or the band character is reflected in the exciton absorption lineshape. In the weak scattering limit the absorption lineshape L(E) for the k = 0 optically accessible exciton at photon energy E is quasi-Lorentzian

$$L(E) = \frac{\hbar/2\pi \tau_0}{(E - E_n(0) - \Delta_0)^2 + (\hbar/2 \tau_0)^2} \tag{6.4}$$

where Δ_0 is a level shift and τ_0 is the scattering lifetime of the k = 0 excitation. The linewidth (6.3 , 4) exhibits a linear temperature dependence at high T. As pointed out by Toyozawa /1958/ interesting complications arise when the k = 0 state is located at the bottom of the lowest exciton band with no density of states below it. Under these circumstances (6.3), which was derived for an average density of final states, is inapplicable. Rather, the lineshape is a distorted Lorentzian, the linewidth being now proportional to D^4/B^3 exhibiting at high temperatures a T^2 temperature dependence. In the second extreme case, i.e., the strong scattering situation, the lineshape is Gaussian

$$L(E) = \frac{1}{\sqrt{2\pi D^2}} \exp[-(E - E(0))^2/2D^2] \tag{6.5}$$

corresponding essentially to a state localized by phonons. The linewidth is determined by the phonon fluctuation D which at sufficiently high temperatures is proportional to $T^{1/2}$.

Three diagnostic criteria are applicable to assess whether an excitonic absorption corresponds to the weak scattering limit or to the strong scattering situation. (1) One can examine the relation between B and D. (2) The exciton absorption lineshapes can be analysed. (3) The temperature dependence of the linewidth is relevant. Regarding pure RGS it should be pointed out that at present the available experimental data are inadequate for the application of criterion (3) and they are difficult to evaluate regarding (2) as the details of the best available spectra (Chap. 3) for Xe, Ar, Kr and Xe are obscured by surface excitons and by the appearance of longitudinal excitons, as well as by the overlap between the $^3P_1$ and $^1P_1$ components for Ne and Ar. Concerning criterion (1) we have to rely on a rough order of magnitude estimate. The exciton bandwidths in k space for the lowest optically accessible n = 1 exciton in RGS vary from 2B = 0.9 eV for Xe to 0.4 eV for Ne (e.g. /Fugol, 1978/). From the estimates of the lattice relaxation energy due to two-centre localization of the diatomic molecule (see Sect. 6.2), which constitute an upper limit to E_{LR}, one can infer from (6.2) that at 50 K, $D \cong 0.4$ eV for Ne and $D \cong 0.07$ eV for Xe. Thus for Xe $D \ll B$ corresponding to the weak scattering limit while $D \cong B$ for Ne, which is still on the verge of weak scattering. The experimental absorption spectra for n = 1 excitons in Ar, Kr and Xe reveal a linewidth (FWHM) of 80 meV /Saile et al., 1977/, while for the n = 1 exciton in Ne the linewidth is 200 meV /Saile and Koch, 1979/. The increase of the linewidth for Ne presumably reflects the transition from the weak scattering limit in Ar, Kr and Xe towards the strong scattering situation. No evidence for an asymmetry of the lineshapes of RGS towards lower energies was observed, while Toyozawa's theory /1958/ for excitation to k = 0 of the lowest exciton band would imply such asymmetry. It should be noted that the n = 1 exciton band having parentage in the $^3P_1$ atomic excitation is not necessarily the lowest exciton state, as the exciton band originating from the $^3P_2$ state may be located below the $^3P_1$ band /Knox, 1959,1963/. Interband exciton scattering from the optically accessible $^3P_1$ band to the lower $^3P_2$ band will result in symmetric line broadening and, provided that the weak scattering limit holds, the lineshape will be Lorentzian and proportional to T.

The local effects which may destroy the band character of the exciton are not restricted to the case of thermal fluctuations. Disorder scattering effects due to either substitutional or positional disorder play a similar role in introducing local effects. These considerations are important for the electronic states of a pure liquid where disorder results in a static spread D_S of the site excitation energy. Provided that $D_S > B$ the electronic excitation is localized. For liquid Xe the linewidth (FWHM) of the n = 1 exciton is ~ 0.3 eV which provides a rough measure of D_S. We expect the exciton bandwidth in the liquid to be lower than in the solid so that $B < 0.45$ eV. Accordingly $D_S \geqslant B$ for liquid Xe. Thus, in the liquid the electronic excitation may well belong to the strong scattering situation.

6.2 Self-Trapping of Excitons

From the discussion of the optical excitation of an exciton coupled to lattice vibrations, we now proceed to consider the subsequent fate of the exciton. The basic question in this context is whether the exciton will be free or localized. In the strong scattering limit the exciton is localized. On the other hand, the condition $D < B$ for weak scattering is not sufficient to ensure that the exciton will remain free on the time scale of its radiative decay /Toyozawa, 1974/. This is apparent as exciton localization lowers the lattice energy by E_{LR} due to nuclear distortions while the energy gain due to exciton motion is B. Thus the nature of the final relaxed state of the lattice depends on the relative magnitudes of B and E_{LR}; the free state is energetically stable when $B \gg E_{LR}$, while a localized state is formed when $B \ll E_{LR}$ /Toyozawa, 1974; Rashba, 1976/. Exciton dynamics in RGS is of considerable interest here, because long-range coupling effects, for example, due to polar modes are absent. In RGS short-range coupling with acoustic phonons prevails. For a system characterized by such a short-range interaction, which can be described in terms of a deformation potential, Toyozawa /1974/ and Rashba /1976,1981,1982/ predicted a discontinuous transition from the free to the localized state when $B \sim E_{LR}$. The localization process due to lattice distortion can be visualized as splitting off a bound state from the exciton band for a finite value of the lattice distortion (Fig. 6.2). When E_{LR} is sufficiently large, the localized state will be characterized by a minimum on the adiabetic potential surface which is separated from the free exciton state by a potential barrier. Toyozawa /1974/ and Rashba /1976/ have described self-trapping in terms of an isotropic continuum model where the lattice distortion corresponds to an elastic dilation. The energy of the electronic excitation is

$$E(\alpha) = B\alpha^2 - E_{LR}\alpha^3 \qquad (6.6)$$

where the lattice relaxation energy is $E_{LR} = E_d^2/2Cr_k^3$ with E_d being the deformation potential while C_1 is the elastic constant, r_k is the nearest neighbour distance, and $0 \leqslant \alpha \leqslant 1$ represents the orbital exponent of a trial Gaussian electronic wave function. $\alpha = 0$ denotes an extended state, while $\alpha = 1$ corresponds to a localized state. The first term in (6.6) accounts for the increase of the kinetic energy with localization while the second term provides a stabilization energy due to dilation of the elastic medium. Although a true minimum of $E(\alpha)$ is not exhibited for $\alpha > 0$, it is obvious from (6.6) that $E(1) < E(0)$ for $E_{LR} > B$. As the continuum approximation breaks down at $\alpha = 1$, an energetically stable localized state with $\alpha = 1$ is formed. The free and the localized states are separated by a maximum of a height

$$H_m = 4B^3 / 27E_{LR}^2 \quad . \qquad (6.7)$$

According to this continuum model two conditions have to be satisfied for the observation of self-trapping in a solid. First, $E_{LR} \geqslant B$ is necessary. Second, dynamic conditions have to be fulfilled which imply that the barrier height (6.7), is comparable to the Debye phonon frequency $H_m \leqslant \hbar\omega_D$, so that nuclear tunnelling from the free state to the localized state is feasible. On the other hand, when $H_m > \hbar\omega_D$ and $H_m > k_B T$, the barrier cannot be overcome and the free state is expected to be stable.

A closer inspection of the dynamic condition for localization has to take into account the crossing of the barrier by tunnelling, thermoactivated tunnelling, over barrier transitions and reverse tunnelling in competition with relaxation of hot localized excitons /Rashba, 1981,1982; Nasu and Toyozawa, 1981/. Approximate expressions are given by a prefactor representing a mean phonon frequency and the transparency Tr. Tunnelling is described by $Tr = \exp(-2S)$ with a classical action S. Thermal activated tunnelling corresponds to $Tr = \exp(-2S) \cdot (1 - T/\Theta)^{-\alpha}$ with a density α of vibrational modes and the Debye temperature Θ /Rashba, 1981,1982/. This treatment and the even more elaborate calculations of Nasu and Toyozawa /1981/ predict high tunnelling rates for the solid rare gases as compared to the radiative lifetimes. According to Nasu and Toyozawa /1981/ an exponential decrease of the localization rate constants is expected for increasing band width B. Luminescence should originate mainly from localized excitons. The probability for localization should decrease in passing from Ne to Xe and the rather spurious contributions from free excitons should be strongest in the case of Xe.

Up to this point the general question of self-trapping was explored on the basis of a continuum model which does not incorporate any details of the microscopic structure. Two stable configurations of the localized state are of particular interest (Fig. 6.3): (1) The one-centre atomic trapped exciton. Here the nearest neighbours displace symmetrically around the excitation. (2) The two-centre trapped excitation. Here two neighbouring atoms displace axially to trap the excitation.

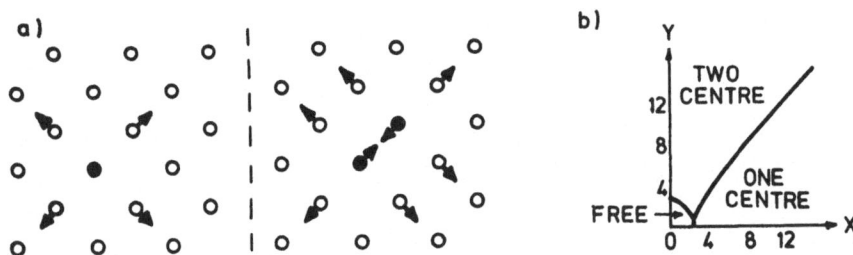

Fig. 6.3a,b. (a) Schematic lattice rearrangement for one centre self-trapping (*left*) and two centre self-trapping (*right*) of excitons in RGS after Song /1969, 1971/ and Jortner /1974/. (b) Map of zones of stable states of excitons for alkali halides in the plane $X = U'/(\alpha|V|)^{1/2}$, $Y = V'/(\alpha|V|)^{1/2}$ (from Song /1969/). For α, V, V' and U' see text.

The question whether the one-centre or the two-centre trapping is favoured should be settled on the basis of energetic and dynamic criteria. Exciton trapping bears a close analogy to hole trapping in alkali halides resulting in the V_k centre formation (e.g. /Das et al., 1964; Song, 1969/). The starting point for a microscopic model is to express the total energy $E_t(Q)$ of the solid as

$$E_t(Q) = E_e + I(Q) + E_L(Q) \qquad (6.8)$$

where E_e is the electronic energy of the free exciton state, $I(Q)$ is the interaction between the exciton and the lattice displacements, while $E_L(Q)$ is the lattice energy. The potential energy depends parametrically on the lattice coordinates Q. Two lines of attack are now feasible. (i) An excitonic model for one-centre and two-centre trapping /Song, 1969/. The exciton lattice coupling is expressed in terms of two contributions, i.e. the change in the diagonal site energies, U', and the change of the off-diagonal pair interactions V' with lattice displacements. The lattice energy is calculated in the harmonic approximation. It was demonstrated by Song /1969/ that large diagonal deformation energies U' result in one-centre trapping, while large off-diagonal deformation energies V' give two-centre trapping (Fig. 6.3). (ii) The molecular model for two-centre trapping /Das et al., 1964/: It can be argued that two-centre localization results in the formation of an ordinary excited diatomic molecule. This point of view is in particular pertinent for RGS where this diatomic homonuclear excited state reveals very similar characteristics to those of the gas phase excimer /Jortner, 1974/. One can then attempt to start from a local approach expressing the electron-lattice coupling $I(Q)$ in terms of the molecular potential curve of the "isolated" diatomic molecule, while the lattice deformation energy $E_L(Q)$ can be calculated along conventional lines provided that potential parameters are available. This approach, which disregards part of the excitonic effects, is also applicable to liquids and to impurity pairs.

The two models are not mutually exclusive but rather complementary. The electronic and geometrical structure of one-centre /Song and Leung, 1981; Song, 1982; Leung et al., 1983/ and of two-centre /Song and Lewis, 1979; Song, 1982; Grosso et al., 1982/ self-trapped excitons have recently been calculated. The results will be discussed in connection with transient absorption experiments in Sect. 6.5.

It should be noted that (6.8) provides the microscopic analogue of the continuum approximation (6.6). In principle, (6.8) gives the potential surface for free, one-centre and two-centre localized excitations and the nature of the potential barriers separating these states. Equation (6.8) can be minimized with respect to nuclear displacements and should provide three energies: $E_t(0)$ corresponding to the free state, $E_t(Q_1)$ for the minimum of the one-centre localized state and $E_t(Q_2)$ for the minimum of the two-centre localized exciton. The energetic criterion for localization is $E_t(Q_1)$ and/or $E_t(Q_2) < E_t(0)$. The dynamic criterion for localization stems from the existence of barriers and considerations of tunnelling probability.

The predominance of the molecular characteristics of the trapped two-centre exciton in RGS /Jortner, 1974/ strongly suggests but does not prove that the molecular approach to the problem of trapping in these materials should provide a feasible starting point. Fugol and Tarasova /1977/ and Fugol /1978/ have considered the energetic stability of trapped excitons in RGS (Fig. 6.4). For two-centre trapping they have estimated E_{LR} from the experimental potential surfaces of the diatomic rare-gas

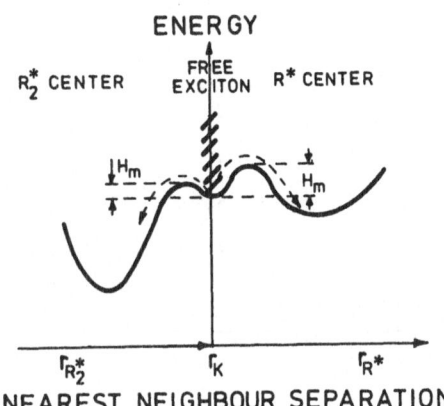

ENERGY

R_2^* CENTER   FREE EXCITON R^* CENTER

H_m H_m

$r_{R_2^*}$ r_k r_{R^*}

NEAREST NEIGHBOUR SEPARATION

Fig. 6.4. Schematic potential curves and lattice distortions describing the localization of free excitons in atomic and excimer centres (after Fugol /1978/)

molecule. The results of their analysis are summarized in Table 6.1. From these estimates they conclude that the two centre trapped exciton, i.e., the diatomic excimer, is energetically stable with respect to the free state in all RGS. They have also estimated E_{LR} for one-centre trapping. These estimates (Table 6.1) reveal that E_{LR} (one centre) < E_{LR} (two centre) for all RGS so that according to the energy criterion the two-centre state is favoured. To assess the dynamic criterion, Fugol and Tarasova /1977/ utilized a hybrid of the isotropic continuum model (6.6) together with the molecular estimates of E_{LR} to derive the barrier humps H_m (Table 6.1). These data predict that the barrier height for two-centre and one-centre localization in Ne and Ar is sufficiently low. In Xe the one-centre localized state is energetically unstable and the barrier height of 20 meV is large for two-centre localization so that the free state should be observed. These predictions agree with the estimate on the tunnelling rates mentioned above. Solid Xe may provide an interesting example for a system where tunnelling between free and localized exciton states occurs.

Table 6.1. Parameters for localized excitons from Fugol /1978/ in eV. $\hbar\omega_D$: Debye energies; 2B: exciton band width; S: Stokes shift between absorption and emission; S*: dissociation energy of the centre; E_{LR}: lattice relaxation energy; Λ: coupling parameter; H_m: barrier height

	excimer centres R_2^*				atomic centres R^*			
	Ne	Ar	Kr	Xe	Ne (Γ 1/2)	Ne (Γ 3/2)	Ar (Γ 3/2)	Xe (Γ 3/2)
$\hbar\omega_D$	0.0064	0.0080	0.0062	0.0055				
2B	0.4	0.7	0.9	0.9				
S	3.7	2.46	1.8	1.3	1.1	0.9	0.5	-0.06
S*	1.85	1.5	0.93	0.4	1.05	0.85	0.32	-0.4
E_{LR}	2.0	1.86	1.38	0.85	1.25	1.05	0.77	0.05
Λ	10	5.3	3.1	1.9	6.25	5.25	2.2	0.11
H_m	0.0003	0.002	0.01	0.02	0.0008	0.001	0.01	5.6

6.3 Maxima in the Molecular Potential Curves

The existence of potential barriers in the total potential energy of an excited solid undergoing structural deformation has been described within the continuum theory for exciton localization (see Sect. 6.2). Starting from the molecular point of view for two-centre excitation trapping one should incorporate the molecular potential curve in I(Q) (6.8). The potential curves for the lowest states of some of these free diatomic excimer rare-gas molecules exhibit barriers /Mulliken, 1964/.

These barriers are very pronounced for the lighter rare gases He_2^* (e.g., /Guberman and Goddard, 1975/) and Ne_2^* (e.g. /Cohen and Schneider, 1974/). They appear also in some branches of potential curves of heavy rare gases. The origin of these maxima is the Pauli exclusion principle and has to be well distinguished from the solid state effects in Fig. 6.4. Nevertheless, these maxima have a similar retarding effect for the formation of excimer centres. Thus it is apparent that the molecular potential curves provide some indispensable ingredients which must be incorporated in a complete theory (see the case of solid Ne discussed in Sect. 6.5).

6.4 Vibrational Relaxation in Excimer Centres

The isotropic continuum theory /Rashba, 1976; Toyozawa, 1974/ implies that the dynamic restriction for the formation of the diatomic species involves tunnelling over the potential barrier. A hidden assumption involved in this description is that vibrational relaxation of the two-centre trapped excitation is extremely fast on the time scale of radiative decay, so that once the barrier is overcome the system instantaneously slides down to form the final relaxed diatomic molecule. Inverse tunnelling has also to be considered /Rashba, 1981,1982/. The vibrational relaxation process essentially corresponds to molecular vibrational relaxation of a diatomic molecule in an RGS /Jortner, 1974/. It is now known that vibrational relaxation of a guest diatomic molecule in a solid can be surprisingly slow /Nitzan et al., 1975/. The vibrational relaxation rate is determined by the energy gap $\Delta E_v = (E_n - E_{n-1})$ between adjacent vibrational states. This interconversion of vibrational energy of the diatomic molecule to lattice phonons ($\hbar\omega_p$) is a multiphonon process of order $N = \Delta E_v/\hbar\omega_p$. The zero temperature rate $\gamma_v(0)$ follows the energy gap law

$$\gamma_v(0) = A \exp(-\delta N) \tag{6.9}$$

where $A = 10^{12}$ s^{-1} is a coupling constant /Jortner, 1976/ and δ is a numerical factor of the order of unity. The rate decreases exponentially with N, and therefore, high-order multiphonon processes are expected to be slow. The exponential dependence of the rate on the energy gap implies that the vibrational relaxation rate can decrease by a few orders of magnitude by going from higher to lower vibrational energy levels of the diatomic molecule. These general considerations imply that vibrational relaxation will be fast within the diatomic centre in Xe, Kr and Ar where the vibrational frequency is low and N small, while for Ne the high-order multiphonon vibrational relaxation will seriously be retarded.

6.5 Localization of Excitons and Vibrational Relaxation in Solid Ne, Ar, Kr and Xe

Generally, two types of luminescence emission bands are observed in RGS, i.e., strong broad bands with a large Stokes shift (1 - 2 eV) and weak sharp bands with some fine structure and much smaller (< 1 eV) Stokes shifts relative to the first exciton absorption band. First we shall discuss the prominent emission bands. In Xe, Kr and Ar most of the luminescence intensity is emitted in broad luminescence bands of nearly Gaussian shape /Jortner et al., 1965/. These bands have been observed in all investigations and a typical set of spectra is shown in Fig. 6.5. The energies are compiled in Table 6.2. These luminescence bands are assigned /Jortner et al., 1965/ to emission from vibrationally relaxed electronically excited homonuclear diatomic rare-gas

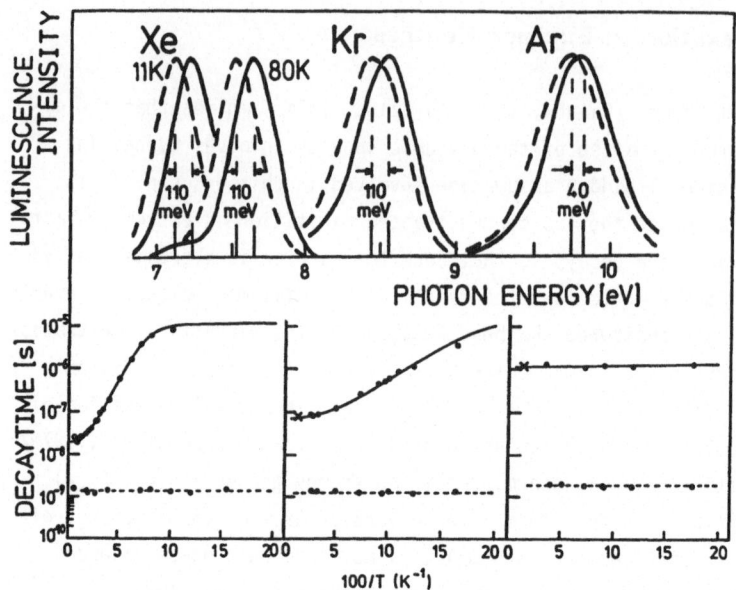

Fig. 6.5. Results of time-resolved luminescence experiments for the main luminescence bands of solid Xe, Kr and Ar. (————) in the intensity curves (*upper panel*)
correspond to the component with short decay times (10^{-9} s region); (— — —) to long
decay times. The emission bands are taken from Dössel et al. /1983/. Decay times
(*lower panel*) for Kr and Ar are taken from Roick /1984/. The decay times for Xe
were determined by Hahn et al. /1977/, Kink et al. /1977/ and Roick /1984/

molecules which correspond to the trapped two-centre excitations. Although the prominent molecular character of these bands in solid Ar, Kr and Xe is well established,
two complicated features are exhibited in the solid state experimental data. Firstly, trivial complications arise due to inhomogeneous broadening of the emission due
to statistical distribution of trapping sites which cause different spectral shifts.
A careful analysis by Heumüller /1978/ of the position and shape of these bands for
different temperatures, as well as preparation and annealing conditions, showed that
in most cases these bands are inhomogeneously broadened. The shifts of these bands
with changing the temperature reported earlier can be ascribed to annealing. Secondly, a non trivial temperature dependence is exhibited in the emission spectrum of
solid Xe which in the temperature range 60 - 130 K exhibits a second prominent emission band peaking at 7.6 eV (Table 6.2 and Fig. 6.6). Heumüller and Creuzburg /1978/
identified four components in the Xe emission band by a deconvolution with Gaussians
(Table 6.2).

Two new techniques were introduced to obtain further detailed information regarding the nature of the emitting molecular species in Ar, Kr and Xe:
(1) *Time-resolved data.* In lifetime measurements for the Xe 7.2 eV emission band,
a short temperature-independent lifetime of some ns and a temperature-dependent long

Table 6.2. The broad emission bands "d" of solid Ar, Kr and Xe (in eV)

Reference	Ar		Kr		Xe			
Jortner et al. /1965/		9.84		8.25			7.19	
Cheshnovsky et al. /1972,1973/ and Gedanken et al. /1973/		9.72		8.38	7.56		7.19	
Basov et al. /1970/				8.42	7.56		7.08	
Creuzburg /1971/			8.66	8.45	7.60	7.24	6.93	6.53
Heumüller /1978/ and Heumüller and Creuzburg /1978	9.80	9.70	8.60	8.45	7.61	7.44	7.22	7.02
Fugol /1978/		9.63		8.33	7.6		7.1	
Hanus et al. /1974/ and Coletti and Hanus /1977		9.76		8.43	7.65		7.21	
Huber et al. /1974/		9.61		8.38				
Nanba et al. /1974/		9.30		8.0			6.97	
Brodmann et al. /1974/ and Ackermann et al. /1976/		9.80		8.25	7.6		7.05	
Tolkiehn /1976/ and Gerick /1977/		9.72			7.6		7.1	
Hahn et al. /1977/							7.2	
Zimmerer /1979/	9.86	9.83	8.46	8.4			7.07	7.03
Kink et al. /1977/					7.6		7.2	
Dössel et al. /1983/	9.79	9.75	8.55	8.44	7.65	7.54	7.23	7.12

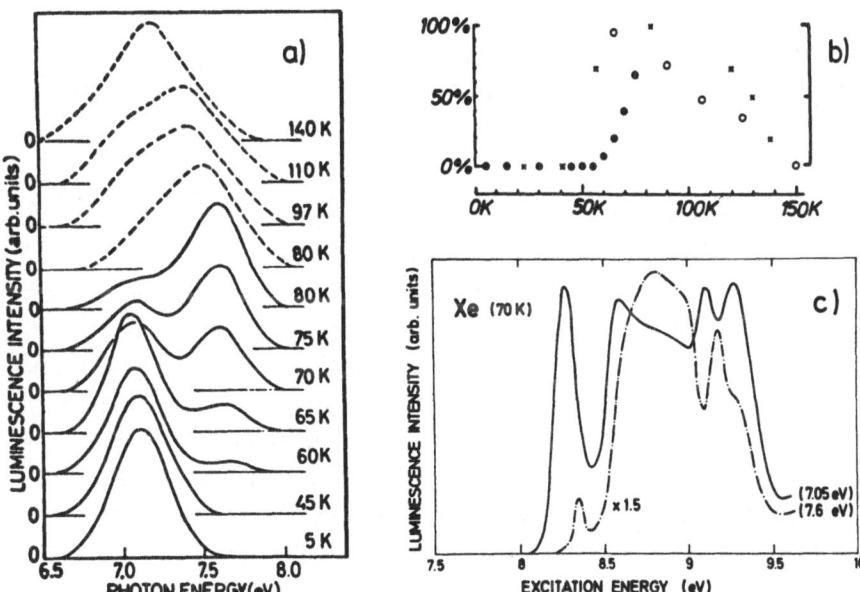

Fig. 6.6a-c. (a) Emission spectra for solid Xe excited by α particles and VUV light /Zimmerer, 1976/. (b) Intensity ratio of the 7.6 eV emission band to the 7.05 eV emission band for different temperatures. (o o o) electron excitation /Basov et al., 1970/; (x x x) α particle excitation /Cheshnovsky et al., 1973c/; (● ● ●) α particle excitation /Zimmerer, 1976/. (c) Luminescence efficiency in the 7.05 eV and 7.06 eV band of solid Xe for a spectrum of excitation energies /Ackermann et al., 1976/

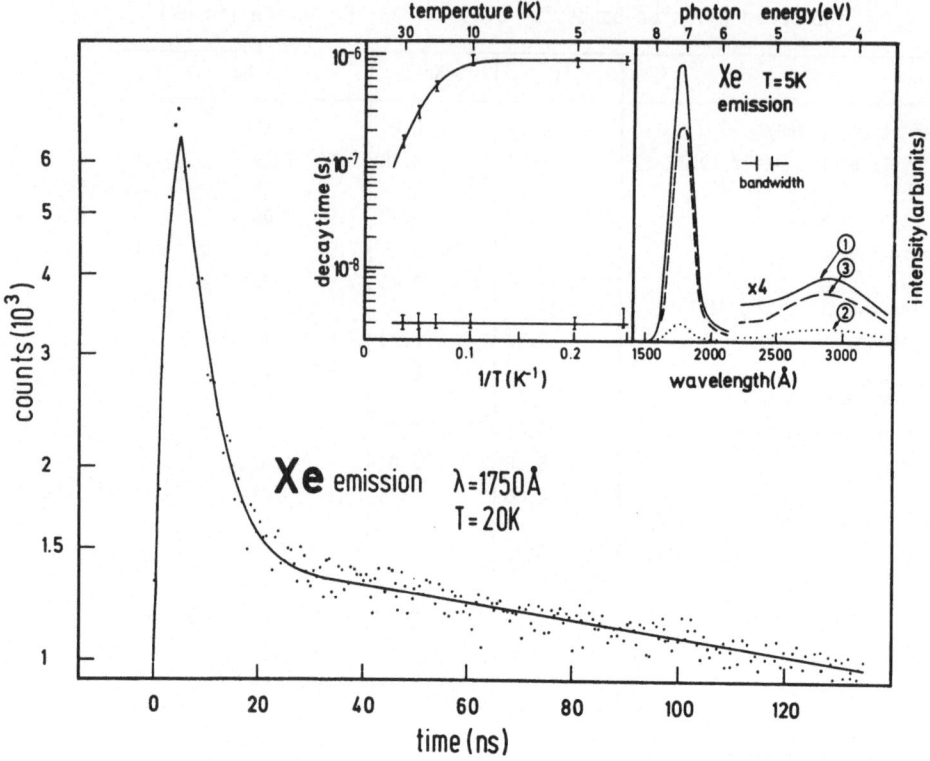

<u>Fig. 6.7.</u> Decay curves for solid Xe. (*Left insert*) temperature dependence of the long and short components; (*right insert*) emission bands for different times after deposition of the sample /Hahn et al., 1977/

lifetime in the μs regime have been reported /Hahn et al., 1977/ (Figs. 6.5,7). In these experiments, the long lifetime is shortened by surface quenching. At low temperature King et al. /1977/ measured a lifetime longer than 10 μs (Fig. 6.5). For Kr and Ar a short component and temperature dependent long components have been observed by Hahn /1978/, Jordan /1978/, Dössel et al. /1983/ and Roick /1984/. Further the emission bands of Xe, Kr and Ar have been separated by these authors into two bands by recording the emission spectra with (i) a time window of some ns just covering the time range of the excitation, and (ii) with a time delay excluding the short component. The long components are found at lower energies whereas the fast components are at higher energies. Dössel et al. /1983/ observed a splitting in both Xe bands (Fig. 6.5). The splittings for each of the Xe bands and for Kr are 110 meV and for Ar a smaller value of 40 meV has been obtained (Fig. 6.5). Time-resolved data have also been reported by Creuzburg and Völkl /1977/, Coletti and Hanus /1977/, Carvalho and Klein /1978/ and Monahan et al. /1977/.

(2) *Transient absorption data.* The self-trapped exciton states are not accessible to usual absorption spectroscopy, because of the limitations imposed by the Franck-

Condon principle. These states have been studied by transient absorption spectroscopy in solid Xe, Kr, Ar and Ne /Suemoto et al., 1977,1978,1979; Dössel et al., 1983/. In solid Kr and Ar transitions from the lowest excited triplet state $^3\Sigma_u^+$ to several higher excited states have been identified (Table 6.3). By comparison with

Table 6.3. Transitions observed in transient absorption spectra. For further data see Dössel et al. /1983/ and Grosso et al. /1982/

	energies (in eV)				
	solid	liquid	solid		
	experimental		calculated		
He	d)	a)		$2s\,^3S \rightarrow 2p\,^3P$	atomic
	0.590	1.145		$\rightarrow b\,^3\Pi_g(0-0)$	
		0.590		$a\,^3\Sigma_u^+ \rightarrow c\,^3\Sigma_g^+(0-0)$	molecular
		1.363		$\rightarrow b\,^3\Pi_g(v\gg 0)$	
		0.646			
Ne	b)	b)		$\rightarrow 2p_2,\,2p_4,\,2p_5$	
	2.29	2.18		$\rightarrow 2p_6,\,2p_7$	atomic
	2.19	2.11		$1s_5 \quad \rightarrow 2p_8,\,2p_9$	
	2.10	2.02		$\rightarrow 2p_{10}$	
	1.90	1.84		$\rightarrow\,^3\Pi_g$	
	1.76	1.67			
Ar			c)	$\rightarrow\,^3\Pi_g$	
	1.38	1.30	1.12 – 1.48	$\rightarrow\,^3\Sigma_g$	
	1.23	1.13	1.02 – 1.21		
Kr	1.6			$^3\Sigma_u$	molecular
	1.21				
	1.05	1.18	1.03 – 1.29	$\rightarrow\,^3\Pi_g$	
	(0.93)	(0.9 – 1.0)	0.95 – 1.07	$\rightarrow\,^3\Sigma_g$	
Xe	<1.1		0.80 – 1.05	$\rightarrow\,^3\Pi_g$	
			0.78 – 0.91	$\rightarrow\,^3\Sigma_g$	

a) Hill, Heybey and Walter /1971/
b) Suemoto and Kanzaki /1979/
c) Song and Lewis /1979/
d) Soley, Leach and Fitzsimmons /1975/

the available calculated molecular potential curves, the spectra for Ar and Kr are correlated to transitions from $^3\Sigma_u$ to $^3\Sigma_g$, $^3\Sigma_g'$ and $^3\Pi_g$ (Fig. 6.8). The transition energies in this Rydberg series of self-trapped excitons have been calculated by Song and Lewis /1979/. These authors used two methods; an effective mass approximation including corrections which were used also for free excitons in RGS by Hermanson /1966/ and a pseudo-potential approach which has been applied successfully to F centres and self-trapped excitons in alkali halides. The results, collected in Table 6.3 agree quite well with the experiment. A similar conclusion has been reached by Grosso et al. /1982/ by applying their envelope function formalism (Table 6.3). The connection between free excitons, self-trapped excitons and the free atoms and excimers has been illustrated in an empirical approach by Dössel et al. /1983/. The only optical experiment reported up to date on solid He /Soley et al., 1975/ used transient absorption to identify the $^3\Sigma_u$ molecular state. These

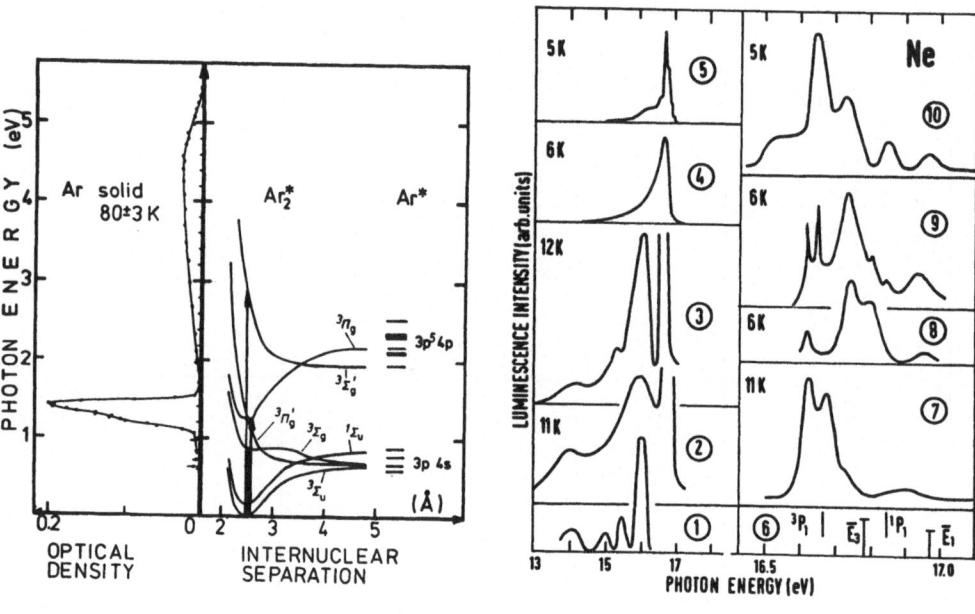

Fig. 6.8   Fig. 6.9

Fig. 6.8. Transient absorption of solid Ar /Suemoto and Kanzaki, 1979/ and calculated potential curves of Ar_2^* molecules /Spiegelman and Malrieu, 1978/. The assignment of the observed features is indicated by vertical arrows

Fig. 6.9. Comparison of luminescence emission spectra for solid Ne for different excitation conditions. In the left set of the spectra (1) - (5) the full emission spectrum is shown. In the right set of the spectra (7) - (10) the high energy part is shown on an expanded scale. (1) calculated R_2^* centre emission /Yakhot, 1975/; (2 and 7) X-ray excitation /Schuberth and Creuzburg, 1975/; (3) β excitation /Packard et al., 1970/; (4) light excitation /Zimmerer, 1978/; (5 and 10) 300 eV e$^-$ excitation /Fugol, 1978/; (6) calculated R^* centre emission /Kunsch and Coletti, 1979/; (8) 5 keV e$^-$ excitation, low current /Coletti and Bonnot, 1978/; (9) 200 eV e$^-$ excitation, high current /Coletti and Hanus, 1977/

experimental data provide further support for the molecular description of the self-trapped exciton in the heavy RGS as well as in solid He.

Proceeding with the survey of the prominent emission bands of RGS we note that the emission spectra of solid Ne are exceptional (Fig. 6.9). They are dominated by a relatively narrow emission band peaking at 16.7 eV which is Stokes shifted by 0.7 eV to lower energies relative to the n = 1 ($^3P_{3/2}$) exciton state. This emission originates from a medium relaxed atomic state /Packard et al., 1970; Gedanken et al., 1973b/ which has been assigned to a one-centre trapped exciton. The splittings observed (Fig. 6.9) in some experiments of this "atomic type" emission into several components have been correlated to the $^1P_1$, $^3P_1$ and $^3P_0$ splitting in the free atom. In this context, the excitation method is crucial. VUV light and X rays do not damage the sample and the penetration depth is large enough to excite mainly the volume. Creation of lattice defects by electrons with energies up to 200 eV is also negligible, but the penetration depth is small and contamination and surface effects will become important. The penetration depth increases with electron energy, but the sample damage will be more severe at higher energies. High currents can cause a local heating of the sample. α particles will induce strong lattice defects due to the large mass which is comparable to the mass of the lattice atoms. In a recent investigation emission from different trapping sites and from desorbing atoms has been established /Coletti et al., 1985/.

The assignment of the prominent emission of solid Ne is confirmed by transient absorption spectroscopy /Suemoto and Kanzaki, 1979/ where besides one transition due to molecular type centres, several transitions due to the atomic-like centres have been observed. The atomic-like transitions start from the lowest atomic state $^3P_2$ ($1s_5$) terminating at the fine structure components of the higher lying p states (Table 6.3).

The medium dilation around atomic centres of solid Ne has attracted the interest of several theoretical investigations /Kunsch and Coletti, 1979; Kusmartsev and Rashba, 1982; Song, 1982; Leung et al., 1983/. These studies derive an increase of the nearest neighbour separation in the relaxed one centre configuration of about 1 Å, i.e., 30% to 50% as compared to the configuration of the ground state. The transient absorption experiments of Suemoto and Kanzaki /1980,1981/ and the interpretation by Song /1982/ show that in addition to this fast bubble formation also up to 3 or 5 vacancies are accumulated at these relaxed centres on a time scale of $10^{-4} - 10^{-6}$ s. The temperature dependence of the growth rate proves that the vacancies are thermally created and that these migrating vacancies are captured by the centre. Further, solid Ne exhibits a medium intensity broad emission (Fig. 6.9) towards lower energies which is due to unrelaxed molecular emission /Schuberth and Creuzburg, 1975; Yakhot, 1975/. This assignment is also borne out by transient absorption experiments.

Next, we survey the weak emission bands from RGS. In solid Ar, Kr and Xe the relative intensities of these emissions (bands a, b, c in Fig. 6.10) are 10^{-2} to 10^{-3} of the major two-centre molecular emission band (band d in Fig. 6.10). The assignment of these weak bands has to consider whether they are indeed intrinsic or whether they are due to structural disorder, i.e., emission from excitons trapped at imperfections, vacancies etc., or at surface states. The experimental results are still controversial concerning the number of peaks, the energy positions (Table 6.4) and the relative and overall intensities. Despite these problems there is now a very detailed experimental investigation of the temperature dependence of the narrow line emission available /Coletti and Bonnot, 1978/ (Fig. 6.10). In particular, the optical excitation data of Kink and Selg /1979/, Roick et al. /1983/ and Roick /1984/ provide good evidence that the weak 8.3 eV emission for solid Xe is intrinsic.

The following classification of the emission bands of RGS will be introduced in the order of decreasing energy:

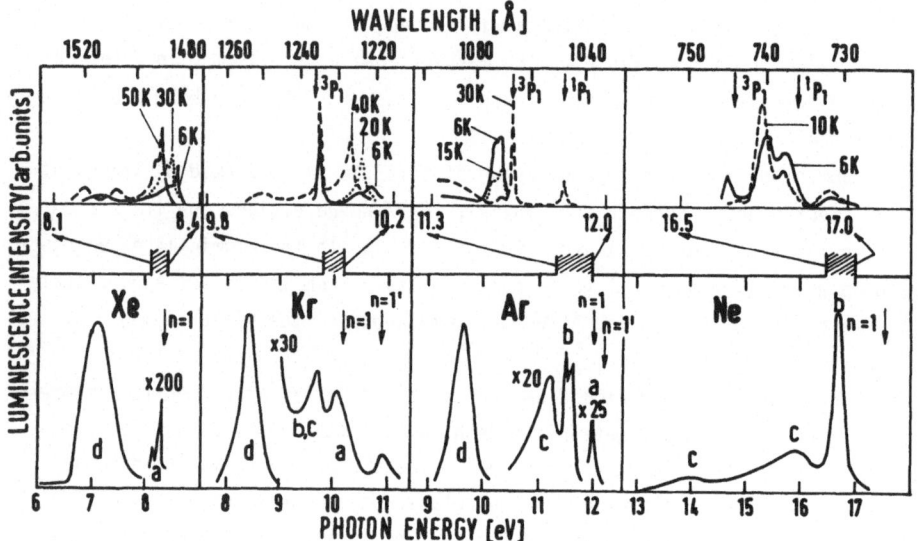

Fig. 6.10. In the lower part an overview of the complete emission spectra of solid Xe, Kr, Ar and Ne are shown with an assignment of the individual features; "a" free excitons; "b" atomic centres; "c" vibrationally excited excimer centres R_2^*; "d" vibrationally relaxed R_2^* centres. The Xe and Ar spectra have been obtained by electron excitation at 5 K /Fugol, 1978/; the spectrum for Kr by light excitation at 5 K /Zimmerer, 1979/; the spectrum for Ne by X-ray excitation at 5 K /Schuberth and Creuzburg, 1975/. In the upper part, the temperature dependence of the emission bands "a" and "b" excited by 5 keV electrons is displayed on an expanded scale /Coletti and Bonnot, 1978/

Table 6.4. The narrow emission bands "a", "b" and "c" of solid Ne, Ar, Kr and Xe (in eV). The assignment is partly tentative

	Ne			Ar			Kr		Xe
	a	b	c	a	b	c	a	b	a
Fugol /1978/	16.95, 16.76, 16.62	16.87, 16.68, 16.56	16.2	12.1	11.64, 11.58	11.37			8.33, 8.18
Hanus et al. /1974/	17.01, 16.66	16.91, 16.64	16.80, 16.79, 16.74, 16.49, 16.1	11.67, 11.53, 10.74	11.61, 10.99, 10.58	11.56, 10.89	10.13, 10.02, 9.69	10.12, 9.92	9.4, 9.2, 9.0, 8.36, 8.34, 8.24, 8.18
Coletti and Hanus /1977/	16.97, 16.79	16.85, 16.67, 16.53, 15.04	16.91, 16.73, 16.18	11.58	11.83, 11.62		10.15	10.65, 10.04	8.35
Coletti and Bonnot /1978/ Kink et al. /1979/	temperature dependence of emission bands see Fig. 6.10								
Packard et al. /1970/	16.7	16.0	14.0						
Brodmann et al. /1974,1976/						11.1, 11.3			8.2
Gerick /1977/					11.85, 11.6				
Schuberth and Creuzburg /1975/	16.9, 14.0	16.8	16.7, 16.65, 16.1						
Heumüller /1978/							10.14	9.99	8.33, 8.26
Kink et al. /1979/									8.35
Fugol et al. /1982/									8.36, 8.35
Roick et al. /1983,1984/					11.62	11.38			8.3

a band. Corresponding to the emission from the free exciton state. This emission can be observed in a system where a free exciton state is stable. Further, in a system where the trapped exciton is energetically and dynamically favoured, competition between exciton localization (Sect. 6.2) and radiative decay of the free exciton will be exhibited. The emission quantum yield Y_f for the emission from the free exciton state is in this case

$$Y_f = \tau_t/\tau_R \qquad (6.10)$$

where τ_R is the pure radiative lifetime of the free exciton and τ_t is the exciton trapping time.

b band. Corresponding to the localized atomic excitation, i.e., the one-centre trapped exciton. The emission peak is red shifted relative to the absorption, because in RGS the short-range one-centre excitonic interaction term (the matrix spectral shift) is repulsive, i.e., U' (see Sect. 6.2) is negative. The emission line will be phonon broadened.

c band. Due to emission from a vibrationally hot diatomic R_2^* molecule. In this case exciton trapping has occurred by the two-centre mechanism forming the vibrationally excited diatomic molecule. However, vibrational relaxation is inefficient or only partially completed on the time scale of radiative decay.

d band. Assigned to the broad prominent Gaussian bands due to emission from a vibrationally relaxed diatomic R_2^* homonuclear molecule, as already discussed above.

We shall now discuss the emission spectra of each RGS in more detail. In Table 6.5 we present a synopsis and a tentative assignment of the emission bands for all RGS. Solid Xe exhibits a weak a band, $Y_f \sim 10^{-3}$, and a dominant d-emission band. In Kr some evidence for weak a , b and c emission is present ($Y_f \sim 10^{-2}$) together with the dominant d band, while Ar reveals very weak a ($Y_f \sim 10^{-4}$) and weak b and c bands ($Y_f \sim 10^{-2}$) together with the strong d band. Finally, for Ne emission d is missing and emission b is strong compared with the broad c bands.

For a detailed interpretation of the molecular type emission spectra (bands d and c) of vibrationally relaxed and non relaxed R_2^* centres, we use input information from gas phase spectra at low and high pressures. The broad band emission d is very similar to the emission observed in the gas phase at high densities. In the gas phase it is well known that excited rare-gas atoms have attracting potential curves and form excimers R_2^*. At high pressure the molecules decay after complete vibrational relaxation by emitting the second continuum (Fig. 6.11). At low pressure, the collision rate is smaller and emission of only partially vibrationally relaxed excimers is observed (first continuum, Fig. 6.11) which corresponds to emission c in the solid phase /Jortner, 1974/.

Table 6.5. Tentative assignment of emission due to free excitons (a), one-centre self-trapped excitons (b), vibrationally excited two-centre self-trapped excitons (c), and vibrationally relaxed two-centre self-trapped excitons (d). The relative intensities of the bands have been estimated from the area of the emission bands and are only reliable within an order of magnitude. A complete set of data by optical excitation is given by Roick et al. /1983,1984/, Roick /1984/ and Gaethke et al. /1984/

	energy				relative intensity			
	a	b	c	d	a	b	c	d
Ne		16.5	13 - 16			1.0	1.0	
Ar	12.1	11.64 , 11.58	11.37	9.8	$1 \cdot 10^{-4}$	$3 \cdot 10^{-2}$	$5 \cdot 10^{-2}$	1.0
Kr	10.15	10.05		9.7	8.4	$5 \cdot 10^{-3}$	$5 \cdot 10^{-3}$	10^{-2}	1.0
Xe	8.35			7.6 , 7.1	10^{-3}			1.0

Ne: Schuberth and Creuzburg /1975/

Ar and Xe: Fugol /1978/

Kr: Zimmerer /1978,1979/

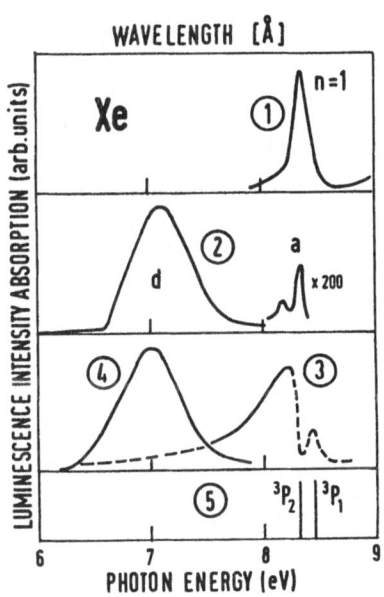

Fig. 6.11. (1) Absorption spectrum of solid Xe /Baldini, 1962/; (2) Emission spectrum of solid Xe /Fugol, 1978/; (3) Xe gas emission at low pressure /Stewart et al., 1970/; (4) Xe gas emission at high pressure /Kashnikov, 1971/; (5) Xe atomic resonance lines (after Fugol /1978/)

The ground state potential curves involved have been reviewed by Barker /1976/. Since the first potential curves for Xe_2^* have been estimated by Mulliken /1970, 1974/, several calculations for R_2^* and R_2^+ molecular states have become available. Gilbert and Wahl /1971/ treated Ne_2^+ and Ar_2^+. Ab initio calculations exist for Ne_2^* and Ne_2^+ /Cohen and Schneider, 1974/. Potential curves for excited states of Ar_2^* have been calculated by Lorentz and Olsen /1972/, Gillen et al. /1976/, Saxon and Liu /1976/ and by Spiegelman and Malrieu /1978/. Furthermore, potential curves have

127

been calculated for Kr_2^+ /Abouaf et al., 1978/, Xe_2^+ /Wadt et al., 1978/ and for Xe_2^* and Xe_2^+ /Ermler et al., 1978/. The equilibrium separations and the dissociation energies have been collected in Table 1.2. Ar_2^* centres in an Ar crystal have been discussed by Song /1971/. In the following Xe will be used as an example, because new potential curves /Ermler et al., 1978/ and detailed additional information concerning the influence of the crystal are available. This is illustrated in the scheme of Fig. 6.12. The influence of the symmetry of the surrounding crystal on

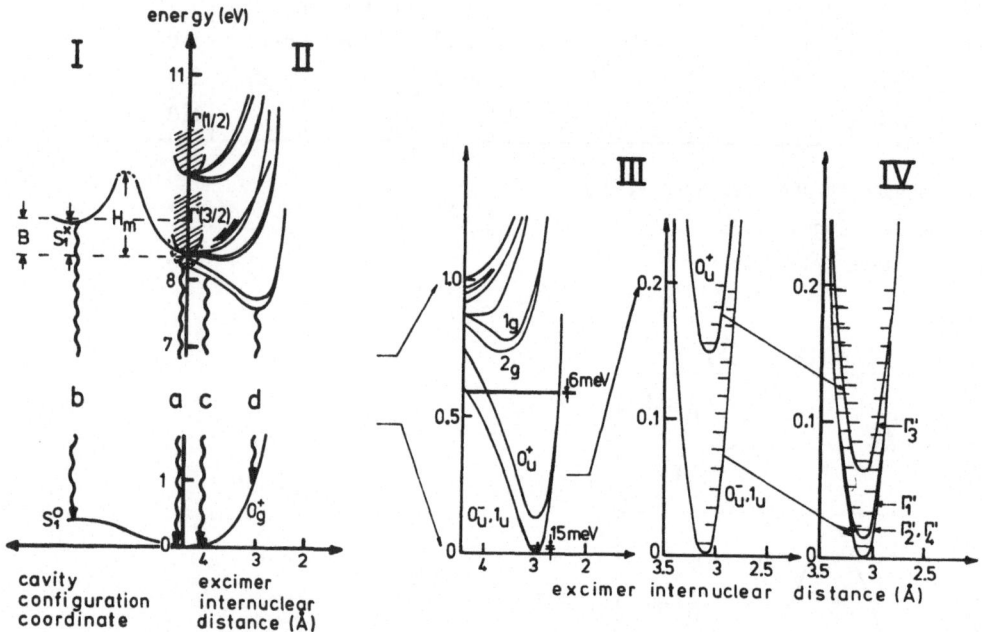

Fig. 6.12. Potential curves for Xe excitons. (I) Atomic centres R^* (after /Fugol, 1978/). For the definition of S_1^0, S_1^+ and H_m see Sect. 6.2, (6.7) and Table 6.1. The dashed region corresponds to free excitons; (II) Molecular centres R_2^* depicted according to the calculation of Ermler et al. /1978/ for the free molecule; (III) Potential curves of the free molecule (II) on enlarged scales; (IV) Experimental results for the potential curves of R_2^* centres in the crystal on the right scale of III. (a, b, c and d) indicate emission of free excitons, of atomic centre R^*, of vibrationally excited R_2^* and of vibrationally relaxed R_2^* centres

the potential curves causes the additional splitting into Γ_1', Γ_2', Γ_3' and Γ_4' /Molchanov, 1972/.

For radiative decay (d bands) the lowest Rydberg states $^3\Sigma_u^+$ and $^1\Sigma_u^+$ are important. In the region of the equilibrium separation, i.e., in the lower vibrational levels, Hund's coupling case "c" has to be applied and $^3\Sigma_u^+$ splits into 1_u and 0_u^-, whereas $^1\Sigma_u^+$ corresponds to 0_u^+ (Fig. 6.12). Vibrational relaxation and electronic relaxation between the singlet (0_u^+) and triplet branches (0_u^-, 1_u) will take place in the excimer centres. In Fig. 6.12,*III* the complications arising from the large variety of states are illustrated. For example, the anharmonicity of the potential curves causes for Xe a reduction of the vibrational spacing from 15 meV to 6 meV. The electronic potential curves of the 0_u^+ and the 0_u^-, 1_u states are separated by about 100 meV which leads to a superposition of the vibrational series. Some of the vibrational levels of different electronic states lie close together and transitions between these electronic states will be accelerated according to the energy gap law. Therefore, emission of singlet and triplet states can be expected with life-times which span the range from 10^{-5} to 10^{-9} s similar to the values observed for the free molecules (Table 6.6). The lifetimes observed in RGS are collected in Table 6.7.

Table 6.6. Radiative lifetime of rare-gas excimer states (not complete)

	$^1\Sigma_u$ (10^{-9} s)	$^3\Sigma_u$ (10^{-6} s)	
Ne_2^*	2.8	11.9	a)
		5.1	b)
		6.62	c)
Ar_2^*	4.2 ± 0.13	3.2 ± 0.3	d)
		3.22	c)
		2.8	e)
Kr_2^*	3.4 ± 0.3	0.264	f)
		0.35	c)
	5.2	0.15	g)
Xe_2^*	6.22 ± 0.8	0.100 ± 2	h)
		0.102 ± 2	i)
		0.101 ± 1	k)
	4.6 ± 0.3	0.99 ± 2	f)

a) Schneider and Cohen /1974/
b) Leichner /1973/
c) Oka, Rao, Redpath and Firestone /1974/
d) Keto, Gleason and Walters /1974/
e) Thonnard and Hurst /1972/
f) Bonifield et al. /1980/
g) Wenck, diploma work Hamburg /1979/
h) Keto et al. /1976/
i) Millet et al. /1978/
k) Wenck et al. /1979/

Table 6.7. Relaxation time constants in RGS. τ_1, τ_2: experimental lifetimes; τ_3: calculated vibrational relaxation time constants of R_2^* centres; ΔE: vibrational energy spacings; $\hbar\omega_p$: maximum of transverse acoustic phonon energies; $N = \Delta E/\hbar\omega_p$: order of vibrational relaxation processes; τ_4: Self-trapping time constant

		τ_1 $\quad$ τ_2 (in ns) $^1\Sigma_u$ $\quad$ $^3\Sigma_u$	temp. (K)	band	τ_3 (ns)	ΔE (meV)	$\hbar\omega_p$ (meV)	N	τ_4 (s) exp.	theor.
Xe	a	1.3 $\quad$ 9000	6	7.2 eV						
	b	20 - 15000	100 -							
	c	2000	6							
	d	30	148		0.005	15 [l]	3.8	4		
	p	58	80	7.6 eV						
	d	26 - $\quad$ 29	136 - 65	7.2 eV						
	d	15 - $\quad$ 16	136 - 65	7.6 eV						
Kr	a	1.2 $\quad$ 3200	6	8.4 eV	0.05	23 [m]	4.3	5	10^{-12}	
	c	300	6							
	d	90	83							
Ar	a	1.8 $\quad$ 1130	6	9.7 eV	1.5	35 [n]	5.9	6	and	
	c	1400	6						$5 \cdot 10^{-10}$	
	f	3.2 $\quad$ 1100	83							
	d	1410	80							
Ne	e	1 $\quad$ 5000	6	molecular	10^{10}	$1 \rightarrow 0$ $\quad$ 67	4.6	15		10^{-12}
	c	500		16.53 eV	10^7	$2 \rightarrow 1$ $\quad$ 61				
	c	2000		16.18 eV	$4 \cdot 10^5$	$3 \rightarrow 2$ $\quad$ 56				
	d	3900		molecular	400	$4 \rightarrow 3$ $\quad$ 51				
	d	560000		atomic	8	$5 \rightarrow 4$ $\quad$ 46	70 [o]			
		g, h			g, h	h			i, r	k

a) Roick et al. /1984/, Roick /1984/
b) Kink et al. /1977/
c) Coletti and Hanus /1977/
d) Suemoto and Kanzaki /1979/
e) Gaethke et al. /1984/
f) Carvalho and Klein /1978/

g) Yakhot /1976/
h) Yakhot et al. /1975/
i) Zimmerer /1976/
k) Martin /1971/
l) Ermler et al. /1978/
m) Jortner /1974/

n) Saxon and Liu /1976/
o) Cohen and Schneider /1974/
p) Monahan et al. /1977/
q) Keto et al. /1979/
r) Roick et al. /1983/

There is general agreement that the d bands in Ar and Kr and the 7.2 eV band in Xe are due to radiative decay of the lowest vibrational level of the 0_u^-, $1_u(^3\Sigma_u^+)$ and 0_u^+ ($^1\Sigma_u^+$) branches of the molecular centres. There is a controversy, however, whether the fine structure indicated by at least two different decay components and a splitting of 40 - 110 meV (see Fig. 6.5) is caused by the two states $^3\Sigma_u^+$, $^1\Sigma_u^+$ or by an additional splitting of $^3\Sigma_u^+$ due to the reduction of the symmetry from $D_{\infty h}$ to D_{2h} for the molecule in the solid (see Fig. 6.12,IV).

The following tentative explanation for the fine structure and temperature dependent lifetimes of the d bands in Ar and Kr and the 7.2 eV band in Xe is consistent with the experimental results and incorporates earlier proposed models. Part IV of the potential curves in Fig. 6.12 is based on this interpretation. The emission at higher energy (see Fig. 6.5) and with the ns lifetimes (Table 6.7) corresponds to the strongly allowed transition from 0_u^+ ($^1\Sigma_u^+$) to the ground state /Hahn et al., 1977/. In the D_{2h} symmetry of the crystal the 0_u^+ state becomes Γ_3' and remains strongly allowed. The transitions with long decay times (Table 6.7) and 40 - 110 meV lower energies are due to the 0_u^-, 1_u ($^3\Sigma_u^+$) states. In the D_{2h} symmetry 0_u^- and 1_u split further into Γ_1', Γ_2', Γ_4' /Molchanov, 1972/ where transitions from Γ_1' to the ground state are strongly forbidden and those from Γ_2', Γ_4' are moderately allowed. The long decay times of Ar, Kr and Xe are attributed to the Γ_1', Γ_2' and Γ_4' sub-levels. The temperature dependence of the transition rate for Xe is explained by a thermal depopulation of Γ_1' to Γ_2', Γ_4' yielding an energy separation from Γ_1' to the centre of Γ_2', Γ_4' of 5 meV in a single phonon model /Hahn et al., 1977/ and of 14 meV with a multiphonon model /Kink et al., 1977/. The splitting of Γ_4' and Γ_2' is expected to be in the µeV region as in alkali halides where it has not been resolved (e.g. /Fischbach et al., 1973/).

One unsettled problem regarding the assignment of the d bands is the appearance of the prominent 7.6 eV emission band of solid Xe around 60 K (Fig. 6.6). Once more a short and a long living component has been identified /Dössel et al., 1983/. Emission from a special site where lattice relaxation around R_2^* is hindered has been proposed by Cheshnovsky et al. /1973c/. This assignment is consistent with the observation of the 7.6 eV emission from Xe_2 impurity pairs in solid Ar /Cheshnovsky et al., 1973c/, but it does not explain the temperature dependence. Alternatively, a phase transition similar to the case of Ne has been discussed by Tolkiehn /1976/. A third explanation has been reintroduced recently by Monahan et al. /1977/ who attribute the 7.6 eV band to emission from self-trapped n = 2 excitons /Molchanov, 1972/, because the energy difference of these two bands is similar to that of the n = 1 and n = 2 excitons. The weak but proven emission at 7.6 eV, when only n = 1 excitons are excited (Fig. 6.6c) is not consistent with this latter explanation for which a change from the n = 1 branch to the n = 2 branch during self-trapping would be required. Heumüller and Creuzburg /1978/ attributed the 7.2 eV band to the 1_u and 0_u^+ states and the 7.6 eV band to the next higher electronic states 2_g and 1_g

(Fig. 6.12,*III*). From the temperature dependence of the relative intensities of the
7.2 eV and 7.6 eV bands, a barrier of 60 meV for the g states has been deduced which
governs the temperature dependent branching ratio for the g and u levels. This model
is inconsistent with the lifetime of this band. Kink et al. /1977,1981/ assigned the
7.6 eV band to the self-trapped n = 1 exciton at an ideal crystal site, whereas the
7.2 eV band is due to an excitonic state localized at a defect site. This assignment
has been supported by a calculation of the energy of the centre with and without a
vacancy, and by the observation of a similar splitting of 110 meV of both bands
/Dössel et al., 1983/.

We shall now proceed to discuss the dynamics of exciton trapping attempting to
utilize the theoretical description of Sects. 6.1 - 4 to account for the following
observations:

1) The efficient trapping of excitons by the two-centre localization mechanism in
   solid Ar, Kr and Xe, where 99% of the excitation results in the formation of
   diatomic excimers.
2) The efficient trapping of the electronic excitation in solid Ne by the one-centre
   trapping rather than by two-centre trapping.
3) The appearance of the free exciton emission in Xe and possibly in Ar but not in
   Ne.
4) The inefficient vibrational relaxation of the R_2^* centre in solid Ne.

Observation (1) implies that the barrier height separating the free and the two-
centre localized states in Ar, Kr and Xe is sufficiently low to allow for effective
tunnelling and stabilization of the diatomic molecule. Observations (2) and (3) imply that for Ne the barrier separating free and one-centre localized states is small
or does not exist at all. This last conclusion is encouraging in relation to the
localization theory provided that we can give a good argument for the inefficiency
of two-centre localization in favour of one-centre localization in solid Ne. The re-
tardation of the formation of the Ne_2^* centre in solid Ne may originate from a molecular effect rather than a solid state effect, resulting from a hump in the molecular
potential curve (Sect. 6.3). Humps appear in the Ne_2^* potential curves calculated by
Cohen and Schneider /1974/. These humps appear at internuclear distances of 2 - 4 Å.
They may reduce the probability to form molecular centres by crossing the barrier
and may force the formation of a cavity. In this way, atomic-like centres are favoured and are stabilized by the molecular potential curve in solid Ne. Using Cohen
and Schneider's potential curves, Kunsch and Coletti /1979/ have calculated the local
structure around the cavity and the resulting transition energies (E_1 and E_3 in Fig.
6.9). For the cavity an increase in the nearest neighbour separation by a factor of
1.37 has been obtained for the first atomic shell surrounding the excited atom. This
deformation extends to distant shells and even for the 16th shell a deformation in
the nearest neighbour separation by 1 percent has been calculated.

Perhaps the most interesting feature of exciton dynamics in RGS is the appearance of the weak but well defined free exciton band a ($Y_f \sim 10^{-3}$) in solid Xe. Xe is the most probable candidate for free exciton emission, as has been pointed out in Sect. 6.2. The energy of the a band of Xe reveals an interesting temperature dependence /Kink and Selg, 1979; Coletti and Bonnot, 1978; Fugol et al., 1982/ which is similar to that of the free exciton band in absorption, providing support for the assignment of this band. The intensity ratio of bands a and d and its temperature dependence (Fig. 6.13) have been explained by a competition of radiative decay and self-trapping by a crossing of the self-trapping barrier /Kink and Selg, 1979/. Recently the self-trapping rate constants of the free excitons in Xe have been determined directly in time-resolved experiments /Roick et al., 1983/. The increase of the self-trapping rate from $2 \cdot 10^9$ s^{-1} at 10 K up to 10^{10} s^{-1} at 40 K has been attributed to thermal activated tunnelling. The self-trapping barrier of 25 meV estimated from these rate constants is in quite good agreement with that derived from the Stokes shift (Table 6.1).

The large oscillator strength $f \approx 0.2$ for the $n = 1$ exciton in solid Xe suggests a description of the absorption and emission features in terms of a polariton picture /Kink and Selg, 1979; Fugol et al., 1982/. Within the polariton model it has been possible to describe the shape of the surprisingly broad absorption bands by a convolution of contributions from the transversal and longitudinal polariton branches /Kink and Selg, 1979; Fugol et al., 1982/. The observed sharpness of the free exciton emission and its position near the transverse polariton energy are an immediate outcome of the polariton nature of this band. Furthermore, a temperature dependent long wavelength tail due to a population of the transverse polariton wing by successive phonon scattering events has been observed /Fugol et al., 1982/.

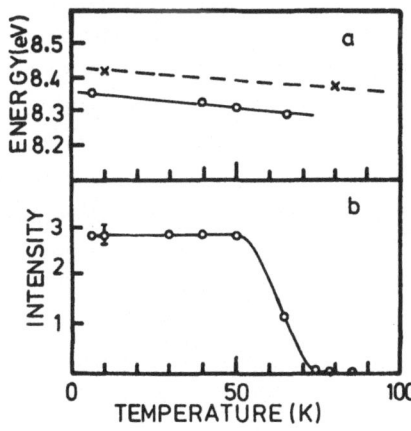

Fig. 6.13a,b. (a) Temperature dependence of the peak position of the resonant emission (————) and of the reflectivity (— — —). (b) Temperature dependence of the intensity of the resonant emission ($\hbar\omega = 8.35$ eV) (after /Kink and Selg, 1980/)

Solid Ar is a good example for the difficulties encountered in band assignments. In a recent study Roick et al. /1984/ reinvestigated the atomic-type emission b and the hot molecular emission c. By varying the energy of the exciting light and by surface covering experiments it became evident that b and c are due to surface contributions. The surface sensitivity of the ratio of atomic-type emission b and molecular-type emission d is caused by an influence of the dimensionality on the self-trapping barriers /Toyozawa and Shinozuka, 1980/. In the three-dimensional case of the bulk the self-trapping barrier for the atomic-type centres (10 meV) is much larger than that for molecular-type centres (2 meV) which prevents atomic-type centres. In the two-dimensional case of the surface both barriers are absent and thus both configurations are populated /Roick et al., 1984/. The restriction of hot molecular emission c to the surface has been explained by the fact that even very small changes in the effective pair potential for an atom at the surface or in the bulk can cause orders of magnitude changes in the vibrational relaxation rates /Roick et al., 1984/.

This leads us to consider the effects of vibrational relaxation in the R_2^* centres (point 4). The intensities of hot luminescence increasing from Kr to Ne /Kink et al., 1981/ reflect the competition between radiative decay times lying roughly in the ns to µs regions for all RGS (Table 6.7) and the decreasing vibrational relaxation rates for lighter RGS. This result can be explained in terms of the energy gap law for vibrational relaxation /Jortner, 1974/ discussed in Sect. 6.4. Relaxation starts from the high vibronic levels of the R_2^* molecular states. In Table 6.7, the largest phonon energies $\hbar\omega_p$ are compared with the vibrational quanta ΔE_v for all RGS /Schwentner, 1978/. The phonon energies are several times smaller than the vibrational quanta and relaxation has to be a multiphonon process. The order $N = \Delta E_v/\hbar\omega_p$ of this process increases from 4 in Xe to 6 in Ar. There is a jump to $N = 18$ in Ne, which retards the vibrational relaxation process of R_2^* /Gedanken et al., 1973b/. The vibrational relaxation of Ne_2^* should be ineffective and the relative contribution from the c band should decrease in the order Ne $\gg$ Ar $\gg$ Kr > Xe. Yakhot /1976/ calculated vibrational relaxation of R_2^* in RGS providing a set of relaxation lifetimes which agree with the radiative yields for the c emission in Ar, Kr and Xe. For solid Ne, Yakhot et al. /1975/, using the ab initio potential curves given by Cohen and Schneider /1974/, considered also the influence of anharmonicity on the vibrational relaxation. From the comparison with the radiative lifetimes (Table 6.7) emission from the 5th to the 3rd vibrational level is predicted. We note, however, that in the calculation of Yakhot et al. /1975/ the close proximity of the 0_u^-, 1_u and 0_u^+ states with the possibility of relaxation between these states has not been considered. Recently Gaethke et al. /1984/ have been able to distinguish between the vibrational relaxation rates in the volume and at the surface.

6.6 Emission from Liquid Rare Gases

6.6.1 Emission from Liquid Ne, Ar, Kr, Xe

The striking similarity between the prominent d-emission bands in solid Ar, Kr and Xe and the emission of the corresponding molecular emission spectra has been illustrated in the previous section. It is interesting to see whether liquid rare gases fit into this picture. Further, liquid helium will also be included in this discussion as sufficient emission data on gaseous and liquid helium are available. Jortner et al. /1965/ compared emission spectra of liquid Xe, Kr and Ar obtained by α particle excitation in all three phases and established excimer formation in the condensed phases. In the course of the search for VUV lasers, emission bands of liquid Xe, Kr and Ar have been investigated with electron excitation by the group of Basov et al. /1970/. Cheshnovsky et al. /1973a/ and Suemoto /1977/ reported emission bands for liquid Kr and Cheshnovsky et al /1972a,b/ for liquid Ar. The only experiment for liquid Ne was performed by Packard et al. /1970/. The emission spectra for liquid He yield information both on radiative decay to the ground state /Stockton et al., 1970,1972; Surko et al., 1970/ and on transitions between excited states of He[*] and He$_2^*$ /Fitzsimmons, 1973/.

In Fig. 6.14 and Table 6.8 we present a collection of spectroscopic data for emission bands of liquid rare gases together with spectra for the gaseous and solid phase. All the emission bands of Xe, Kr, Ar and Ne in the liquid phase are attributed to the decay of excimer-like molecular centres R_2^* from the lowest excited bonding molecular states $^1\Sigma_u^+$ (0_u^+) and $^3\Sigma_u^+$ (1_u) to the ground state $^1\Sigma_g^+$ (0_g^+) (Fig. 6.12). For

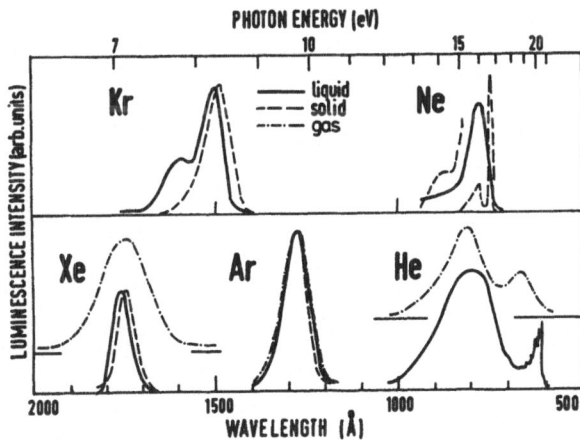

Fig. 6.14. Compilation of emission bands in liquid rare gases together with solid and gas phase spectra. (Xe gas) Jortner et al. /1965/; (Xe solid and liquid) Basov et al. /1970/; (Kr) Cheshnovsky et al. /1973c/; (Ar) Cheshnovsky et al. /1972a/; (Ne) Packard et al. /1970/; (He) Stockton et al. /1972/

Table 6.8. Emission bands in liquid rare gases

Xe a)	Kr a)	Ar b)	Ne c)	He			
				to ground state d)		between excited states e)	
				Å	eV	Å	eV
1767 Å	1472 Å	1295 Å	774 Å	800	15.5 $A^1\Sigma_u^+(v'=0) \rightarrow X^1\Sigma_g^+$	10400	1.192 $d^3\Sigma_u^+ \rightarrow c^3\Sigma_g^+$
7.02 eV	8.42 eV	9.57 eV	16.02 eV	(600)	(20.66)	9182	1.350 $c^3\Sigma_g^+ \rightarrow a^3\Sigma_u^+$
			(900 Å)	604	20.52	6400	1.937 $d^3\Sigma_u^+ \rightarrow b^3\Pi_g$
				608	20.39	9136	1.357 $c^1\Sigma_g^+ \rightarrow A^1\Sigma_u^+$
				613	20.23	6600	1.878 $D^1\Sigma_u^+ \rightarrow B^1\Pi_g$
				619	20.02	10911	1.136 $2p^3P \rightarrow 2s^3S$
				627	19.77 $A^1\Sigma_u^+(v'=16) \rightarrow X^1\Sigma_g^+$	7270	1.705 $3s^3S \rightarrow 2p^3P$
				636	19.49	7060	1.756 $3s^1S \rightarrow 2p^1P$
				648	19.13		
				665	18.64		
				686	18.07		
				707	17.53		

a) Basov et al. /1970/
b) Jortner et al. /1965/
c) Packard, Reif and Surko /1970/
d) Stockton, Keto and Fitzsimmons /1972/
e) Keto et al. /1974b/

Xe and Ar the emission bands in the three phases are almost identical. The differences are smaller as, for example, the shifts due to temperature broadening /Chesh-
novsky et al., 1973/ or due to different preparation conditions observed in the
solid phase. For Kr, besides the maximum at 8.42 eV (1472 Å) which is common in the
solid and liquid phase, a shoulder appears on the low energy tail at 7.85 eV (1580 Å).
This shoulder has been attributed to a Kr-Xe* molecule /Cheshnovsky et al., 1973/
which is formed due to the presence of Xe impurities. Thus liquid Ar, Kr and Xe exhibit only the d-type molecular emission band.

The emission spectrum for liquid Ne differs dramatically from that of solid Ne
(Fig. 6.9). The b-type atomic emission line observed in solid Ne at 16.69 eV (743 Å)
is missing in liquid Ne. The absence of the atomic emission in liquid Ne suggests
that either this centre is not formed in the liquid or that it does not live long
enough to radiate. In liquid Ne atomic transitions from the $^3P_2$ state have been
found in transient absorption spectra, but the lifetime of the $^3P_2$ state is much
shorter in liquid than in solid Ne. The radiative lifetimes in both phases are expected to be of the same order. Therefore, the shorter lifetime in liquid Ne is attributed to quenching of Ne* centres by Ne$_2^*$ excimer formation. In the solid phase,
the barrier in the molecular potential curves is sufficiently high to suppress Ne$_2^*$
formation, whereas at the higher temperatures of the liquid these humps may perhaps
be crossed. The prominent emission band of liquid Ne peaks at 16.0 eV (774 Å), coincides with the c emission in the solid and is attributed to a c band in the liquid.

Additional information regarding these molecular centres in liquids can be derived from lifetimes. In liquid Xe, Kr, Ar and Ne lifetimes have been investigated
by Kubota et al./1978a/, Carvalho and Klein /1978/ and Suemoto et al. /1979/. A fast
component of some ns and a slow component have been identified (Table 6.9). These
are attributed to allowed transitions in the R$_2^*$ centres from $^1\Sigma_u^+$ (0_u^+) and transitions from $^3\Sigma_u^+$ (1_u) which become more forbidden going from Ne, Ar to Xe by an ad-
mixing of the $^1\Pi_u$ character due to increasing spin-orbit coupling. These data support the notion that the R$_2^*$ molecular centre is practically identical in the gas
and condensed phases. The insensitivity of the R$_2^*$ emission to medium effects can be
reconciled with a large medium distortion around this large radius Rydberg state in
the liquid, as is the case in the solid.

We conclude that the features of excited-state dynamics in pure liquid Ne, Ar,
Kr and Xe are characterized by three features. First, excited-state trapping via
two centre localization occurs in all liquids. A combination of the molecular approach for the description of the R$_2^*$ potential surface together with a solid state
(continuum or microscopic) approach for lattice relaxation is required for a theoretical description of exciton trapping in liquids. Second, medium-induced vibrational relaxation of R$_2^*$ in liquid Ar, Kr and Xe is efficient on the time scale of
the radiative decay while a third feature in Ne emission is exhibited from vibrationally excited states of Ne$_2^*$, which is in accord with the energy gap law for vibrational relaxation.

137

Table 6.9. Experimental lifetimes τ_1, τ_2 (in 10^{-9} s) and intensity ratios A_1/A_2 for liquid rare gases

liquid	τ_1 $^1\Sigma_u^+$, 0_u^+ a)	τ_2 $^3\Sigma_u^+$, 1_u a)	 b)	A_1/A_2 a)	$\tau_1 A_1/\tau_2 A_2$ a)
Ne			2900		
Ar	5.0 ± 0.2	860 ± 30	1110	7.8	0.45
Kr	2.1 ± 0.3	80 ± 3	110	0.9	0.02
Xe	2.2 ± 0.3	27 ± 1		0.6	0.05

a) Kubota, Hishida and Raun /1978a/
b) Suemoto and Kanzaki /1979/

6.6.2 Emission from Liquid He

We turn now to the emission properties of liquid He considering first emission to the electronic ground state. In liquid He a broad emission band centred around 15.5 eV (800 Å) (Fig. 6.14, Table 6.8) has been reported by Stockton et al. /1970, 1972/ and by Surko et al. /1970/. The conversion of excitation energy to VUV light is very efficient. Absolute measurements show that about 30% of the energy of 200 KeV electrons is emitted in the 800 Å band. On the basis of these results a VUV lamp has been proposed yielding 3 Watts (10^{10} photons/sÅ) at 800 Å /Fitzsimmons, 1973/. The 800 Å band corresponds to transitions from low vibrational levels (v' = 0) of the lowest excited singlet state $A^1\Sigma_u^+$ to the repulsive ground state $^1\Sigma_g^+$ (Fig. 6.15). In addition, a series of sharp maxima is observed near 20 eV (600 Å) (Fig. 6.14, Table 6.8). The limitation of the spectrum to higher energies is due to reabsorption in the liquid He sample. The emission in this region has been attributed by Stockton et al. /1970,1972/ to the decay of the v' = 16 vibrational level in the $A^1\Sigma_u^+$ state. The fine structure is caused by the projection of the Franck-Condon distribution in the v' = 16 level onto the strongly repulsive ground-state potential. The broad band emission and the fine structure have been observed also in the gas phase. The rapid and selective population of this high vibrational level needs to be explained. v' = 16 and v' = 17 are the two highest quasi-bound vibrational levels of $A^1\Sigma_u^+$. The v' = 16 level is within $1.2 \cdot 10^{-2}$ eV resonant with the $2s^1S$ metastable atomic state (Fig. 6.15) which is produced at high rates. This radiative recombination process involves inverse predissociation /Herzberg, 1950/. What is remarkable is that this inverse predissociation occurs in the liquid and that the v' = 16 , 17 quasi-bound vibrational states are not dumped by vibrational relaxation on the time scale of their radiative decay.

138

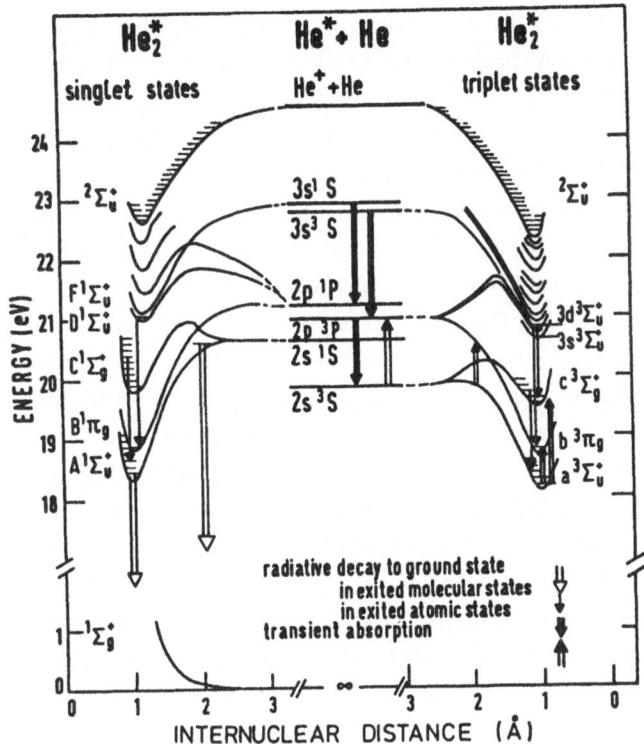

Fig. 6.15. Observed transitions in liquid He. The potential curves have been adopted from Ginter and Battino /1970/, except for the singlet states above $C^1\Sigma_g^+$: Guberman and Goddard /1975/. Transitions to the ground state: Stockton et al. /1972/; radiative decay between excited states: Dennis et al. /1969/; transient absorption: Soley and Fitzsimmons /1974/

The metastable $2s\,^3S$ atomic state decays non-radiatively by a molecular mechanism similar to the decay of the $2s\,^1S$ atomic state (Fig. 6.15). The lifetime of 15 μs obtained for the state by transient absorption spectroscopy /Keto et al., 1972,1974/ is considerably too short for the radiative decay of this state. The decay of the $2s\,^3S$ state gives rise to a buildup of a new transient absorption at 1.92 μm, which is explained by a $a^3\Sigma_u^+(v) \rightarrow b^3\Pi_g(v')$ transition involving high initial and final vibrational states. Thus, the $2s\,^3S$ state decays non-radiatively by the formation of vibrationally excited $a^3\Sigma_u^+$ molecular state. Again, vibrational relaxation is slow in the liquid. Of considerable interest are the lifetimes of those states which are involved in transitions to the ground state. For the $A^1\Sigma_u^+$ state of He_2^* in liquid He experimental lifetimes are not available. The high efficiency of this transition shows that the lifetime is determined mainly by radiative decay. The $A^1\Sigma_u^+$ is the lowest state for allowed transitions to the ground state. But the role of the even lower $a^3\Sigma_u^+$ molecular state and of the lowest atomic state $2s\,^3S$ (Fig. 6.15) have also to be considered. Both states are metastable, and non-radiative decay of these states has been reported. However, interesting dynamic effects of these metastable states were observed. The intensity of the VUV luminescence which is emitted within 10^{-6} s after α-particle excitation, depends on the sample temperature in the region of 0.2 K to 4 K /Moss and Hereford, 1963; Fischbach et al., 1969; Roberts and Hereford, 1973/. The explanation of the temperature dependence requires that part of the luminescence intensity (15%) is due to annihilation processes of metastables. These decay processes are influenced by the temperature via the transport properties of liquid He as will be shown below. In addition, Surko and Reif /1968/ found energy transport by neutral excitations in superfluid He between 0.3 K and 0.6 K when the sample was irradiated with α particles. At low temperatures, the neutral excitations travel over distances greater than 1 cm without appreciable attenuation. At higher temperatures, the range is reduced. The neutral excitations carry enough energy to produce electrons and ions at the liquid He surface and at metal surfaces. Due to the amount of energy and the long lifetime, these excitations are attributed to metastable atoms and molecules. These studies have been extended by Mitchell and Rayfield /1971/, Calvani et al. /1972,1973/ and Arrighini et al. /1974/. The lifetimes of these metastable states have been investigated by Keto et al. /1972,1974/ using transient absorption spectroscopy. Metastable states with a density of about 10^{12} to 10^{13} per cm^3 have been prepared by electron beam excitation (160 keV). In these experiments, the beam was pulsed with a pulse length of ≈ 3 ms and rise and fall times of 10 ns. The decay of the population has been probed by observing, during and after the excitation, an absorption spectrum for transitions from $2s^3S \rightarrow 2p^3P$ at 1.145 eV (1.083 μm) (metastable atoms) and from $a^3\Sigma_u^+(v = 0) \rightarrow b^3\Pi_g(v' = 0)$ at 0.59 eV (2.1 μm) (metastable molecules) (see Fig. 6.15). The initial density of metastable states has been changed by varying the electron beam current between 0.1 μA and 4 μA.

In the discussion of these lifetimes we will first be concerned with processes determining the lifetimes of molecules in the $a^3\Sigma_u^+$ state. Two experimental facts are pertinent. First, for a fixed temperature and fixed initial concentration, the inverse of the concentration of the $a^3\Sigma_u^+$ state increases linearly with time for the time range of 1 μs up to 100 μs after excitation. Second, the steady state concentration increases with the square root of the beam current.

These two observations show that the dominating loss process for metastable molecules in liquid He is a bilinear collision process between pairs of metastable states:

$$He_2^*(a^3\Sigma_u^+) + He_2^*(a^3\Sigma_u^+) \xrightarrow{\alpha(T)} He + He + He_2^* \; . \tag{6.11}$$

The time dependent concentration M of metastable states is described by second-order kinetics

$$1/M = 1/M_0 + \alpha(T) \cdot t \tag{6.12}$$

140

with the steady state concentration M_0 and the bilinear reaction rate $\alpha(T)$. M_0 is given by the number I_0 of initially produced metastable molecules $M_0 = [I_0/\alpha(T)]^{1/2}$. To provide a quantitative estimate we note that the lifetime of the $a^3\Sigma_u^+$ state due to the bilinear quenching is about 1 ms at a beam current of 1 μA and the lower limit of the radiative lifetime is at least as long as 0.1 s. A most important experimental observation is that the bilinear reaction rate $\alpha(T)$ increases with decreasing temperature. $\alpha(T)$ is inversely proportional over the higher part of the temperature range to the number density of rotons given by $e^{-\Delta/T}$ with $\Delta = 8.6$ K /Keto et al., 1972,1974/. This interesting temperature dependence reveals that the reaction rate is diffusion limited and determined by scattering with rotons. The energy gap of rotons is pressure dependent. The change of $\alpha(T,P)$ with external pressure P has been measured /Keto et al., 1974/. The observed decrease of $\alpha(T,P)$ with pressure agrees with the predictions based on the pressure dependence of Δ.

Some direct implications of the annihilation process (6.11) were recorded. The afterglow observed in the VUV emission of the $A^1\Sigma_u^+$ state is due to repopulation of this state in the annihilation process of two metastable $a^3\Sigma_u^+$ states as follows from the similarity in the rate constants for annihilation $\alpha(T,P)$ and for repopulation. The destruction of metastable molecules feeds the $A^1\Sigma_u^+$ channel as well as higher-lying states shown in Fig. 6.15 /Keto et al., 1974/.

Similar results for triplet-triplet annihilation have been obtained by Soley et al. /1975/ in the only optical experiment reported for solid He. In solid He the molecular absorption band for the $a^3\Sigma_u^+ \rightarrow b^3\Pi_g$ transition (Fig. 6.15, Table 6.3) is similar to that in the liquid phase. Also the time dependence of the $A^1\Sigma_u^+$ afterglow, i.e., the bilinear reaction rate $\alpha(T)$ is the same. The common reaction rate is surprising, because α is determined by diffusion of metastable molecules and the transport properties for other species, like positive and negative ions, are about four orders of magnitude smaller in the solid than in the liquid phase. Thus, mass transport in the solid is excluded. Short-range electron exchange interaction /Dexter, 1953; Jortner et al., 1964/ provides an attractive mechanism for triplet migration in solid He. If this is indeed the excitation transport mechanism, the triplet-triplet annihilation in solid He bears a close analogy to triplet-triplet annihilation in aromatic crystals /Wolf, 1967/.

From the foregoing discussion it is apparent that in liquid He as in other liquid rare gases it is justified to use the gas phase molecular potential surface for He_2^* for the study of energetics and dynamics in liquid He. An interesting feature of the potential curves for the He molecule is that at least up to the $F^1\Sigma_u^+$ singlet and $3d^3\Sigma_u$ triplet states, no antibonding states exist. The potential curves show humps instead of antibonding states, a feature which has been discussed since the early work by Mulliken /1964/ and by Ginter and Battino /1970/ and has been analysed in detail by Guberman and Goddard /1975/. This feature of the He_2 potential curves (Fig. 6.15) differs qualitatively from the general characteristics of the potential

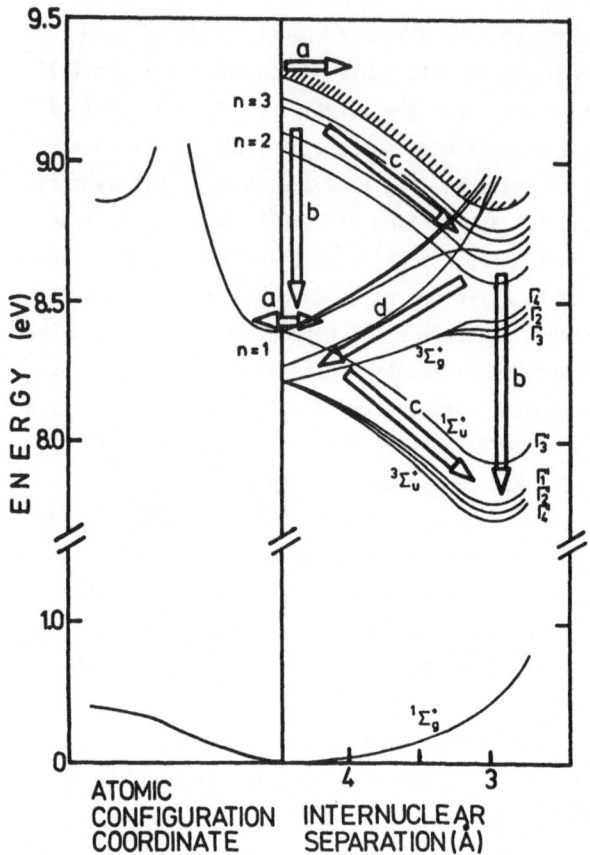

curves for the heavy rare gases (Fig. 6.16). For all heavier rare-gas diatomics, there are several branches of antibonding potential curves which start from the lowest excited state and which cross higher lying self-trapped exciton states near the minima at the equilibrium internuclear distances of these R_2^* states (Fig. 6.16). This qualitative difference between He and heavy rare gases influences the dynamics in highly excited states of the corresponding liquids. In the condensed heavy rare gases the highly excited R_2^* centres originating from high electronic states are expected to decay non-radiatively by predissociation, as the crossing points of the bonding and antibonding potential curves lie for the heavier rare gases near the minima of the bonding potential curves. Therefore, predissociation is possible even after vibrational relaxation of the diatomic centres. This non-radiative relaxation follows the antibonding potential curves which bridge the gaps between the bonding states. The radiationless processes governing these configurational changes are expected to be very efficient. Thus in condensed rare gases, excited non-selectively by α, X-ray or electron excitation, this cascading results in the formation of the n = 1 exciton state on a time scale which is short compared to the radiative decay

of high electronically excited states. Accordingly, no emission was observed from
high excited states in heavy RGS and liquids. Thus, for liquid He transitions between excited states (Fig. 6.15) are expected and were identified in the long wavelength emission spectrum (Fig. 6.17).

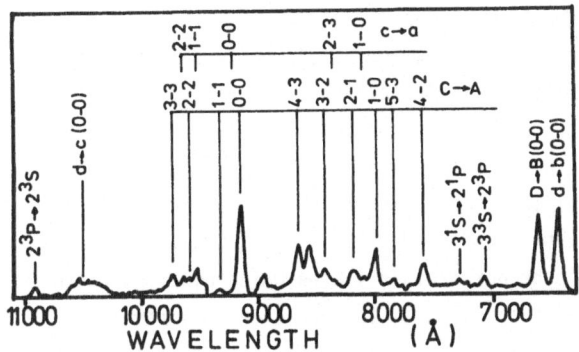

Fig. 6.17. Emission spectrum
of liquid He in the long wavelength region (after /Keto et
al., 1974/)

Radiative transitions for liquid He are observed in the spectral range from
6000 - 11000 Å (2 - 1 eV) /Dennis et al., 1969; Keto et al., 1974/ and (Fig. 6.17).
Similar to the gas phase, three atomic transitions are observed: $3s^1S \rightarrow 2p^1P$,
$3s^3S \rightarrow 2p^3P$ and $2p^3P \rightarrow 2s^3S$ (Figs. 6.15,17 and Table 6.8) Higher states are not
expected, because of the Hornbeck and Molnar /1951/ process

$$He^* + He \rightarrow He_2^+ + e \tag{6.13}$$

which quenches He^* atoms in states above the reaction threshold at 23.0 eV. The
transition $2p^1P \rightarrow 2s^1S$ expected at 20581 Å (0.60 eV) with a radiative transition
time of 510 ns /Wiese et al., 1969/ has not been observed in the liquid, probably
due to the fast (1.8 ns) radiative depopulation of the $2p^1P$ state to the ground
state. The most prominent structures in the emission spectrum (Fig. 6.17) correspond to molecular transitions. As is shown in Fig. 6.15 and Table 6.8, transitions
between $C^1\Sigma_g^+ \rightarrow A^1\Sigma_u^+$, $D^1\Sigma_u^+ \rightarrow B^1\Pi_g$ in singlet states and $c^3\Sigma_g^+ \rightarrow a^3\Sigma_u^+$, $d^3\Sigma_u^+ \rightarrow b^3\Pi_g$ and
$d^3\Sigma_u^+ \rightarrow c^3\Sigma_g^+$ in triplet states have been identified. All the observed molecular
emissions originate from Σ states. Though the emission spectrum demonstrates that
the $b^3\Pi_g$ and $B^1\Pi_g$ states are populated effectively in the course of the radiative
cascade, no emission from these Π states has been detected in careful experiments
/Dennis et al., 1969/. Thus, it is evident that Π states are strongly quenched by
non-radiative processes. The details of the emission spectrum show that the emitting
states are not in thermal equilibrium with the sourrounding liquid. In $C^1\Sigma_g^+$, emission from a series of vibrational levels v' = 0 , 1 , 2 , 3 , 4 and in $c^3\Sigma_g^+$, emission
from v' = 0 , 1 , 2 is observed. Even a rotational fine structure in the D → B, d → b

and d → c transitions can be identified from the broadening of the lines as in the gas phase. The population of higher vibrational levels of D and d states /Keto, 1974/ and a long lifetime of the rotational excitations in the $a^3\Sigma_u^+$ state /Hill et al., 1971/ has been proven in transient absorption spectra. A connection of the radiationless quenching rate with the symmetry of the states is obvious. The strong quenching of Π states has been mentioned. The Σ states can be divided into two sub-classes according to symmetry and quenching probability. In all Σ states, the vibrational excitations have a long lifetime. However, the rotational excitations are strongly quenched in the pσ orbitals (C,c), while they have a long lifetime of 10^{-3} s and more in the sσ orbitals (D,d). The presence of strong non-radiative quenching processes for all higher excited states leading to a population of the lowest emitting state $A^1\Sigma_u^+$ simply follows from intensity considerations. More than 99% of the overall emitted intensity belongs to the $A^1\Sigma_u^+ \to {}^1\Sigma_g^+$ transition. That part of the population of each of the higher excited states which contributes to the emission spectrum in the visible is about three to four orders of magnitude smaller than that in the $A^1\Sigma_u^+ \to {}^1\Sigma_g^+$ radiative decay /Keto et al., 1974/. The non-radiative relaxation cascade is much more efficient than the radiative. This follows also from the time dependence of the emission in the visible region /Keto, 1974/. The intensity drops within the response time of the experimental set-up (20 μs) by one order of magnitude. The further slow decrease is due to repopulation by the annihilation of $^3\Sigma_u^+$ metastables. The efficiency of the non-radiative cascade can be increased even further by some orders of magnitude by applying static external pressure to liquid He /Soley and Fitzsimmons, 1974/. The intensity for all higher excited states is quenched exponentially with increasing pressure (Fig. 6.18) and only the $A^1\Sigma_u^+ \to {}^1\Sigma_g^+$ transition is quenched very little (a factor of two up to 25 atm). The slopes of intensity versus pressure curves on a logarithmic intensity scale are characteristic for the symmetry of the upper state. For transitions starting from p-type states C , c , $2s^3P$, a common slope is observed which corresponds to a very strong increase in quenching efficiency with pressure (about two orders of magnitude at 15 atm). For the s-type states D , d , $3s^3S$ and A the slopes are much smaller. They depend on both the initial and the final state (Fig. 6.18), as shown by the example d → c and d → b with different slopes.

   In summarizing the results concerning the balance between radiative and non-radiative relaxation in excited states of liquid He, we can state that the non-radiative processes are by some orders of magnitude faster. Many of the radiationless processes will be faster than 10^{-9} s, because the times for competing radiative processes like $2p^3P \to 2s^3S$ or $3s^3S \to 2p^3P$ are $9.8 \cdot 10^{-8}$ s and $3.6 \cdot 10^{-8}$ s, respectively /Wiese et al., 1969/. On the other hand, higher vibrational and rotational excitations of molecular states have a lifetime up to milliseconds.

   The energy shifts between the gas and liquid phases contain salient information about the nuclear configurational changes around excited states in liquid He. Hill

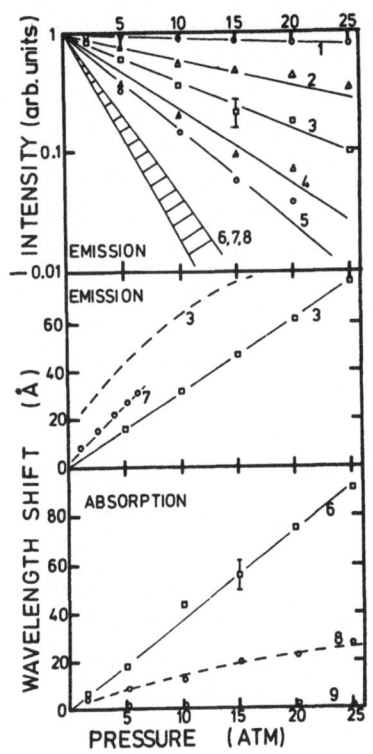

Fig. 6.18. (*Upper panel*) decrease of the emission intensity with external pressure in liquid He; (*Middle panel*) wavelength shifts in the emission spectra from liquid He with external pressure; (*Lower panel*) wavelength shifts in transient absorption spectra from liquid He with external pressure. Experimental points are connected with (———). (– – –) show the results of calculations by Hickman et al. /1975/. The numbers refer to the following transitions:

(1) $A^1\Sigma \rightarrow X^1\Sigma$;   (2) $d^3\Sigma \rightarrow c^3\Sigma$;

(3) $3^3S \rightarrow 2^3P$;   (4) $d^3\Sigma \rightarrow b^3\Pi$;

(5) $D^1\Sigma \rightarrow B^1\Pi$;   (6) $c^3\Sigma \leftrightarrow a^3\Sigma$;

(7) $C^1\Sigma \rightarrow A^1\Sigma$;   (8) $2^3P \leftrightarrow 2^3S$;

(9) $b^3\Pi \leftarrow a^3\Sigma$

et al. /1971/ compared the transition energies in emission and in transient absorption of liquid He with gas phase values. They observed shifts to the blue and to the red. All the shifts relative to the gas phase values are very small, i.e., of the order of 0.01 eV. For a discussion of these shifts it is important to realize that the 3 Å average interatomic spacing in liquid He is substantially smaller than the diameters of the extended Rydberg states of excited atoms and molecules with diameters of 10 - 20 Å. The small energy shifts are consistent with dramatically large solvent dilatation around the electronically excited state. The large electron-He repulsive interaction in the excited state results in a large medium expansion leading to a formation of a "bubble" in analogy with the bubble formed around a free electron in liquid He. These bubbles around free electrons in liquid He have been studied extensively from the experimental and theoretical point of view and the reader is referred to the literature (see, for example, /Schwarz, 1975; Cole, 1974/. The origin of both the electron bubble and the bubble around the excited species is the repulsive interaction due to the Pauli exclusion principle. The diameter of the bubble corresponds to the balance between these repulsive forces and attractive terms which are given in a simplified picture by the surface tension of the bubble. A sensitive test for this model and its applicability for the more complicated case of electrons in bound excited states follows from the introduction of an additional attractive term given by external pressure. Due to external pressure, the diameter of the bubble should be reduced. The influence of pressure on the transition energies and linewidths can be compared with the predictions of model calculations. Soley and Fitzsimmons /1974/ determined pressure shifts of the transitions in absorption and in emission up to pressures of 25 atm. For the emission bands a linear shift with pressure of roughly 3.2 Å/atm to the blue has been found for all molecular and atomic transitions (Fig. 6.18). In absorption the shifts in transition energies versus pressure show different slopes (Fig. 6.18).

The calculation of the spectral shifts and line broadening requires the dependence of the energy of both states, i.e., the upper one and the lower one, on the bubble radius with the pressures as parameter in the calculations. These potential curves are nothing but adiabatic potential surfaces. The shapes of the potential curves and the diameters corresponding to the minima in the potential curves are expected to be different for the upper and lower state, because of the larger extension of higher excited states and because of the different symmetries of the states. The symmetry of p-type states may cause even non-spherical bubbles. In the adiabatic approximation, the transitions take place in a time which is short compared to the relaxation time of the bubble and the observed energies correspond to vertical transitions. In transient absorption and in emission the transitions for the same pair of states start from bubbles with different sizes and shapes which cause the observed difference in transition energies. Hansen and Pollock /1972/ calculated the pair distribution of ground state He atoms in the vicinity of an excited metastable $2s^3S$ He atom from the Percus-Yevick theory. Hickman and Lane /1971/ and Hickman et al. /1975/ adopted the approach of Jortner et al. /1965/ which has been developed for free electron bubbles and modified it to account for bubble formation around electronic excitations. The energy $<E>$ of the bubble is divided into two parts

$$<E> = E_{cavity} + E_{atom} .$$
(6.14)

E_{cavity} describes the energy necessary to form the cavity and contains three terms corresponding to the surface tension, the volume pressure and a kinetic energy which is due to the density gradient at the edge of the cavity. E_{atom} describes the electronic perturbation due to the surrounding liquid. Hickman et al /1975/ treated the problem for metastable atoms. E_{atom} has been obtained from a pseudopotential for the system He^{++} plus two electrons interacting with one another and with the rest of the bulk liquid by a variational method. The calculated shifts with external pressure for the $2s^3S \rightarrow 2s^3P$ absorption line agree quite well with the experimental results (Fig. 6.18). For the corresponding emission line the initial state has p symmetry and the bubble will be non-spherical. Therefore, the treatment causes more problems, but the observed shift has been explained by the earlier calculation of

Hickman and Lane /1971/. In the case of n = 2 states ($2s^3S$, $2s^3P$) the cavity radius of 6 Å is rather small. The model describes very well the increase of the bubble radius with the increasing of the main quantum number. In the case of n = 3 a bubble radius of 11 Å has been calculated for the $3s^3S$ state and of 13 Å for $3s^1S$. The calculated energy shifts for the $3s^3S \rightarrow 2s^3P$ emission line (Fig. 6.18) are in good agreement with the experimental results. The experimental pressure shifts of the $3s^1S \rightarrow 2s^1P$ transition are less accurate in the experiment due to the strong quenching. Beside the shifts also the observed broadening of the lines with increasing pressure has been explained by the bubble model. The method can be applied, in principle, to molecular centres, but the treatment is more complicated due to the additional interactions inherent in a system of two He atoms with four electrons. On the basis of the bubble model Steets et al. /1974/ made an attempt to explain the strong quenching of higher excited states with external pressure. They included a more detailed description of the density of the ground state He atoms just at the border of the bubble. Adopting an exponential, pressure dependent, distribution function, they have been able to rationalize the experimental slopes of the pressure dependence (Fig. 6.18) by processes where only a few (1 - 3) liquid atoms are involved. However, the nature of the quenching process is still obscure.

6.7 Emission at High Excitation Densities

The broad bandwidth emission continua from noble gas excimers, extending from 60 nm in the case of He to 172 nm for Xe favour these gases for short wavelength lasers. The bound excimer states decay radiatively to the strongly repulsive part of the ground state potential curve. The repulsive ground state ensures an extremely fast depopulation by dissociation (10^{-13}s) thus facilitating population inversion. The large width of the emission bands will allow for a broad tuneability region. The particular problems encountered in the construction of the short wavelength laser follow from an λ^{-5} scaling law for the pumping power per unit volume which is required to produce a gain coefficient of unity per unit length. The idea to exploit the high density in the condensed phases for obtaining a sufficient high density of excimers led to the first successful operation of an excimer laser by Basov et al. /1970/. Amplified stimulated emission from electron beam excited liquid Xe at 178 nm has been observed. This project has been discontinued because of the low efficiency which has been attributed mainly to reabsorption of the light. Subsequently, the gas phase excimer lasers were developed /Rhodes, 1984/. Attempts to obtain stimulated emission in the region below 180 nm from Ar, Kr and Xe crystals have recently been reported /Schwentner et al., 1982/. Most of the emphasis has been concentrated on Ar crystals which radiate around 126 nm. Several experimental observations indicate that stimulated emission from Ar crystals has been achieved.

The short-lived dipole-allowed transition of a $^1\Sigma_u^+$ excimer state to the ground state is the only laser active band because the cross section for stimulated emission from the close-lying $^3\Sigma_u^+$ state is two orders of magnitude smaller. The cross section σ_s for stimulated emission follows from

$$\sigma_s = \frac{1}{8\pi n_r^2} \cdot \frac{\lambda^2}{\tau_r \Delta\nu} \qquad (6.15)$$

with the index of refraction n_r, the lifetime τ_r and the line width $\Delta\nu$. The optical data yield a value for solid Ar of $\sigma = 0.7 \cdot 10^{-17}$ cm^2 which is close to the gas-phase value of $1 \cdot 10^{-17}$ cm^2.

The light flux from spontaneous emission in a strongly excited crystal will also induce stimulated emission. The intensity $I(t)$ due to spontaneous emission and amplified spontaneous emission within the cone $d\Omega$ into the amplifying direction is given by

$$I(t) = \frac{dh}{\sigma_s \tau_r} \{ \exp[n(t) \sigma_s L] - 1 \} d\Omega \qquad (6.16)$$

with the excimer density $n(t)$, the penetration depth d, the length L and the height h of the crystal. The gain α due to amplification is given by

$$\alpha = n(t) \sigma_s L \quad . \qquad (6.17)$$

This gain is obtained for a single pass through the excited crystal. In order to obtain a gain of ≈ 1 in a single pass for a crystal with a length of 1 cm it is necessary to create a density of about $n = 10^{17}$ excimers/cm^3.

The growth even of single crystals from noble gases is well known. The demands for laser crystals are less severe since only optically clear but free standing crystals with a volume of about 1 cm^3 and a front surface of 2×1 cm^2 are necessary. The crystals were excited by an electron gun delivering electron pulses of up to 10 J (600 keV, 5 kA, 3 ns). The penetration depths of the electrons are about 0.24 , 0.12 and 0.08 mm in solid Ar, Kr and Xe, respectively. About 0.22 J have been deposited in the crystal. This energy is enough to create the necessary density for amplification of the order of 10^{17} excimers/cm^3. Amplification has been proved by the following observations. a) A decrease by 10% of the FWHM of the $^1\Sigma$ band for an increase of the deposited energy power. b) Most of the $^1\Sigma$ contribution is emitted synchronously with the exciting Febetron pulse and much faster than a convolution of the exciting pulse with τ_r would predict. c) The intensity in the $^1\Sigma$ contribution increases progressively in contrast to the $^3\Sigma$ contribution which increases linearly or even slightly sub-linearly with deposited energy. d) The measured absolute photon fluxes correspond to $8 \cdot 10^{20}$ photons/s into $d\Omega = 10^{-2}$ sterad or to a density of $n = 5 \cdot 10^{16}$ cm^{-3} of $^1\Sigma$ excimers. A peak power of 1300 Watts due to amplified spontaneous emission has been detected in 10^{-2} sterad. The peak power in all directions amounts to 10^6 Watts. e) With a laser resonator a divergence of the amplified radiation of ~ 1 mrad has been obtained. The observations a) to e) yield a gain of $0.3 < \alpha < 1$ per single pass /Schwentner et al., 1982/. Fig. 6.19 shows a comparison of the efficiences for an Ar gas laser and a solid-state laser starting with the electron gun.

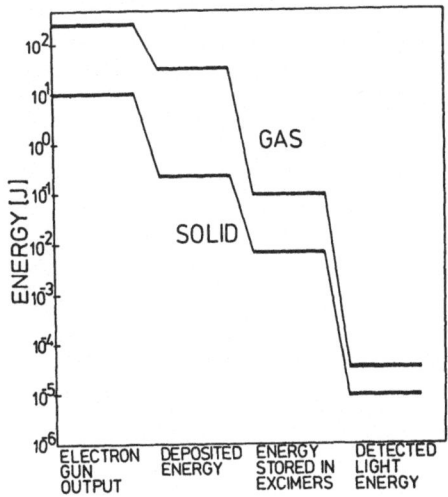

Fig. 6.19. Approximate conversion efficiencies for Ar gas and Ar crystals (from Schwentner et al. /1982/)

Evidently the crystal can be more efficient concerning the conversion of deposited energy into photons /Schwentner et al., 1982/. An alternative way for high density excitation of condensed rare gases is pursued by Muller et al. /1982/, Böttcher and Schmidt /1984/ and by Kessler et al. /1985a/. They achieved excitation of rare gases by multiphoton absorption of an ArF and KrF excimer laser beam.

At the high excitation densities necessary for laser applications a dense cloud of free electrons, excitons and excimer centres is produced. In Kr and Xe crystals a flat continuum emission is observed under high excitation conditions extending from the infrared, through the visible and down to the ultraviolet spectral region /Schwentner et al., 1982; Kessler et al., 1985b/. This emission is only present during the excitation time of $3 \cdot 10^{-9}$ s. In Ar crystals this continuum emission is absent or at least one order of magnitude smaller than in Kr. This emission is due to free-free Bremsstrahlung and free-bound recombination radiation. The emission coefficient ε_ν for this radiation does not depend on the photon energy and is given by /Maecker and Peters, 1954/:

$$\varepsilon_\nu (\text{W s sr}^{-1}\ \text{m}^{-3}) = 5.44 \cdot 10^{-52}\ \frac{N_e \cdot N_i}{\sqrt{T}} \tag{6.18}$$

with the electron and ion densities N_e, N_i in (m^{-3}) and the temperature T in (K).

The measured absolute fluxes prove that in Xe and Kr crystals during the excitation pulse a dense plasma of electrons with a density of 10^{18} electron/cm^3 and a distribution of kinetic energies corresponding to a temperature of 5000 - 8000 K is maintained. The crystal lattice remains at its low temperature of 10 - 60 K. This radiation is similar to the plasma radiation observed in high pressure arcs. The black spot and the arrows in Fig. 6.20 illustrate the characteristics of this plasma in rare-gas crystals in comparison with other plasma sources. This plasma in rare-gas crystals presents an extra ordinary case of an electron temperature of several thousand K and an electron density comparable to general plasma sources but with the ions fixed at the lattice sites at very low temperatures. The scattering properties of these hot electrons and the interaction with the free excitons and excimers will be an interesting field in the future.

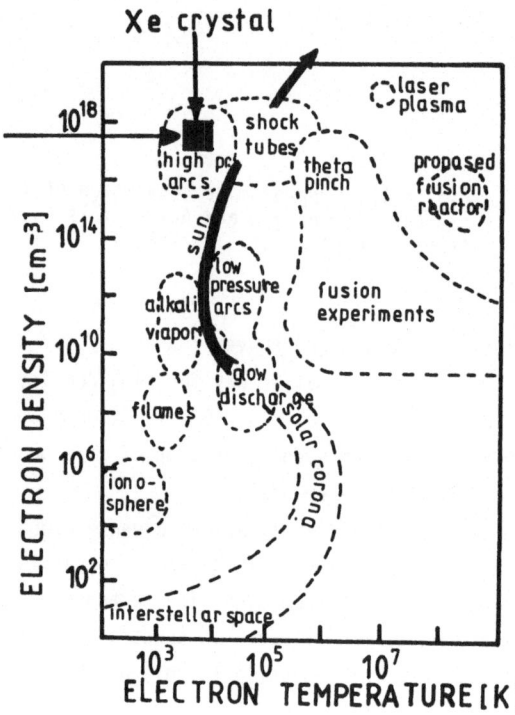

Fig. 6.20. Electron densities and electron temperatures for typical plasmas. The arrows and the black dot indicate the plasma parameter for high excited Xe crystals (from Kessler et al. /1985b/)

6.8 Localized Excitation and Lattice Relaxation of Impurity States

The cavity model for the localization of an electronic excitation in liquid He implies a considerable nuclear relaxation around the electronically excited state. In
this case, the short-range electron-He repulsive interaction leads to a rearrangement into a new equilibrium configuration, which is characterized by an appreciable
increase of the average separation between the electronically excited species and
its nearest He atoms. Apart from the name "cavity model", which emphasizes large
displacements, such medium relaxation phenomena are common in impurity centres in
solids and liquids. Optical excitation of these localized impurity states often
leads to a high degree of phonon excitation in the electronically excited state. The
electronically-vibrationally excited state, populated by vertical transitions, will
subsequently relax by phonon emission to a new equilibrium nuclear configuration.
A proper description of excitation and relaxation within a localized impurity centre
is described in terms of the configurational coordinate model sketched in Fig. 6.21
which portrays the nuclear dependence of the adiabatic potential surfaces. $U_g(Q)$

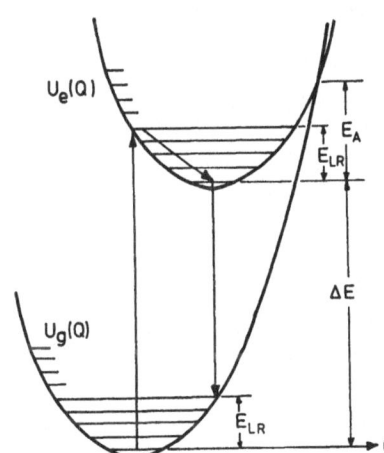

Fig. 6.21. Configuration coordinate diagram
for an impurity centre. U_g is the energy of
the ground state (for absorption) or the
energy of any state below the state with
energy U_e (for electronic relaxation) and U_e
is the energy of an excited state versus the
configuration coordinate Q. E_{LR} is the lattice rearrangement energy and ΔE is the energy difference of the minima of the potential
surfaces. E_A is the activation energy given
by the crossing of the potential surfaces

and $U_e(Q)$ correspond to the ground electronic state and to the electronically excited state, respectively, while Q designates the (dimensionless) normal nuclear
coordinates. In the harmonic approximation these are

$$U_g(Q) = \frac{1}{2} \sum_j \hbar\omega_j \cdot Q_j^2 \tag{6.19}$$

$$U_e(Q) = U_g(Q) - \sum_j C_j Q_j + \Delta E \tag{6.20}$$

where ΔE is the electronic energy gap, ω_j the phonon frequencies while the coeffi-

151

cients C_j represent the phonon coupling constants, so that the displacement of the equilibrium configuration of the jth mode is $\Delta_j = C_j/\omega_j$. The lattice rearrangement energy of the centre is

$$E_{LR} = \frac{1}{2}\sum_j C_j^2/\omega_j = \frac{1}{2}\sum_j \omega_j \Delta_j^2 \qquad (6.21)$$

being due to nuclear distortions while the Stokes shift between emission and absorption is given by $E_S = 2E_{LR}$. We have already used the notion of lattice distortion energy in relation to exciton trapping, and now the same concepts are employed for the simpler case of a localized centre. The quantum energy levels on the two potential surfaces are $|gv\rangle = |g\rangle|v\rangle$ and $|ev'\rangle = |e\rangle|v'\rangle$, where $|g\rangle$ and $|e\rangle$ are electronic wave functions, while $|v\rangle$ and $|v'\rangle$ denote vibrational states for the ground and excited electronic configurations. The corresponding energies are E_{gv} and $E_{ev'}$. The absorption lineshape is

$$L(E) = \langle\Sigma|\langle ev'|\mu|gv\rangle|^2 \; \delta(E + E_{gv} - E_{ev'})\rangle_T \qquad (6.22)$$

where $\langle\;\rangle_T$ denotes thermal averaging over the ground state vibrational states and μ is the transition dipole operator. The theory of optical lineshapes is well-known /Kubo and Toyozawa, 1955; Markham, 1959; Gold, 1961; Pryce, 1966; Maradudin, 1966/. It is useful to introduce the Huang-Rhys coupling constant

$$G = \frac{1}{2}\sum_j \Delta_j^2 \quad . \qquad (6.23)$$

Two limiting situations are of interest to us:

1) *Weak coupling limit* $(G \ll 1)$. The electron phonon coupling is weak, i.e., the displacement of the equilibrium configuration in the excited state is small. The optical spectrum consists of a zero phonon line followed by one-phonon and multi-phonon contributions at higher energies

$$L(E) = \exp\left[-\frac{1}{2}\sum_j \Delta_j^2 \coth\left(\frac{\hbar\omega_j}{2kT}\right)\right] \delta(E - \Delta E) + \text{higher order terms} \; . \qquad (6.24)$$

The Debye-Waller factor appearing in the exponent in (6.24) determines the relative intensity of the zero phonon line to the rest of the band. This situation is realized, for example, by N_2 excited impurity states in Ne /Gürtler and Koch, 1980/.

2) *The strong coupling limit* $(G \gg 1)$. The distortion of the equilibrium configuration is large. The absorption lineshape can be expressed in terms of a Gaussian

$$L(E) = \frac{1}{\sqrt{2\pi D^2}} \exp\left[-\frac{(E - \Delta E - E_{LR})^2}{D^2}\right] \qquad (6.25)$$

$$D^2 = \hbar \sum_j \omega_j \Delta_j^2 \coth\left(\frac{\hbar\omega_j}{2kT}\right) \quad . \tag{6.26}$$

In the limit of low temperatures

$$D^2 = \hbar \sum_j \omega_j \Delta_j^2 \qquad k_BT \ll \hbar\omega_j \tag{6.27}$$

while in the high temperature limit

$$D^2 = 2k_BT \cdot E_{LR} \quad . \tag{6.28}$$

The latter expressions describe the local thermal fluctuations of the phonon field around the impurity which were utilized in Toyozawa's work /1958/ for the description of the modulation of exciton motion [see (6.2 - 5)]. Adopting a mean phonon frequency $\langle\omega\rangle$ the optical absorption lineshape in the strong coupling limit is a Gaussian, (6.25), with a line width according to

$$H = H_0 \left(\coth\frac{\hbar\langle\omega\rangle}{2kT}\right)^{1/2} \tag{6.29}$$

where H_0 is the line width at $T = 0$.

Quantitative data concerning the lineshape of impurity states in RGS originate from Ophir's /1970/ study of the $n = 1$ (3/2) and $n = 2$ (3/2) lineshapes and their temperature dependence for Xe/Ar (Fig. 6.22). Ophir found that the lineshape for the $n = 1$ (3/2) state is Gaussian, so that the strong coupling situation prevails. The half line width H varies from 93 meV $\pm$ 1 meV at 33 K to 139 meV $\pm$ 1 meV at 80 K (Fig. 6.23). The temperature dependence of the line width can well be fitted by the multiphonon theory in the strong coupling limit given by (6.29). The experimental data for H(T) of the $n = 1$ (3/2) state yield $H_0 = 82$ meV and $\hbar\langle\omega\rangle = 8.2$ meV. The electron phonon coupling strength is given by $G = \eta(H_0/\hbar\langle\omega\rangle)^2$, where $\eta = 0.18$ is a (T independent) correction factor relating the half line width to the second moment of the absorption line (see, for example, /Pryce, 1966/). Thus $G = 18$ which, indeed, corresponds to the strong coupling situation. The Stokes shift E_S between absorption and emission is given by /Kubo and Toyozawa, 1955/ $E_S = -2\hbar \langle\omega\rangle G = -0.3$ eV resulting in a large red Stokes shift in accordance with the emission data for Xe in Ar ($E_S = -0.39$ eV) /Cheshnovsky et al., 1972b; Hahn and Schwentner, 1980/. The simple model cannot quantitatively account for the Stokes shift as it assumes that the phonon frequencies in the two electronic states are identical. Experimental evidence for frequency changes between the two electronic states is obtained for the temperature dependence of the energy maximum of the $n = 1$ absorption peaks of Xe impurity in Ar and Kr (Fig. 6.23). Multiphonon theory /Pryce, 1966/ predicts that the first moment of the absorption band shifts as

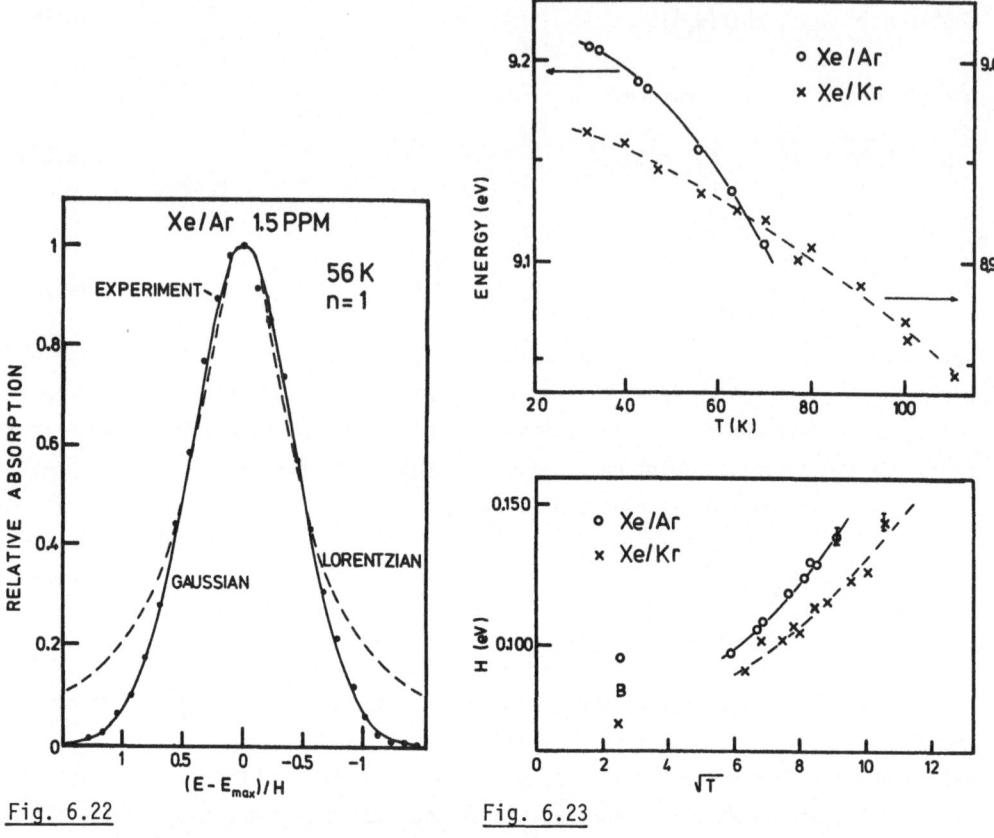

Fig. 6.22

Fig. 6.23

Fig. 6.22. Experimental absorption lineshape (points) of the n = 1 exciton of Xe guest atoms in an Ar matrix at 56 K and fits with a Gaussian (———) and Lorentzian (— — —) lineshape. The centre of the line E_{max} corresponds to 9.156 eV and the half-width (FWHM) H is 119 meV (after /Ophir, 1970/)

Fig. 6.23. Temperature dependence of the position of the n = 1 excitons of Xe guest atoms in Ar (o o o) and Xe guest atoms in Kr matrix (× × ×) (*upper part*) and of the corresponding halfwidth H (*lower part*). The result of Baldini /1965/ is shown by B. The solid (Ar matrix) and dashed (Kr matrix) lines correspond to fits with (6.29) (*lower part*) and (6.30) (*upper part*) (after /Ophir, 1970/)

$$E_{max} = E_{max}(0) + \hbar\Delta\omega \coth(\frac{\hbar<\omega>_g}{2k_BT})$$
(6.30)

where $<\omega>_g$ is a characteristic ground-state frequency and $\Delta\omega$ is the frequency change. This formula provides a semiquantitative fit of the experimental data (Fig. 6.23).

In the strong coupling situation a highly excited phonon state is produced by optical excitation. The optically excited "doorway state" is $\mu|gv>$ which is not an eigenstate of the nuclear potential in the electronically excited configuration and will exhibit time evolution. The non-equilibrium "doorway state" will exhibit intrastate phonon relaxation to the equilibrium configuration. Dissipation of the vibrational energy is expected to occur on a time scale of the reciprocal Debye frequency $\sim (\omega_D)^{-1}$ or, alternatively, it is determined by anharmonic interactions. The time scale for lattice relaxation around a single impurity state is $\approx 10^{-13}$ s.

6.9 Emission from Impurity States in Rare-Gas Solids

Experimental studies of optical emission from electronically excited states of impurity centres in RGS were conducted for dilute rare-gas alloys. The emitting impurity centres involved a single guest atom or a pair of guest atoms embedded in a host RGS (e.g. /Cheshnovsky et al., 1972a,b , 1973a-c; Gedanken et al., 1973a-e; Nanba et al., 1974; Fugol and Belov, 1975; Fugol, 1978; Möller, 1976; Ackerman, 1976; Hahn and Schwentner, 1980; Hahn et al., 1982/). These studies are of interest because of three reasons. Firstly, the large Stokes shift between absorption and emission provides information on substantial medium phonon relaxation around the electronically excited impurity centres. Secondly, identification of the emitting states of atomic impurity centres can be conducted by the assignment of a unique atomic parentage to those medium relaxed electronic excitations. Thus, medium perturbations are much smaller in emission than in absorption. Thirdly, the nature of the emitting atomic impurity states resulting from photoselective excitation /Hahn and Schwentner, 1980; Hahn et al., 1982/ provide central information on electronic relaxation processes within the impurity centre which is coupled to the phonon field. In what follows we shall proceed to discuss the implications of emission spectroscopy to elucidate intrastate medium relaxation processes, while interstate electronic relaxation phenomena will be discussed in Sect. 6.12.

In early studies of emission from atomic impurity states in RGS /Cheshnovsky et al., 1972a,1973a-c; Gedanken et al., 1973a-e/ α-particle excitation and later electron excitation (e.g. /Fugol and Belov, 1975/) and X-ray excitation (e.g. /Schuberth and Creuzburg, 1978/) has been used (see also Table 6.10). The spectra of Xe/Ar at 80 K excited by α particles /Cheshnovsky et al., 1972/ were rather simple, exhibiting only the impurity emission from the atomic Xe $1s_5$ state, the molecular emission from Xe_2^* due to impurity dimers and the Ar_2^* emission of the matrix (Fig. 6.24). Subsequent studies of emission induced by α excitation of Xe/Ar alloys at 4 - 20 K /Gedanken et al., 1973a-e/ revealed two distinct trapping sites for the emitting atomic impurity and for the excimer (Table 6.10). The major drawback of the studies, which involve non-specific excitation, were surpassed by the utilization of photoselective excitation methods by Hahn and Schwentner /1980/ and Hahn et al. /1982/ who were able to identify emission from the $1s_2$, $2s_3$ and $1s_4$ states of the Xe impurity (Table 6.10,11).

Table 6.10 Emission bands in rare-gas alloys except for the Ne matrix and the assignment ($^3p_1 \cong {}^1s_4$, $^3p_2 \cong {}^1s_5$, $^1p_1 \cong {}^1s_2$) given by the authors. s means emission from special sites

	electrons a)	X rays b)	α particles c)	vacuum UV light d)	e)	f)	g)	Stokes shift g)
Kr* in Ar	9.26 Ar Kr*	9.33 Ar Kr* 8.90 Ar Kr*	10.92 s ³p₁ 10.33 ³p₁ 9.32 Ar Kr*				10.95 ¹p₁ 10.5 s 10.4 ³p₁ 9.9 s 9.3 s	0.39 0.39
Kr*₂ in Ar	8.42 8.30	8.53	8.55				8.55	
Xe* in Ar	8.63 8.56 s 8.52 8.46 s 8.40 s 8.38 ³p₁ 8.28 ³p₂ 7.5 Ar Xe*	9.88 ¹p₁ 8.36 ³p₁ 8.28 ³p₂	8.67 s ³p₁ 8.38 ³p₁	8.32	9.80 9.61 ¹p₁ 9.40 8.70 8.44 ³p₁, ³p₂ 8.27	9.9 ¹p₁ 8.50 8.36	9.7 ¹p₁ n=2 8.83 ³p₁ 8.42 s 8.32 s	0.25 0.39
Xe*₂ in Ar	7.08	7.24 7.04	7.56 7.18	7.56 7.21	7.56 7.21	7.54 7.25		
Xe* in Kr	8.40 ³p₁ 8.30 ³p₂ 7.95 Kr Xe* 7.4 Kr Xe*	8.75 Kr Xe* 8.10 7.87 Kr Xe* 7.71 Kr Xe*	8.49 ³p₁ 7.94 Kr Xe*					
Xe*₂ in Kr	7.1	7.65 7.45 7.26 7.06	7.70 7.20					

a) Fugol et al. /1975/ and Fugol /1978/
b) Heumüller and Creuzburg /1978/ and Heumüller /1978/
c) Gedanken, Raz and Jortner /1973/
d) Nanba, Nagasawa and Ueta /1974/
e) Möller /1976/
f) Ackermann /1976/
g) Hahn, Haensel and Schwentner /1982/

Table 6.11. Energies of emission bands (E) and comparison with the corresponding states in absorption (A) and in the free atom (G). Energies in eV

		emission bands in Ne matrix (E)			atomic levels (G)			absorption bands in Ne matrix (A)		(A)-(E) Stokes shift	(G)-(E) Matrix shift
		a)	b)	c)				a)			
Xe/Ne	I	8.61	8.43	8.34 / 8.47 / 8.61	6s	$1s_5$ / $1s_4$	8.31 / 8.44	n = 1	9.06	0.45	- 0.16 / - 0.17
	II	9.80	9.56	9.77	6s'	$1s_3$ / $1s_2$	9.45 / 9.57	n = 1'	10.05	0.25	- 0.11 / - 0.20
	III	10.78			5d	$3d_2$	10.40	n = 2	11.32	0.54	- 0.38
Kr/Ne	I	10.10	10.03	10.02 / 10.16	5s	$1s_5$ / $1s_4$	9.91 / 10.03	n = 1	10.6	0.5	- 0.11 / - 0.13
	II	10.78			5s'	$1s_3$ / $1s_2$	10.51 / 10.64	n = 1'	11.22	0.44	- 0.14
	III	12.12			4d	$3d_5$	12.04	n = 2	13.35	1.23	- 0.08
Ar/Ne	I	11.70	11.61	11.63 / 11.71	4s	$1s_5$ / $1s_4$	11.54 / 11.62	n = 1	12.51	0.81	- 0.07 / - 0.09
	II	11.92		11.81 / 11.92	4s'	$1s_3$ / $1s_2$	11.72 / 11.83	n = 1'	12.74	0.82	- 0.09 / - 0.09
	III	13.93			3d	$3d_5$	13.86	n = 2	14.82	0.89	- 0.07

a) Hahn and Schwentner /1980/
b) Gedanken, Raz and Jortner /1973/
c) Schuberth and Creuzburg /1978/

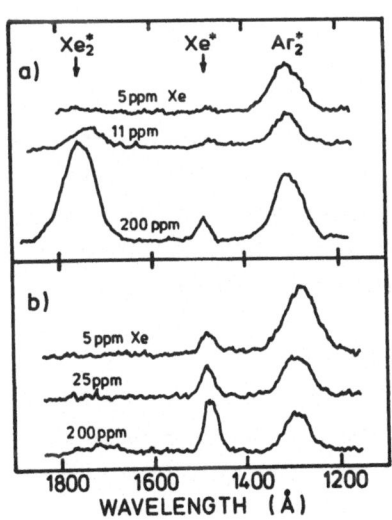

Fig. 6.24a,b. Emission spectra of Xe-doped liquid argon at 87 K (a) and Xe-doped solid argon at 80 K (b) (after Cheshnovsky et al. /1972b/)

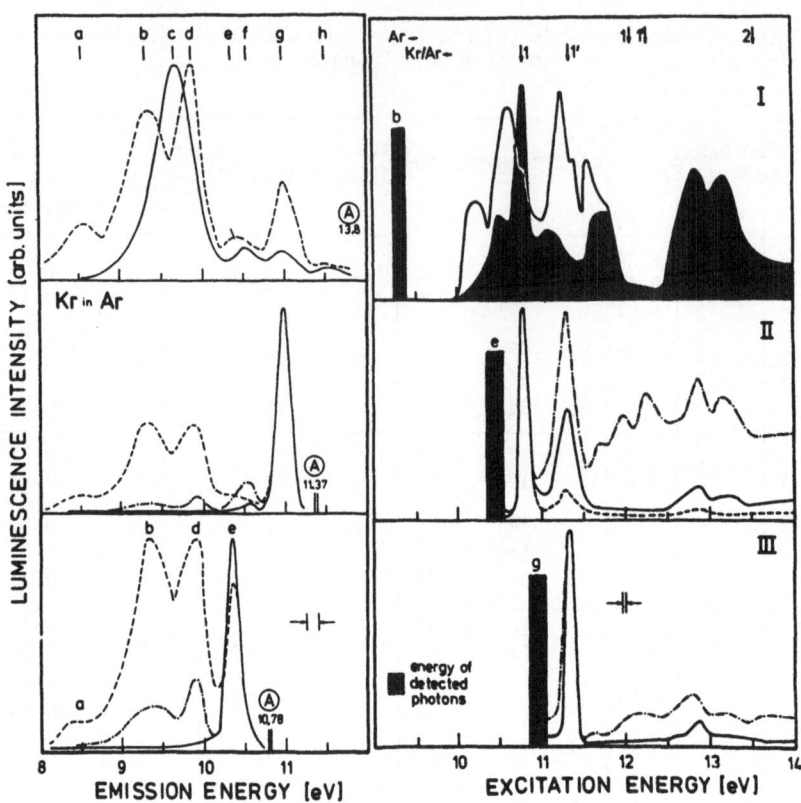

Fig. 6.25. (*Lefthand panels*) Emission spectra of Kr guest atoms in an Ar matrix for several excitation energies A of 13.8 eV, 11.37 3V and 10.78 eV and for several Kr concentrations: (− − −) 3% Kr; (−·−·) 0.3% Kr; (———) 0.03% Kr. (*Righthand panels*) I, II and III are excitation spectra of the emission bands b (site with energy lower than the relaxed Kr n = 1 exciton), e (relaxed Kr n = 1 exciton) and g (relaxed n' = 1 exciton), respectively. The identification of the emission bands is shown on the lefthand panels. (−·−·) 3% Kr; (———) 0.3% Kr; (− − −) 0.03% Kr (from Hahn et al. /1982/)

The system Kr in Ar serves as an example for the complications caused by the presence of single guest atoms and pairs of larger clusters of guest atoms as well as for multiple trapping sites. Emission spectra for three optical excitation energies and three concentrations are shown in Fig. 6.25 /Hahn et al., 1982/. For the higher doping concentrations the presence of pairs of guest atoms can be expected in the sample /Kreitman and Barnett, 1965/. These pairs tend to form Kr_2^* excimers in the excited state and a molecular emission band is observed for Kr. In Fig. 6.25 it is labelled a (Table 6.10). The emission bands c and h in Fig. 6.25 appear only for excitation energies which are high enough to excite intrinsic states of the Ar matrix (upper part of Fig. 6.25, $\hbar\omega$ = 13.8 eV). The bands c and h correspond to the excimer emission of the host Ar_2^* from the vibrational ground state and from the excited vibrational levels, respectively. The excitation spectra II and III in Fig. 6.25 clearly show that maximum e corresponds to the decay of the n = 1 exciton of Kr atoms in the Ar matrix and maximum g to the n' = 1 excitons. The remaining emission bands b and d (Fig. 6.25) have been observed also after particle excitation /Cheshnovsky et al., 1973b; Fugol and Tarasova, 1977/. These have been ascribed to (KrAr)* excimers. Emission has also been observed for Kr impurities in liquid Ar /Cheshnovsky et al., 1973a-c/, which supports the assignment of heteronuclear excimer emission. The assignment of the b and d bands in Fig. 6.25 to the KrAr* excimer is yet uncertain in view of the new results of Hahn et al. /1982/ using photoselective optical excitation. They found, firstly, that emission bands b and d can be excited by photon energies below the threshold of absorption of Kr atoms in an Ar matrix (curve I in Fig. 6.25); secondly, at low Kr concentration, the emission bands b and d disappear when the sample is annealed for fifteen minutes at 15 - 20 K (solid curves in Fig. 6.25). Hahn et al. /1982/ suggested the existence of special sites for a number of Kr atoms in the Ar matrix leading to a smaller threshold for absorption and to lower emission energies. The number of these sites is reduced by annealing. Nowak and Fricke /1985/ presented new and convincing results concerning the heteronuclear excimers ArKr*, ArXe* and KrXe*.

Solid Ne, which is transparent up to 17.4 eV, provides an excellent matrix for the study of the emission spectra of Ar, Kr and Xe impurity atoms. This host solid has another advantage, namely that heteronuclear excimer states between the electronically excited guest and the host atom are not formed. The emission spectra of dilute Ne alloys resulting from photoselective excitation are presented in Fig. 6.26, while the energies of the emission bands of Xe, Kr and Ar in solid Ne are summarized in Table 6.11. Each guest atom exhibits three emission bands which were compared with the radiative transitions in the free atom /Hahn and Schwentner, 1980/. The emission bands can be correlated with the three lowest dipole allowed transitions of the free atom, which are characterized by a red matrix-induced Stokes shift of some tenth of an eV (Table 6.11). The states associated with these dipole-allowed transitions for Ar and Kr are the fine structure components $1s_4$, $1s_2$ and $3d_5$ of the states ns, ns' and (n - 1)d (Table 6.11). For Xe the next higher state $3d_2$ is observed instead of the $3d_5$ state. These assignments are confirmed by a comparison of the radiative lifetimes of the states in the matrix and in the free atom (/Hahn and Schwentner, 1980/ and Table 6.12). The influence of the Ne matrix on the radiative lifetimes of the guest atoms was accounted for in terms of the reduction of the macroscopic radiation field due to the dielectric constant n of the sample and the difference in the effective radiation field E_{eff} at the site of the excited species from the macroscopic field E_0 due to the polarizability of the species /Blair et al., 1972; Fowler, 1968/. The radiative lifetime is roughly

$$\tau^r(\text{crystal}) \approx \tau^r(\text{gas}) / [n \cdot (E_{eff}/E_0)^2] \quad . \tag{6.31}$$

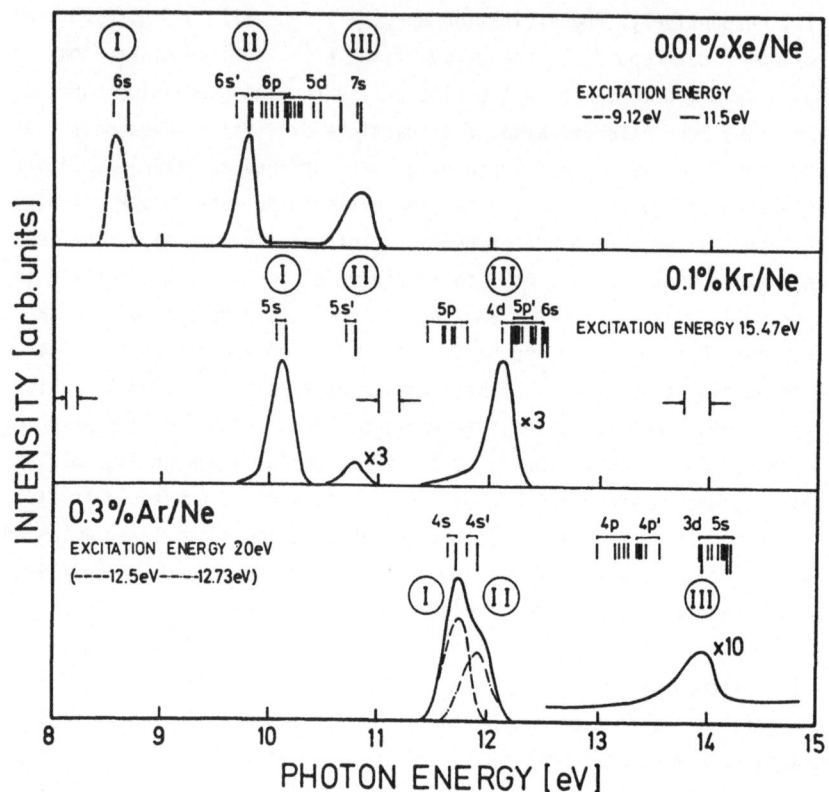

Fig. 6.26. Emission spectra for Xe, Kr and Ar atoms in the Ne matrix with emission bands I, II and III. The atomic states are also shown. Long bars correspond to dipole-allowed, and short bars to dipole-forbidden transitions. The absolute positions of the atomic transitions have been shifted by 0.23 eV, 0.14 eV and 0.09 eV for Xe, Kr and Ar, respectively (from Hahn and Schwentner /1980/)

For a guest centre in a matrix E_{eff} has been approximated by mean values of the dielectric constant of the host and the guest /Person, 1958; Dubost and Charneau, 1976/. This treatment did not account for medium effects on the electronic states of the impurity. It is apparent that the medium effects on the radiative lifetimes are small.

In Tables 6.10,11 the assignments of the impurity emission bands in dilute RGS alloys are summarized. Emission bands originating from the following states were observed. (1) In all cases atomic-type emission is exhibited, for example from the $1s_4$, $1s_2$ and $3d_5$ states. (2) In addition, emission from special trapping sites for atomic impurity states has been observed. (3) Impurity pairs involving the emission for the lowest excited states of excimers. (4) Impurity pairs in special trapping sites. (5) Heteronuclear excimers, such as Xe^*Kr and possibly Kr^*Ar.

Next we shall discuss the spectroscopic implications of intrastate relaxation. As is evident from the data summarized in Tables 6.10,11, the emission from single

Table 6.12. Radiative lifetimes of free atoms and of guest atoms in Ne matrix
(in 10^{-9} s)

		radiative lifetime in Ne matrix	atomic radiative lifetime	
Xe	6s , $1s_4$	2.4 ± 0.2	3.46 ± 0.09	
	6s', $1s_2$	3.5 ± 0.2	3.44 ± 0.07	a)
	5d , $3d_2$	1.3 ± 0.2	1.40 ± 0.07	
Kr	5s , $1s_4$	2.5 ± 0.2	3.18 ± 0.12	a)
	5s', $1s_2$	3.1 ± 0.5	3.11 ± 0.12	
	4d , $3d_5$	13.0 ± 2	44.5	b)
Ar	4s , $1s_4$	5.8 ± 0.3	8.4	
	4s', $1s_2$	1.2 ± 0.2	2.0	c)
	3d , $3d_5$	420 ±20	?	

a) Matthias et al. /1977/
b) Gruzdev and Loginov /1975/
c) Wiese, Smith, Miles /1969/

impurity states reveals a Stokes shift of ~ 0.2 eV - 1.2 eV relative to the corresponding exciton state seen in the absorption spectrum. This large Stokes shift originates from phonon relaxation around the excited impurity /Cheshnovsky et al., 1972b;
Jortner, 1974/. The observation of these large Stokes shifts is consistent with the
notion that the electronic excitation of impurity states in RGS corresponds to the
strong electron-phonon coupling limit (Sect. 6.8), being in good agreement with the
results of the analysis of the line broadening data. The experimental Stokes shift
for the Xe/Ar system E_s = -0.39 eV at 6 K deduced from the emission studies is consistent with the value of E_s = -0.3 eV for the Stokes shift deduced /Ophir, 1970/
from the temperature dependence of the line broadening in this system (Sect. 6.8).

The medium dilation around electronically excited impurity states in RGS causing
the large Stokes shifts bears a close physical analogy to the bubble formation around
electronically excited states in liquid He (Sect. 6.7). The changes in the equilibrium configuration coordinates for the ground and the excited states can be described
semiquantitatively for Xe in Ne. Messing et al. /1977e,f/ measured the dependence
of the energy of the atomic absorption bands of Xe atoms for an increasing density
of surrounding Ne atoms. They scanned a range of densities from pure Xe up to Xe in
liquid Ne. In the middle part of Fig. 6.27 the energy shifts of the Xe transitions
observed in absorption are shown as a function of the mean Xe-Ne nearest neighbour
separation together with the energies of the peaks of the excitation spectra of Xe
in solid Ne /Hahn and Schwentner, 1980/ which continue the trends of the density de-

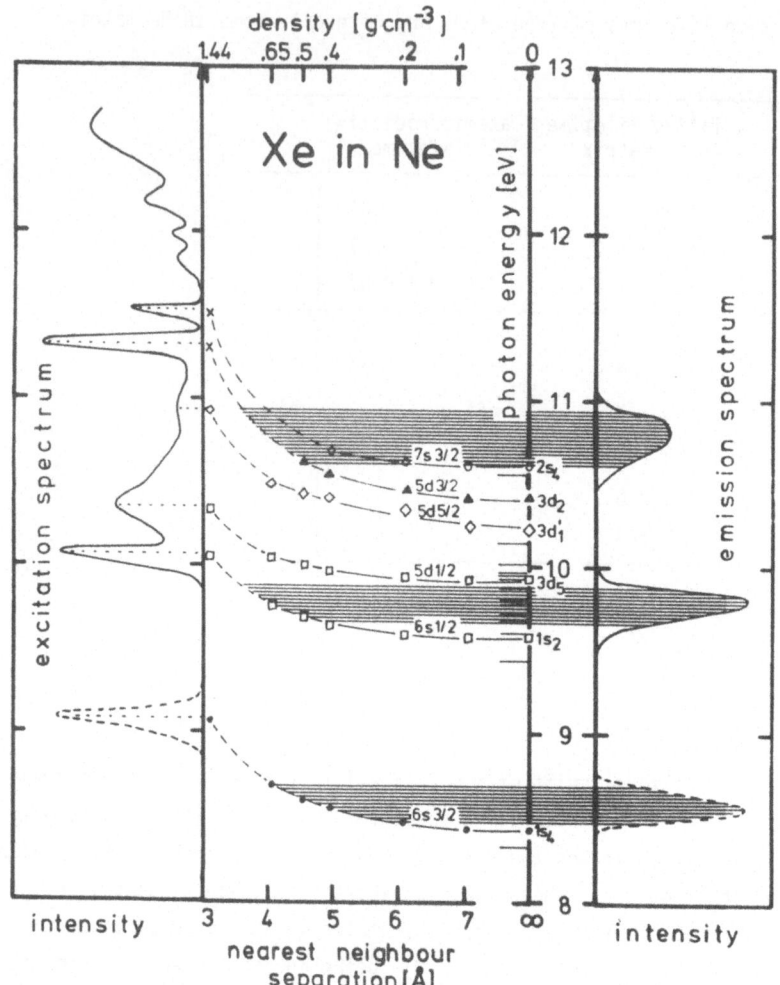

Fig. 6.27. Dependence of the energy position of the lowest states of Xe atoms on the density of surrounding Ne atoms. The density is expressed by the mean nearest neighbour separation between a Xe atom and surrounding Ne atoms. The Ne density ranges from 0 through liquid to the solid phase. For solid Ne (density 1.44 gcm^{-3} $\hat{=}$ 3.13 Å) the maxima of the excitation spectra (*lefthand side*) have been included. The density around relaxed Xe atoms follows from a comparison with the emission spectra (*righthand side, horizontal lines*) (from Hahn and Schwentner /1980/)

pendence for the different states. A comparison of the energy positions of the emission bands (righthand side of Fig. 6.27) with the absorption maxima shows the larger Xe-Ne nearest neighbour separation due to medium dilation occurring before radiative decay. This guest-host separation increases from 3.2 Å in the ground state configuration to values between 4 Å to 5 Å in the medium relaxed excited state.

162

The large medium relaxation around impurity states brings the impurity emission close ($\approx 0.1 - 0.2$ eV in all cases) to the energy of the corresponding atomic emission in the gas phase, as is apparent from Tables 6.10,11. The proper theoretical description of such medium-relaxed electronically excited impurity states should involve a tight-binding approach rather than a Wannier picture. The effects of medium dilation around impurity states are exhibited also for excited impurity pairs, as matrix effects on the emission spectra of the excimers are negligibly small.

Finally, we would like to comment on the time scale for intrastate medium relaxation around excited atomic impurity states. The rise time of the emission for the medium-relaxed impurity excited states is shorter than 10^{-11} s /Hahn and Schwentner, 1980/. This result is compatible with the theoretical estimates of Sect. 6.8 for the medium relaxation occurring on the time scale of $\approx 10^{-13}$ s. We thus expect that the quantum yield for hot luminescence originating from the radiative decay of electronically excited impurity states prior to medium relaxation will be $10^{-4} - 10^{-5}$. Such low quantum yields of hot luminescence induced by photoselective excitation are amenable to experimental observation and are expected to provide pertinent information concerning the dynamics of intrastate medium relaxation.

6.10 Emission from Impurity States in Liquid Rare Gases

Emission from impurity states of guest rare-gas atoms in host liquid rare gases was excited by α particles /Cheshnovsky et al., 1972b , 1973a; Jortner, 1974; Raz et al., 1976/ and by multiphoton absorption /Muller et al., 1982/. The spectroscopic interpretation of the emission spectra of impurity states in liquid rare gases (Table 6.13) is somewhat simpler than in the corresponding solids, as complications due to multiple trapping sites are not exhibited in the liquid.

Table 6.13. Emission spectra of liquid rare-gas alloys (doping concentration $10 - 10^4$ ppm) at 85 K (Xe/Ar , Kr/Ar) and 120 K (Xe/Kr) after Cheshnovsky et al. /1972 , 1973/ and Raz et al. /1976/

Assignment	Xe/Ar	Xe/Kr	Kr/Ar
Ar_2^*	9.69		9.69
$Xe\ ^3P_1$	8.32		
Kr_2^*		8.32	8.38
$Xe\ Kr^*$		7.85	
$Ar\ Kr^*$			9.18
Xe_2^*	7.12	7.12	

The only case of a genuine atomic impurity emission in a liquid recorded up to date involves the emission from the $^3P_1$ state of Xe in liquid Ar shown in Fig. 6.24 /Cheshnovsky et al., 1972b/. The Xe emission is Stokes shifted by 0.4 eV to the red relative to the impurity absorption in the liquid, being very close to the corresponding atomic emission in the gas phase. These features clearly indicate the effects of medium dilation, i.e., cavity formation around the excited impurity state in the liquid /Cheshnovsky et al., 1972a/.

From Fig. 6.24 it is apparent that the molecular impurity emission at low Xe concentrations is much more prominent in the liquid than in the solid. This is not surprising, as formation of Xe_2^* impurity molecules will occur in the liquid by the diffusion controlled "collision" process

$$Xe(^3P_2) + Xe(^1S_0) \rightarrow Xe_2^*(^3\Sigma_u) \quad . \tag{6.32}$$

This process provides an example of the applicability of diffusion controlled kinetics to reactions involving electronically excited atoms or molecules in simple dense fluids.

Emission spectra of Xe/Kr liquid alloys are portrayed in Fig. 6.28. They reveal three prominent emission bands. The Kr_2^* host emission peaked at 1470 Å, the Xe_2^* impurity pair emission at 1740 Å and an additional band peaked at 1580 Å. This additional band appears at the lowest accessible Xe impurity concentration (10 ppm); its relative intensity reaches a maximum at 100 ppm Xe and starts decreasing in favour of the Xe_2^* 1740 Å emission band at higher Xe concentrations (Fig. 6.28). This 1580 Å band is assigned /Cheshnovsky et al., 1973a/ to emission from the heteronuclear XeKr* diatomic molecule. The Xe_2^* molecule in this system is formed via the reaction

$$XeKr^* + Xe(^1S_0) \xrightarrow{k_F} Xe_2^*(^3\Sigma_u) + Kr(^1S_0) \quad . \tag{6.33}$$

This mechanism requires that the intensity ratio between the 1740 Å emission and the 1580 Å emission is proportional to the Xe impurity concentration. Such a relation is indeed obeyed, as is evident from Fig. 6.29 which results in the rate constant $k_F = 2\cdot10^{-13}$ cm^3s^{-1} for the reaction. This magnitude of the rate constant is compatible with a diffusion controlled process which is characterized by a diffusion coefficient $D \cong 10^{-6}$ cm^2s^{-1}.

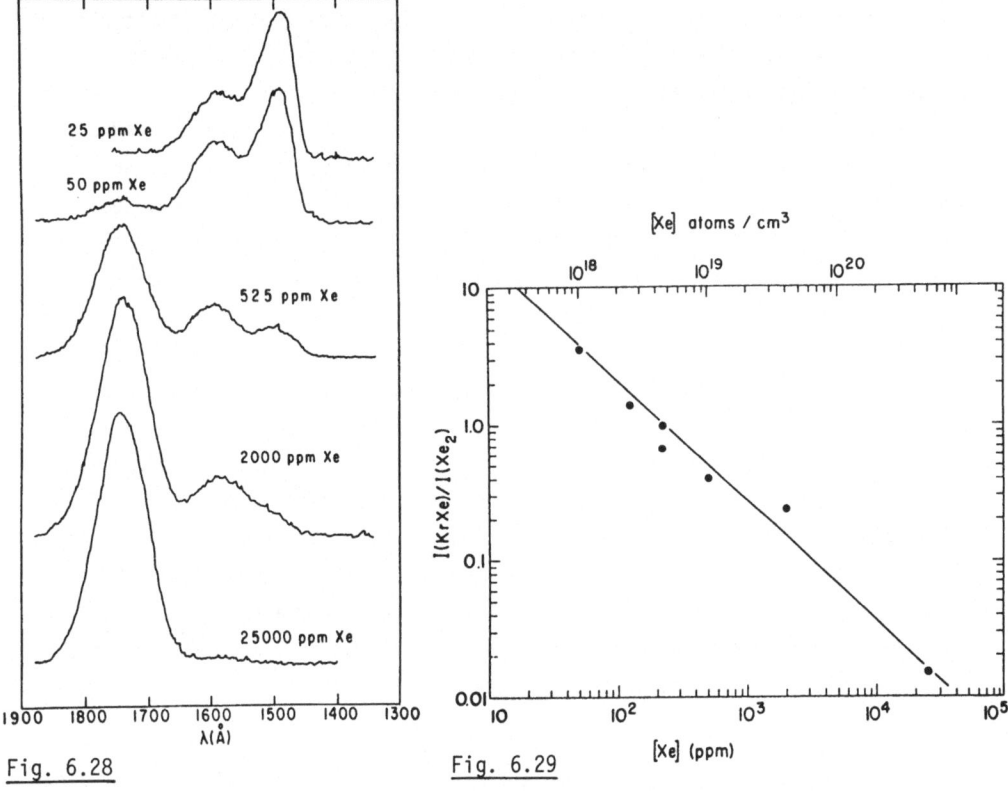

Fig. 6.28   Fig. 6.29

Fig. 6.28. Emission spectra of liquid Kr doped with Xe at various concentrations
(T = 120 K) (from Cheshnovsky et al. /1973a/)

Fig. 6.29. Dependence of the intensity ratio of the heteronuclear excited KrXe*
molecule emission to the excited homonuclear diatomic Xe$_2^*$ emission upon the concentration in liquid Kr (from Cheshnovsky et al. /1973a/)

6.11 Electronic Relaxation

A central feature of excited-state dynamics involves interstate radiationless electronic relaxation between different electronic states. In this relaxation process
electronic energy is converted into vibrational energy (e.g. /Kubo and Toyozawa,
1955; Riseberg and Moos, 1968; Englman and Jortner, 1970; Auzel, 1978/). For atomic
impurity states in RGS electronic energy of the impurity is dissipated by acoustic
phonons. Such a non-radiative process can be described in terms of relaxation between two potential surfaces corresponding to two different electronic configurations (Fig. 6.21). The coupling which induces the non-radiative processes is given
by the non-adiabatic interstate interaction originating from nuclear kinetic energy.
The electronic relaxation process is invisioned as a transition from a phonon mani-

fold $|ev'>$ of the initial excited electronic configuration $|e>$ to a set of phonon states $|gv>$ corresponding to the final electronic ground state $|g>$. The microscopic rate constant W_{ev} for the $|ev'> \rightarrow |gv>$ transition can be expressed in terms of the non-adiabatic formalism /Kubo and Toyozawa, 1955; Perlin, 1963; Englman and Jortner, 1970/ provided that the residual coupling is weak relative to the characteristic phonon frequency $<\omega>$ of the system. The non-adiabatic microscopic rate is expressed within the framework of the Condon approximation

$$W_{ev'} = \frac{2\pi}{\hbar} |V_{eg}|^2 \sum_{v} |<v'v>|^2 \delta(E_{gv} - E_{ev'})$$

(6.34)

where V_{eg} is the non-adiabatic electronic coupling, $<v'v>$ is the nuclear overlap integral, while E_{gv} and $E_{ev'}$ are the energies of the vibronic states. A basic assumption underlying most of the studies of electronic relaxation is that medium-induced vibrational relaxation and excitation within each electronic manifold are fast on the time scale of the electronic process. Thus, for example, in the case of impurity states in RGS one can assert that medium relaxation around electronically excited states is fast ($\approx 10^{-13}$ s, Sect. 6.8) on the time scale of interstate electronic relaxation. Under these circumstances the transition probability W_{eg} for electronic relaxation can be expressed in terms of a thermal average of the microscopic rates

$$W_{eg} = \Sigma_{v'} \exp(-E_{ev'}/k_BT) \cdot W_{ev'} / \Sigma_{v'} \exp(-E_{ev'}/k_BT) \quad .$$

(6.35)

Before proceeding to discuss any approximate calculations of the rate (6.35) two general limiting cases can be distinguished:

a) The low temperature limit is realized when the thermal energy is considerably lower than all the characteristic vibrational (phonon) frequencies of the system. The low temperature rate $W(T \rightarrow 0) = W_{eo}$ is temperature independent over the range $k_BT \ll \hbar<\omega>$. The temperature independent rate exhibits the effects of nuclear tunnelling from the zero-point state of the initial nuclear configuration.

b) The high temperature limit is encountered when the thermal energy considerably exceeds the nuclear vibrational frequency. The high temperature rate

$$W \propto \exp(-E_A/k_BT)$$

(6.36)

assumes an activated rate expression with the activation energy E_A being given by the potential energy of the initial electronic state at the lowest intersection point of the two potential surfaces (see Fig. 6.21).

Considerable effort was devoted towards the derivation of explicit expressions. Most of these studies (e.g. /Kubo and Toyozawa, 1955; Perlin, 1963; Englman and

Jortner, 1970; Auzel, 1978/ treat the nuclear motion in the harmonic approximation considering only linear electron-phonon coupling corresponding to configurational distortion of the origins of the potential surfaces. The physical picture is analogous to the treatment of phonon broadening of impurity absorption lineshapes surveyed in Sect. 6.8. However, while optical excitation corresponds to nearly vertical transitions at finite photon energy, the non-radiative process corresponds to a horizontal transition in the limit of zero energy.

Within the framework of the simple harmonic picture, the non-radiative rate can be expressed as a product of a non-adiabatic electronic coupling term $|V_{eg}|^2$ and a thermally averaged nuclear Franck-Condon overlap term. The latter contribution can approximately be expressed in terms of the electronic energy gap ΔE (see Fig. 6.21) between the minima of the two potential surfaces, the phonon characteristic frequency $\hbar<\omega>$ and the electron phonon coupling strength G (6.23). Although some general but cumbersome expressions were derived /Kubo and Toyozawa, 1955; Webman, 1976/ in considering phonon dispersion, a coarse graining procedure over phonon frequencies, taking only the characteristic phonon frequency $\hbar<\omega>$ into account, leads to great simplifications. The approximate general rate can be expressed in the form /Huang and Rhys, 1950/

$$W = A \exp[-G(2\bar{v}+1)] \cdot I_p \left\{ 2G\left[\bar{v}(\bar{v}+1)\right]^{1/2}\right\} \left[\frac{\bar{v}+1}{\bar{v}}\right]^{P/2} \tag{6.37}$$

where the reduced energy gap is

$$P = |\Delta E| / \hbar<\omega> \quad . \tag{6.38}$$

$I_p\{ \}$ is the modified Bessel function of order P and v represents the thermal population of phonon states

$$\bar{v} = [\exp(\hbar<\omega> / kT) - 1]^{-1} \tag{6.39}$$

while the pre-exponential factor is

$$A = \frac{2\pi / V_{eg}^2}{\hbar^2<\omega>} \quad . \tag{6.40}$$

For exothermic processes, $\Delta E < 0$, the low temperature rate for nuclear tunnelling is obtained for (6.37) in the form

$$W(T \to 0) = A \exp(-G) \frac{G^P}{P!} \tag{6.41}$$

while in the high temperature limit (6.37) results in the activated rate

$$W = A \left(\frac{\hbar<\omega>}{4\pi G k_B T}\right)^{1/2} \cdot \exp\left[-\frac{(\Delta E - G\hbar<\omega>)^2}{4G\hbar<\omega>k_B T}\right] \quad . \tag{6.42}$$

For the intermediate temperature region (6.37) has to be applied.

Of considerable interest is the low temperature rate which, for reasonable values of P > 3G, can be expressed by the use of Stirlings's approximations

$$W(T \to 0) = \frac{A}{(2\pi P)^{1/2}} \exp(-G) \cdot \exp(-\gamma P) \qquad (6.43)$$

$$\gamma = \ln\left(\frac{P}{G}\right) - 1 \quad . \qquad (6.44)$$

Equation (6.43) expresses the energy gap law for electronical relaxation. From (6.43) it is apparent that the following general features of the physical system are expected to affect the low-temperature electronic relaxation rate within the impurity centre:

1) The pre-exponential factor A. For atomic impurity states in RGS A was roughly estimated to be $A \sim 10^{14}$ s^{-1} /Webman, 1976/.

2) The electronic energy gap ΔE. The increase of P with ΔE (6.38) results in an exponential decrease of W.

3) The electron-phonon coupling G. When $P \simeq G$ we expect that $\gamma \simeq -1$ and the energy gap dependence of W is suppressed. Efficient electronic relaxation can occur for large energy gaps provided that $G \sim P$ /Webman, 1976/.

Webman /1976/ has conducted model calculations for the rates of electronic relaxation in impurity centres in RGS /Jortner, 1974/ which rest on a Wannier-type description of the electronic excited states. Such an approach indicated that non-radiative cascading from high excited states, characterized by large main quantum numbers, is efficient. However, this treatment is too crude for a detailed theoretical understanding of electronic relaxation within low lying electronically excited impurity states in RGS, which are essentially of atomic parentage.

6.12 Electronic Relaxation Within Impurity Centres in RGS

The dynamics of excited states of impurity centres in RGS can be investigated by measurements of emission spectra, excitation spectra, emission quantum yields, transient absorption and decay lifetimes. The central experimental information involves the total lifetime τ_I of the Ith electronic state where the non-radiative decay rate $W_I = \Sigma_F W_{IF}$ involves the contribution from all non-radiative decay channels, while the corresponding radiative decay time τ_I^r incorporates radiative decay to all lower lying F states, so that $(\tau_I^r)^{-1} = \Sigma_F (\tau_{IF}^r)^{-1}$. The experimental lifetime is given by the well-known relation

$$(\tau_I)^{-1} = W_I + (\tau_I^r)^{-1} = \Sigma_F W_{IF} + \Sigma_F (\tau_{IF}^r)^{-1} \quad . \tag{6.45}$$

The partial quantum yield for the radiative $I \rightarrow F$ transition is:

$$Y_{I \rightarrow F}^r = (\tau_I)/(\tau_{IF}^r) \tag{6.46}$$

while the quantum yield for the non radiative $I \rightarrow F$ process is

$$Y_{I \rightarrow F}^{nr} = W_{IF} \cdot \tau_I \quad . \tag{6.47}$$

Major experimental effort should be directed towards (i) the elucidation of the nature of the various radiative and non-radiative decay channels, (ii) the quantitative determination of the total decay rates τ_I^{-1}, W_I and $(\tau_I^r)^{-1}$, (iii) the determination of partial radiative and non radiative decay rates $(\tau_{IF}^r)^{-1}$ and W_{IF}, and (iv) determination of quantum yields $Y_{I \rightarrow F}^r$ and $Y_{I \rightarrow F}^{nr}$ for radiative and non-radiative channels, which is complementary to (iii). The theory of electronic relaxation of localized states in insulators (Sect. 6.11) should provide answers to the following questions regarding excited-state dynamics of impurity states in RGS:

a) Which states exhibit appreciable emission? Effective radiative decay of a given state I requires that $W_I \lesssim (\tau_I^r)^{-1}$.

b) What are the rates of non-radiative cascading from highly excited states which terminate in the emitting states?

c) What quantitative information can be obtained regarding non-radiative decay rates between specific electronic states?

d) What information can be obtained concerning details of electronic relaxation processes, such as temperature dependence of individual non-adiabatic rates?

We shall start with a survey of the experimental information. The early studies /Cheshnovsky et al., 1972a,b; Gedanken et al., 1973a-e; Raz et al., 1976/ using α excitation of dilute rare-gas alloys, established the occurrence of emission from low lying excited states. In all cases the $^3P_1$ state of the guest emits. A notable exception was Xe in Ne where emission from the $^1P_1$ state, was observed. These results were qualitatively rationalized in terms of the energy gap law for electronic relaxation /Jortner, 1974; Raz et al., 1976/. Much more detailed information was obtained by Hahn and Schwentner /1980/ using photoselective excitation. The most extensive and informative data were obtained for the emission of Xe, Kr and Ar in solid Ne. Typical experimental results for the excitation spectra, the emission spectra and time-resolved decay are presented in Figs. 6.30,31. We have already discussed the assignment of these impurity emission peaks (Table 6.11) and will now consider the electronic relaxation processes. Excitation of the

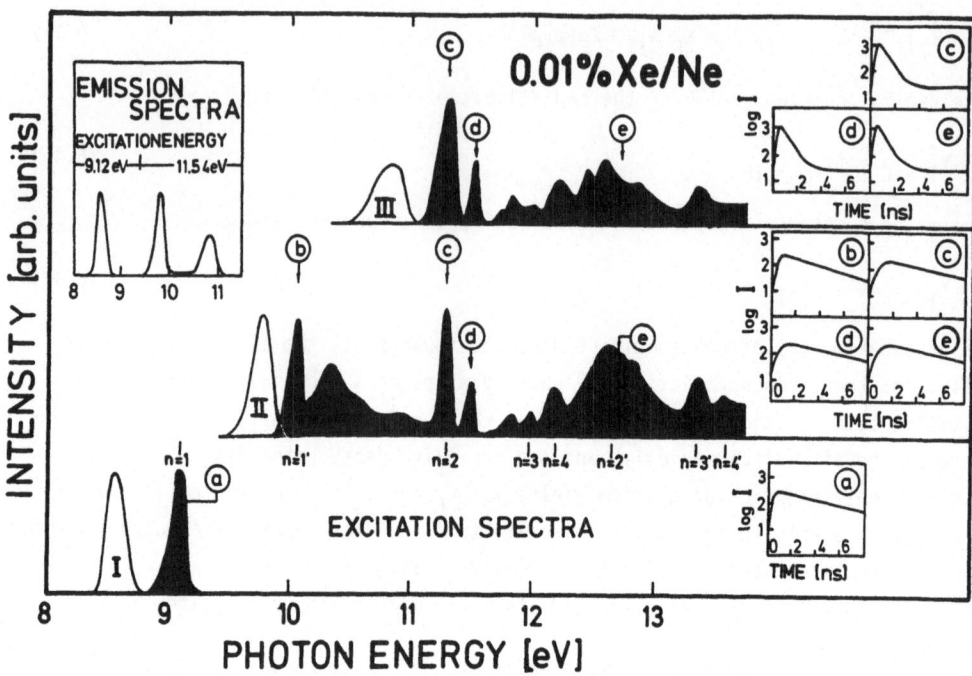

<u>Fig. 6.30.</u> Excitation spectra (emission intensity versus excitation energy) for Xe in Ne (hatched curves) for each of the three emission bands I, II and III (see also Fig. 6.26). The emission bands are shown for each of the corresponding excitation spectrum at lower photon energies. Prominent maxima in the excitation spectra are marked by (a, b, c, d, e). Further, the positions of excitonic states n = 1, 2, 3, 4 and n' = 1', 2', 3', 4' are marked. The inserts at the righthand side provide a survey of the time dependence of the intensity in the emission bands at prominent excitation energies. The insert on the left shows the complete emission spectrum (from Hahn and Schwentner /1980/)

n = 1 impurity state of Xe, Kr and Ar in solid Ne results in medium relaxation leading to the formation of the $1s_4(^3p_1)$ impurity state. On the basis of the experimental data (Table 6.12) for the radiative decay, it is apparent that the major contribution to the lifetime of the $1s_4$ state originates from the radiative decay channel to the ground state which is much faster than the other two competing radiationless processes, i.e., radiationless relaxation either to the only lower lying excited state $1s_5(^3P_2)$ or to the ground state. The radiationless relaxation to the ground state is hindered by the very large ($\simeq 10$ eV) energy gap. The separation from $1s_4$ to the $1s_5$ state is $0.1 - 0.2$ eV (see Figs. 6.32,33) requiring a radiationless transition of the order of P = 10 - 30. Provided that the electron-phonon coupling is not too large the radiationless $1s_4 \rightarrow 1s_5$ decay will be effectively blocked by the appreciable energy gap. In a similar manner the energy gap law determines the relaxation of the n = 1' impurity states of Xe and of Ar in solid Ne. Excitation of

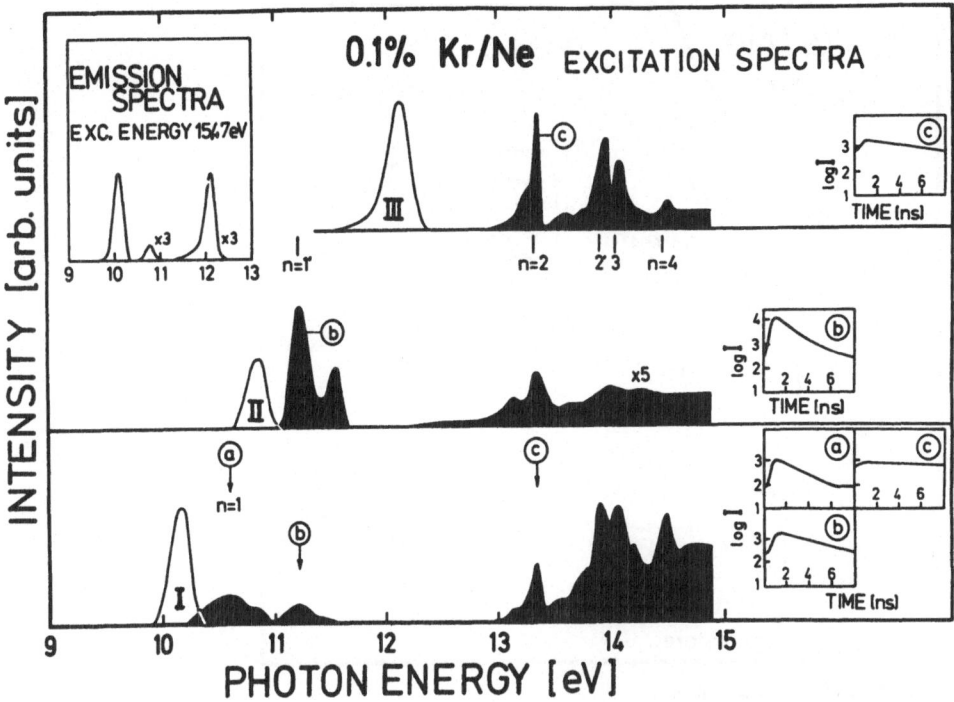

Fig. 6.31. Excitation spectra (emission intensity versus excitation energy) for Kr in Ne similar to Fig. 6.30 (from Hahn and Schwentner /1980/)

n = 1' excitons leads to the formation of the medium relaxed $1s_2(^1p_1)$ state. The experimental decay lifetimes (Table 6.12) are again dominated by the pure radiative decay channel, so that radiationless electronic relaxation for $1s_2(^1p_1)$ to the $1s_3(^3p_0)$, $1s_4(^3p_1)$ and $1s_5(^3p_2)$ states are not exhibited on the time scale of 10^{-6} s.

The energy gaps of 0.2 eV for Ar and 1.1 eV for Xe, due to spin-orbit coupling, are sufficiently large to prohibit electronic relaxation. For Kr in solid Ne an interesting complication due to impurity-impurity energy transfer arises. In this system there is an accidental energetic overlap between the medium relaxed $1s_2(^1p_1)$ state and the n = $1(^3p_1)$ state. The rise and decay times of the emission bands of the $1s_2$ and the $1s_4$ states after excitation of n = 1' excitons indicate that a resonant dipole-dipole energy transfer takes place from an excited Kr atom in the $1s_2$ state to a nearby Kr atom in the ground state /Hahn and Schwentner, 1980/. The Förster-Dexter radius for this transfer is 21 Å.

Next, we consider the fate of high energy impurity excitations in solid Ne. There are several characteristic differences for the n = 2 exciton relaxation cascades for Xe, Kr and Ar guest atoms in a Ne matrix. The origin is an increasing spread on an energy scale of the (n+1)s, the (n-1)d, the np and the ns' states in going

171

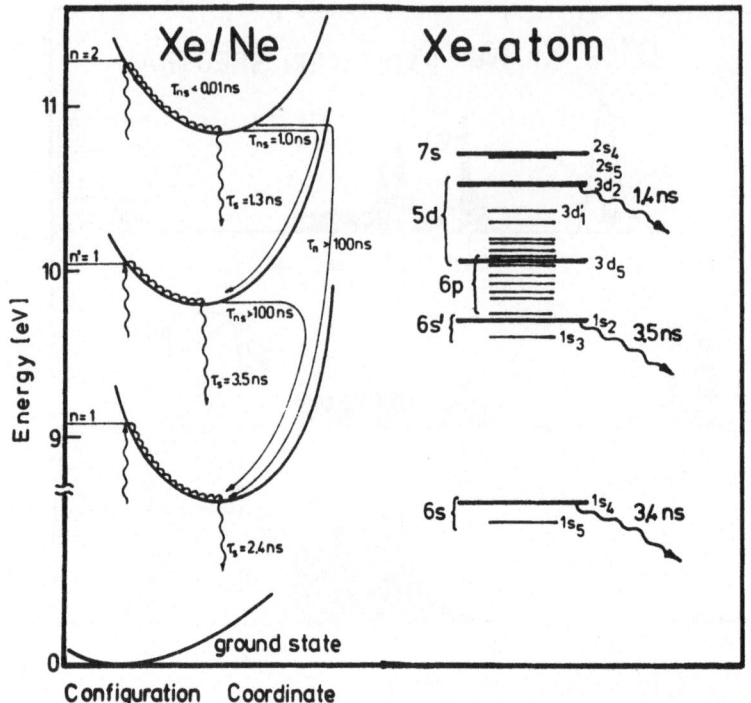

Fig. 6.32. (*Left part*) experimental radiative (∿➤) and non-radiative (——➤) electronic relaxation pathways and time constants (in ns) for Xe in Ne in a configuration coordinate diagram. The excitation channels n = 1, n' = 1 and n = 2 and the
matrix relaxation (∿➤) are included. (*Right part*) atomic energy levels (Paschen
notation) for Xe with some lifetimes for radiative transitions to the ground state
(∿➤). Long bars correspond to states with allowed transitions, and short bars to
forbidden transitions to the ground state (from Hahn and Schwentner /1980/)

from Xe to Ar. This spread results in groups of close-lying levels which are separated by increasing gaps (Figs. 6.32,33). Excitation of n = 2 excitons in Xe leads
to a population of the medium relaxed $2s_4$ and $3d_2$ states (Fig. 6.32). The $3d_2$ state
is separated from the next state $3d_1'$ by a gap of 0.18 eV. It was experimentally
established that the $3d_2$ state decays radiatively with a time constant of $1.3 \cdot 10^{-9}$ s
as well as non radiatively to $3d_1'$ with a time constant of $1.0 \cdot 10^{-9}$ s. We note that
in this case radiationless processes of the order of P = 30 can have time constants
as fast as 10^{-9} s. This observation is incompatible with the considerations based
on the energy gap law and can be rationalized in terms of the cancellation effect
which is exhibited when P ≃ G in (6.43,44). The energy interval between the $3d_1'$ and
the $1s_2$ states is so densely populated with other states (Fig. 6.32) that the radiationless processes will be extremely fast. Radiative transitions, i.e., the dipole allowed decay from the $3d_5$ state, are not observed. In the $1s_2$ state the radiationless relaxation cascade terminates, as has been previously noted. No relaxation to $1s_3$, $1s_4$ or $1s_5$ is observed.

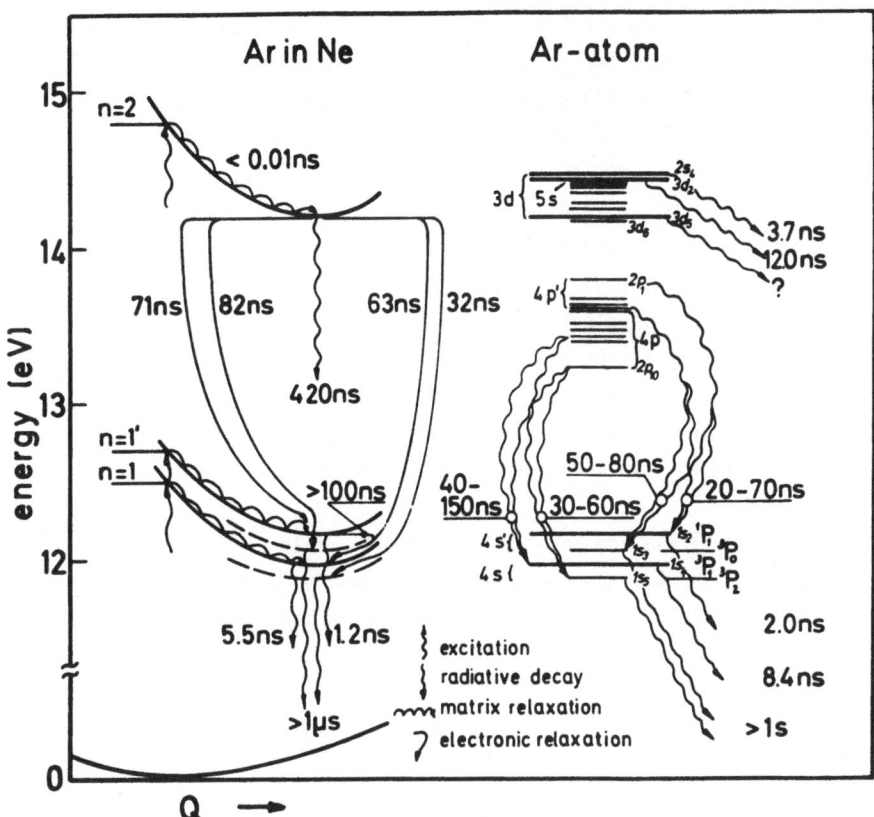

Fig. 6.33. Relaxation cascade for Ar in Ne similar to the results shown in Fig. 6.32. Important radiative transitions between excited atomic transitions are included (〰➤) (from Hahn and Schwentner /1980/)

The relaxation cascade resulting from the n = 2 excitation of Ar in solid Ne (Fig. 6.33) results in an initial population of the medium relaxed $2s_4$ state which relaxes effectively to the lowest 3d states ($3d_5$ and $3d_6$). The gap between the 3d and (4p, 4p') states is sufficiently large to make non-radiative decay rather slow. Branching occurs between radiative decay of the $3d_5$ state ($420 \cdot 10^{-9}$ s) and slow non-radiative decay to 4p' and 4p states. The (4p', 4p) states are separated by a large energy gap from the 4s and 4s' levels. Therefore, both the 4p' and 4p states are depopulated radiatively to the 4s as well as the 4s' state with similar time constants. The intensity is distributed between radiative channels from the fine structure components of the 4p' to 4s' and the 4p to 4s states with the branching ratios given in Fig. 6.33.

The relaxation cascade of Kr in solid Ne (Fig. 6.31) is similar to that of the Ar impurity state /Hahn and Schwentner, 1980/. Compared to the energy scheme for

Ar (Fig. 6.33) the main quantum numbers n have to be increased by one and the group of 5p' levels is shifted into the range of the 4d levels. The energy of the n = 2 exciton does not result in radiative decay from the medium-relaxed $2s_4$ or from the $3d_2$ states. Thus, electronic relaxation from $2s_4$ and $3d_2$ into the lower 4d states ($3d_5$ and $3d_6$) is fast on the time scale of 10^{-9} s. Subsequently, 4d → 5p electronic relaxation competes with radiative decay of the lower 4d states. The energy gap between the 5p state and the 5s' and 5s states is so large that radiationless processes do not occur. Further, radiative decay from the p symmetric 5p state to the p symmetric ground state is dipole forbidden. Therefore, the 5p states are depopulated by radiative decay to the 5s and 5s' levels. The transitions to the 5s states which are singlet-singlet transitions with transition times of $22 \cdot 10^{-9}$ s are favoured compared to 5s' transitions ($750 \cdot 10^{-9}$ s). In these radiative depopulation channels to the 5s states a branching to the fine structure components $1s_4$ and $1s_5$ with equal probability takes place.

The results for the relaxation cascades of impurity states in solid Ne are collected in Table 6.14. The radiative population efficiencies of the fine structure components derived in this relaxation cascade agree with the intensities in emission

Table 6.14. Experimental relaxation times between excited guest atom states in Ne matrix (in 10^{-9} s); n.o. means not observed

	$n = 2 \rightarrow 1s_2$	$n = 2 \rightarrow 1s_3$	$n = 2 \rightarrow 1s_4$	$n = 2 \rightarrow 1s_5$	$1s_2 \rightarrow \begin{matrix} 1s_3 \\ 1s_4 \\ 1s_5 \end{matrix}$	$1s_4 \rightarrow 1s_5$
Xe	1.0	n.o.	n.o.	n.o.	n.o.	n.o.
Kr	750	?	22	22	n.o.	n.o.
Ar	71	82	63	32	n.o.	n.o.

bands found by Schuberth and Creuzburg /1978/ with X-ray excitation. A quantitative theoretical description of these relaxation cascades is not yet available and will be of considerable interest.

Results for atomic impurity states in heavy RGS have been collected in Table 6.10. Obviously, further experimental and theoretical work concerning the electronic relaxation processes is required. An interesting new result in this field involves the study of the temperature dependence of the electronic relaxation from the $1s_4({}^3P_1)$ state of Kr in Ar which is produced by medium relaxation around the n = 1 $({}^3P_1)$ state /Hahn et al., 1982/. The $1s_4$ state is depopulated by radiative decay to the ground state and by non radiative decay to the $1s_5({}^3P_2)$ state. In Fig. 6.34 the temperature

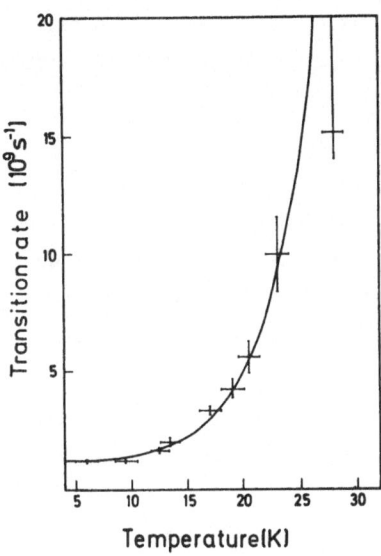

Fig. 6.34. Temperature dependence of the transition rate of Kr n = 1 excitons in an Ar matrix. Crosses, experimental points; solid line, fit with (6.37) (from Hahn et al. /1982/)

dependence of the decay rate of the $1s_4(^3p_1)$ state is shown, which exhibits the temperature independent nuclear tunnelling at low temperatures and temperature activated processes at higher temperatures. Analysis of the data in terms of (6.37) can be performed taking ΔE = 82 meV and $\hbar\omega_p$ = 4 meV, i.e., P = 20.

6.13 Electronic Energy Transfer

6.13.1 Basic Processes

Some of the most interesting features of the dynamics of electronic excitations in molecular crystals involve energy transfer processes /Knox, 1963/. There have been extensive studies of migration of singlet and triplet Frenkel-type excitons in organic crystals (e.g. /Rice and Jortner, 1967; Wolf, 1967; Powell and Soos, 1975; Kepler, 1976; Silbey, 1976/) which led to estimates of the exciton diffusion length and the exciton diffusion coefficient in these molecular crystals. Studies of eletronic energy transfer in pure RGS are of interest to establish the possibility of exciton migration on an ultrashort time scale prior to exciton self-trapping and in relation to energy transfer from self-trapped excitons. The effects of substitutional disorder on energy transfer are manifested for transfer to atomic and molecular impurity states, while some consequences of structural disorder are exhibited for energy transfer in liquid rare gases. The following electronic energy transfer processes in condensed rare gases are of interest (see, for example, /Jortner, 1974; Raz et al., 1976; Fugol, 1978; Schwentner, 1978 , 1980; Schwentner et al., 1984/):

A) Migration of "free" excitons. The term "free" excitons refers to exciton states prior to self-trapping. In this context two limiting cases of energy transfer should be considered:

A1) Coherent exciton transport. The limiting situation for weak exciton-lattice coupling involves coherent transfer characterized by the exciton group velocity, so that the displacement of the exciton at time t is proportional to t. This idealized state of affairs is not realized in real life as exciton-scattering mechanisms due to phonons, structural imperfections, etc., prevail even in pure crystals. Coherent exciton transport should be envisioned in terms of a mean free path Λ, which considerably exceeds the lattice spacing a, i.e., $\Lambda \gg a$.

A2) Incoherent exciton transport. When the exciton scattering by phonons, disorder, etc., is strong, the exciton mean free path is comparable to the lattice spacing $\Lambda \simeq a$. Exciton transport is now characterized by a diffusion coefficient D which is related to the diffusion length ℓ by

$$\ell = (D\tau)^{1/2} \tag{6.48}$$

where τ is the exciton lifetime. For RGS we identify τ with the lifetime ($\tau \approx 10^{-12}$ s) of the free exciton prior to self-trapping (Sect. 6.2). Experimental evidence for incoherent, strong scattering diffusive type exciton transport will be discussed in Sects. 6.13.2, 3.

B) Electronic energy transfer from self-trapped excitons in RGS. On a long time scale relative to the self-trapping time, the exciton energy becomes localized (Sect. 6.2). Energy migration involving polaron-type motion /Holstein, 1959/ e.g., $R_2^* + R \rightarrow R + R_2^*$ in solid Ar, Kr and Xe, would be accompanied by large nuclear configurational changes. It is unfavourable in view of the large binding energies of the R_2^+ diatomic molecules, and therefore this exciton transport mechanism can safely be neglected. Electronic energy transfer from self-trapped excitons to impurity states in the bulk and to boundaries can occur via:

B1) Transfer from two-centre trapped excitons in solid Ar, Kr and Xe where the energy donor is the diatomic R_2^* molecule and the energy acceptor is an impurity centre.

B2) Transfer from a one-centre trapped exciton in solid Ne to an impurity centre.

Mechanisms (B1) and (B2) involve Förster-Dexter /Förster, 1948; Dexter, 1953/ pairwise electronic energy transfer, where the self-trapped exciton acts as the energy donor which transfers energy to randomly distributed impurity acceptors.

C) Electronic energy transfer between impurity states. Two types of processes can be distinguished in this context:

C1) Transfer between impurities of the same type. This energy transfer process which has been frequently observed in liquids since the pioneering work of Förster

/1948/ is not commonly exhibited in doped RGS. The large Stokes shift between absorption and emission of atomic impurity states in RGS (poor spectral overlap; see Sect. 6.9) leads in general to a low efficiency of energy transfer from an excited impurity excitation to another impurity of the same type. An interesting case in this category involves energy transfer between different electronic states of impurities of the same type. It is exhibited in the transfer from the medium relaxed $n = 1'$ state of Kr to the $n = 1$ state of another Kr impurity in solid Ne /Hahn and Schwentner, 1980/ which has been discussed in Sect. 6.12.

C2) Transfer between impurities of different types, which involves the conventional Förster-Dexter mechanism. Examples for processes B and C will be discussed in Sects. 6.13.3 , 4.

The physical consequences of donor-acceptor electronic energy transfer in RGS can be classified in terms of two types of transitions:

a) Bound-bound transitions where the energy acceptor is produced in a bound excited state which is located below the ionization energy of this acceptor. The excited state of the acceptor can subsequently decay either radiatively or non radiatively.

b) Bound continuum transitions. When the energy of the donor exceeds the solid state ionization potential of the energy acceptor, the energy transfer process will result in the ionization of the acceptor. Such ionization processes in RGS bear a close analogy to Penning ionization in the gas phase. From the foregoing classification it is apparent that two general techniques can be utilized to interrogate the dynamics of energy transfer in RGS. The consequences of transfer resulting in bound-bound transitions can be explored by the techniques of emission spectroscopy. On the other hand, energy transfer resulting in ionization can be investigated by photoemission studies monitoring the electronic yield and the photoelectron energy distribution.

6.13.2 Dynamics of "Free" Excitons in Rare-Gas Solids

To explore electronic energy transfer by "free" excitons one has to utilize sensitive interrogation methods in order to probe processes occurring on the 10^{-12} s time scale. Photoemission yield and energy distribution measurements provide an adequate tool.

In pure RGS two separate photoemission mechanisms are exhibited.

1) In the excitonic energy range $E_{n=1} \leqslant E < E_{Th}$ only extrinsic photoemission can occur. Here extrinsic means that the photoemission process is mediated by energy transfer of the exciton to a boundary from where a photoelectron is released.

2) Above the threshold, i.e., $E > E_{Th}$, the usual intrinsic photoemission takes place (see Sects. 3.1 , 7.3 , 4). Extrinsic photoemission yield spectra from pure RGS were

studied by O'Brien and Teegarden /1966/, Schwentner et al. /1973/, Koch et al.
/1974a,b/, Steinberger et al. /1974/, Ophir et al. /1975a,b/, Pudewill et al. /1976a,b/,
Hasnain et al. /1977a-c/ and Schwenter et al. /1981/. These studies encompassed all
RGS. An example for the pronounced structure in the photoemission yield curves associated with the exciton states is depicted for solid Kr in Fig. 6.35. Yield spectra
are shown for a number of film thicknesses /Schwentner et al., 1981/. Direct emission at the surface is forbidden energetically. Koch et al. /1974/, Ophir et al.
/1975/ and Pudewill et al. /1975/ have previously assigned similar extrinsic photoemission in pure solid Ne, Ar and Xe to the exciton diffusion to the gold substrate
(the emitter electrode), followed by electron ejection from the electrode. Several
alternative interpretations, such as impurity effects, non linear optical processes,
ionization of excitons near the surface and long-range energy transfer from a trapped

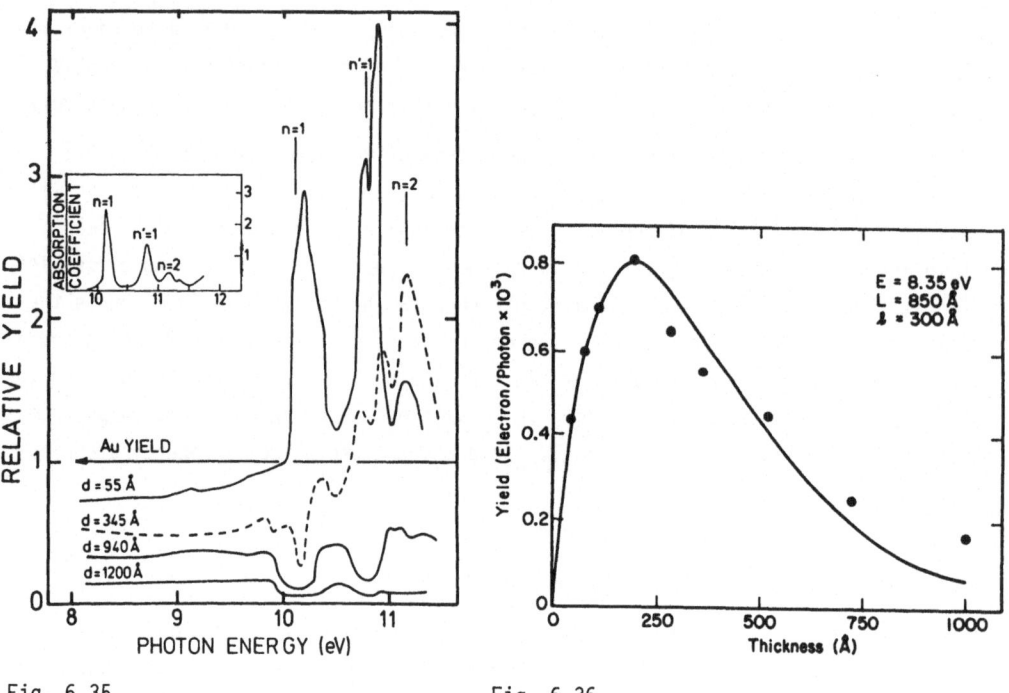

Fig. 6.35   Fig. 6.36

Fig. 6.35. Photoelectron yield from Kr layers with thickness d due to energy transfer to the gold substrate normalized to the photoelectron yield of the substrate.
The n = 1, n' = 1 and n = 2 exciton bands of Kr are marked. In the insert, the absorption coefficient for Kr versus photon energy is given (from Schwentner et al.
/1981/)

Fig. 6.36. Thickness dependence of the photoelectron yield from the n = 1 exciton
of Xe due to energy transfer to the gold substrate. Points represent experimental
data, the curve shows a fit with a diffusion length of 300 Å and an electron escape
depth of 850 Å as parameters (from Ophir et al. /1975a/)

excited diatomic molecule to the gold surface could be eliminated. The experimental data for pure Xe, which were corrected for reflection effects and direct emission from the substrate, were analysed in terms of an exciton diffusion model, involving competition between energy transfer to the substrate and decay of mobile excitons /Ophir et al., 1975a/. The photoemission yield Y is expressed as $Y = F(\ell,L,\alpha,d)$ where L is the escape length of the electrons (Chap. 7); L is measured independently, α is the absorption coefficient, and d is the film thickness, while the diffusion length ℓ is given by (6.48). From the thickness dependence of Y (Fig. 6.36) at several photon energies, a diffusion length of $\ell = 300$ Å was determined for solid Xe. Similar yield measurements and the same kind of analysis have been reported for Ne /Pudewill et al., 1976/.

Returning to the case of solid Kr (Fig. 6.35, Schwentner et al. /1981/) we note that the rare-gas film is transparent for energies below 9.8 eV. The small photoelectron yield represents the escape probability for electrons from the Au substrate. The strong increase of Y for photon energies in the Kr exciton region above 10 eV is due to the energy transfer to the substrate. The absolute value of the yield indicates an efficiency for electron emission due to energy transfer of 0.3 - 0.5 electrons per exciton compared with the direct photoyield efficiency of 0.03 - 0.05 electrons per photon. A similar high efficiency is observed in the Penning ionization at the metal surface by metastable rare-gas atoms (e.g. /Dunning et al., 1975/). These large efficiencies agree with calculations for the interaction of a metal surface with a dipole in front of it. The interaction mechanism in this case is the coupling of the dipole near field to surface plasmon modes /Chance et al., 1976/. It is interesting to note that the quantitative analysis of the extrinsic photoemission from solid Kr /Schwentner et al., 1981/ shows a pronounced energy dependence of the diffusion length (Fig. 6.37). The diffusion length rises from $\ell \simeq 30$ Å in the low energy tail of the n = 1 exciton to $\ell \simeq 300$ Å for the n = 2 exciton. It is intriguing to inquire whether this monotonic increase of ℓ does reflect a transition from incoherent strong scattering motion at low energies to coherent motion at higher energies.

An independent source of information concerning the dynamics of free excitons in RGS was obtained from exciton-induced impurity photoionization. For dilute impurity states three intrinsic photoemission mechanisms can occur: (1) Direct excitation of the impurity above its threshold for $E_{Th}^i < E < E_{n=1}$. (2) Impurity photoemission resulting from excitation of the host excitonic series for $E_{n=1} \leqslant E < E_{Th}$. (3) Intrinsic photoemission from the host matrix at $E > E_{Th}$.

Exciton-induced impurity photoionization in RGS was first studied by Ophir et al. /1974/ for C_6H_6 in solid Xe and subsequently extended to dilute rare-gas impurities in RGS by Saile et al. /1974/, Ophir et al. /1975b/ and Pudewill et al. /1976/. Benzene in solid Ne, Ar and Kr matrices were studied in photoemission yield experiments

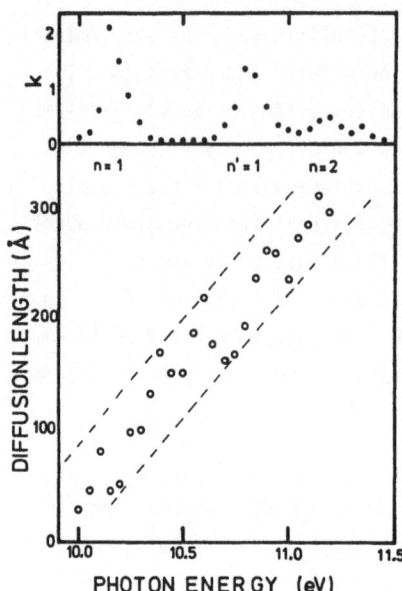

Fig. 6.37. Diffusion length of Kr excitons versus exciton energy derived from Fig. 6.35. In the upper panel, the absorption coefficient in the exciton bands is shown (from Schwentner et al. /1981/)

by Hasnain et al. /1977/, while photoelectron energy distribution measurements for C_6H_6 in solid Ar, Kr and Xe have been reported by Schwentner et al. /1978/. In the following we limit our discussion to the photoemission results obtained for rare-gas impurities in rare-gas matrices.

The photoelectron yield spectra for Xe atoms in Ar matrices are shown as an example in Fig. 6.38. At photon energies below 12 eV the matrix is transparent and electrons from the Au substrate and from the Xe guest atoms are emitted. Both contributions are small at all Xe concentrations up to 1%. At photon energies between 12 eV and 13.9 eV light is absorbed by excitons of the Ar matrix. Hence the yield from the Au substrate and from direct absorption by Xe guest atoms is reduced. Nevertheless the yield increases. The observed larger yield in this range originates from the creation of host excitons (n = 1, 1', 2, ...). Since these excitons lie below the vacuum level of the matrix, direct photoemission is impossible. The explanation for the photoelectron yield and its strong increase with Xe concentration is the energy transfer of Ar excitons to Xe guest atoms leading to ionization of the Xe atoms. The small maxima in pure Ar are due to energy transfer of Ar excitons to the Au substrate. Above hν = 13.9 eV direct photoemission of the Ar matrix causes a further steep increase of the yield. In Table 6.15 the matrix exciton energies and the energy of the emission bands are compared for various systems together with the threshold energy E_{Th}^i necessary for electron emission from the guest levels. The occurrence of energy transfer is marked. The essential result from these experiments is that in all RGS matrices electron emission starts when the energy of primarily excited excitons just exceeds the threshold for photoemission of the guest atom.

Table 6.15. Comparison of energies which are transferable by the matrix either in the self-trapped exciton states ($E_{R_2}^*$) or in the free exciton states n = 1 ($E_{n=1}$) and n = 1' ($E_{n=1'}$) with the lowest excitation energies ($E_{n=1}^i$) and ionization energies E_{Th}^i of the guest atom. E_{Th}: ionization energy of the matrix. + and - means observation or absence of energy transfer in electron emission spectra (energies in eV)

matrix	guest	E_{Th} e)	E_{Th}^i	$E_{n=1}^i$ f)	$E_{R_2}^*$	$E_{n=1}$	$E_{n=1'}$	
Ne		20.3			16.80	17.36	17.50	
	Xe		11.60	9.08	+	+	+	c) d)
	Kr		13.48	10.62		+	+	c)
	Ar		15.05	12.5		+	+	c)
Ar		13.8			9.80 ± 0.44	12.06	12.24	
	Xe		10.2	9.22	-	+	+	b) d)
	Kr		12.2	10.79	-	-	+	b)
Kr		11.9			8.45 ± 0.32	10.17	10.86	
	Xe		10.3	9.1	-	-	+	a) b)

a) Ophir, Raz and Jortner /1974/
b) Ophir et al. /1975/
c) Pudewill et al. /1976b/
d) Schwentner and Koch /1976/
e) Schwentner et al. /1975/
f) Baldini /1965/

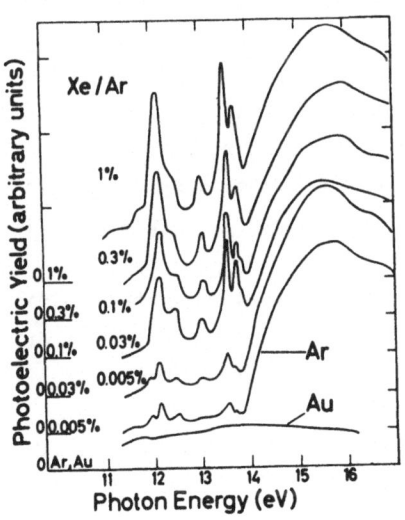

Fig. 6.38. Dependence of the photoelectric yield of Xe doped Ar films on the Xe concentration. The film thickness is 60 Å. The spectra are not corrected for the hot electron contribution from the gold substrate and the reflectivity. Photoelectric yield curves for the gold substrate and for pure Ar films are also shown (from Ophir et al., /1975b/)

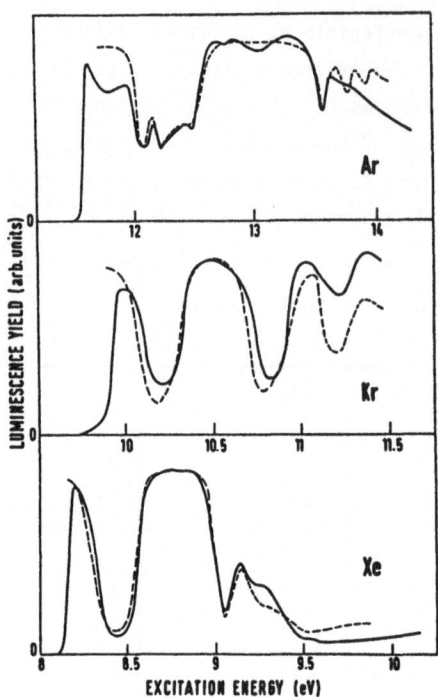

Fig. 6.39. Luminescence efficiency of the R_2^* bands for solid Ar, Kr and Xe at 5 K for excitation energies in the exciton region. (— — —) represent results from model calculations (from Zimmerer /1978/)

For Ar and Kr matrices the luminescence emission bands lie below this threshold and energy transfer from trapped excitons cannot lead to photoemission. From the observation of electron emission due to energy transfer in these systems (see Table 6.15) we conclude that free excitons are responsible also for the energy transfer process. The high yield (Fig. 6.38) for Xe in Ar proves that at high concentrations this is the dominant transfer process (see also Sect. 6.13.2). By fitting the concentration and thickness dependence, diffusion lengths have been derived. We note that in a Ne matrix both the energy of the free exciton and of the self-trapped excitons exceed the threshold energy for guest ionization /Pudewill et al., 1976/.

Some related additional information regarding exciton dynamics was obtained from energy transfer to boundaries of RGS which are contaminated by impurity molecules. Such processes, which are important for luminescence quenching, result in minima in the energy dependence of the luminescence efficiency of RGS (Fig. 6.39) which coincide with the exciton energies /Nagasawa and Nanba, 1974; Nanba and Nagasawa, 1974; Ackerman et al., 1976a,b; Gericke, 1977/.

Unfortunately, an unambiguous interpretation of these results cannot be provided /Ackerman et al., 1976/ as the energy dependence of the luminescence efficiency can be accounted for both in terms of diffusion of free excitons and by the Förster-Dexter transfer from localized self-trapped excitons.

Erosion experiments /Schou et al., 1984; Reimann et al., 1984; Coletti et al., 1984/ are closely related to energy transfer to surfaces. For solid rare gases extremely high erosion yields Y_{er} of the order of $Y_{er} \approx 10$ ejected atoms per electron for keV electrons /Børgesen et al., 1982/ and up to $Y_{er} \approx 10^3$ for MeV charged particles /Johnson and Inokuti, 1983/ have been reported. One model explains the high erosion yield by a thermal spike in the vicinity of the track of the penetrating high energy particle because about 2 eV of the kinetic energy of electrons produced by the stopping of the particle are converted to nuclear motion in the dissociative recombination of localized holes with these electrons (Fig. 6.1). Many atoms are evaporated at the crossing point of the particle track with the surface due to this thermal spike /Johnson and Inokuti, 1983/. An alternative model /Børgesen et al., 1982/ explains the yield by an efficient transport of energy to the sample surface, which has been deposited in form of excitons along the track of the particle. The transport is attributed to the diffusion of free excitons, thus involving the diffusion length. The excitons are trapped as excimer centres at the surface. They decay radiatively to the repulsive ground state and one atom for each excimer is ejected with a large energy. The kinetic energy of the ejected atom is given by the repulsive energy of the ground state of about 1 eV.

In Table 6.16 we have summarized the available results for the diffusion lengths

Table 6.16. Diffusion lengths for energy transfer to guest atoms and to boundaries (in Å)

	excitons	photoelectron emission		luminescence	
		guest atom	boundary	guest atom	boundary
Ne	n = 1, 1', 2	2500 ± 500 a)	observed a)		
Ar	n = 1, 1' n = 2, 2', 3	120 b)	observed b)		50 i) 100 i) 50 g)
Kr	n = 1 n = 1' n = 2	observed b)	10 - 100 150 - 250 d) 150 - 350	300 f)	200 - 250 f) 200 g) h)
Xe	n = 1, 2	170 c)	300 e)	25 - 260 f)	150 - 1000 i) 500 f)

a) Pudewill et al. /1976b/
b) Ophir et al. /1975b/
c) Ophir, Raz and Jortner /1974/ corrected for electron escape depth of a),
   Ophir /1976/
d) Schwentner, Rudolf and Martens /1981/
e) Ophir et al. /1975a/
f) Ackerman /1976/
g) Ackerman et al. /1976a/
h) Ackerman et al. /1976b/
i) Gerick /1977/

of "free" excitons in RGS. From these results it is apparent that the diffusion
length ℓ considerably exceeds the lattice constant. To obtain a rough estimate for
the diffusion coefficient of free excitons in solid Ar, Kr and Xe, we take $\tau \approx$
10^{-12} s which results in $D \approx 0.2 - 1 \ cm^2 s^{-1}$ (6.48). The question of whether the exciton transfer process is diffusive or coherent has not yet been answered definitely.
However, the experimental results support the notion that exciton motion is diffusive and hence all authors have analysed their experiments within a diffusion model.
The excitons can be scattered by phonons and crystal imperfections such as grain
boundaries. An extensive discussion by Fugol /1978/ arrived at the conclusion that
the diffusion lengths derived from the experiments reported so far are limited by
scattering of the free excitons by crystal imperfections.

6.13.3 Competition Between Relaxation and Energy Transfer

Photoelectron energy-distribution measurements (EDC's) on RGS have been used to study
exciton dynamics, energy transfer and relaxation processes /Schwentner and Koch,
1976; Schwentner et al., 1978/. Detailed information concerning the competition of
relaxation and energy transfer has been derived from these experiments. From yield
spectra primarily threshold energies for the various processes can be derived. From
the analysis of the kinetic energy of emitted electrons together with the known
vacuum level, the transferred energy is determined directly in EDC measurements.
EDC's from thin films of solid Ar and Ne each doped with 1% Xe have been presented
by Schwentner and Koch /1976/ for several photon energies. In Fig. 6.40 an overview
of EDC's for 1% Xe in Ar is given. On the righthand side the counting rate is plotted
versus the kinetic energy of the emitted electrons, where zero corresponds to the
vacuum level. The spectra are shifted upwards proportional to the increase in photon
energy. In this display, the diagonal lines connect those structures in different
EDC's which are due to the same initial state (see insert). The crosses in the lefthand curve show the total number of electrons emitted at a given photon energy. They
are compared with a total yield curve of 1% Xe in Ar. At the lowest photon energies
(hυ = 11 eV and 11.5 eV) only Xe atoms can be excited in the film and the EDC's correspond to the upper Xe $5p_{3/2}$ level. This state and its spin-orbit partner Xe $5p_{1/2}$
are seen in all spectra up to hυ = 19 eV. At these larger photon energies mainly
electrons from the Ar 3p valence band contribute to the EDC. In the photon energy
region, where the n = 1, 1', 2 and 2' states of Ar are excited, the total emission
rate (lefthand side) shows strong maxima due to energy transfer to the guest atoms,
as has been described before on the basis of yield spectra (see Fig. 6.38). The
EDC's in the excitonic region are shown on an enlarged scale (Fig. 6.41) using the
same kind of display. Upon excitation of the n = 1 and n' = 1 exciton of the Ar matrix, the EDC's show the same structure at the same energy position as if the Xe
atoms would have been excited directly by photons. The energy is transferred by Ar

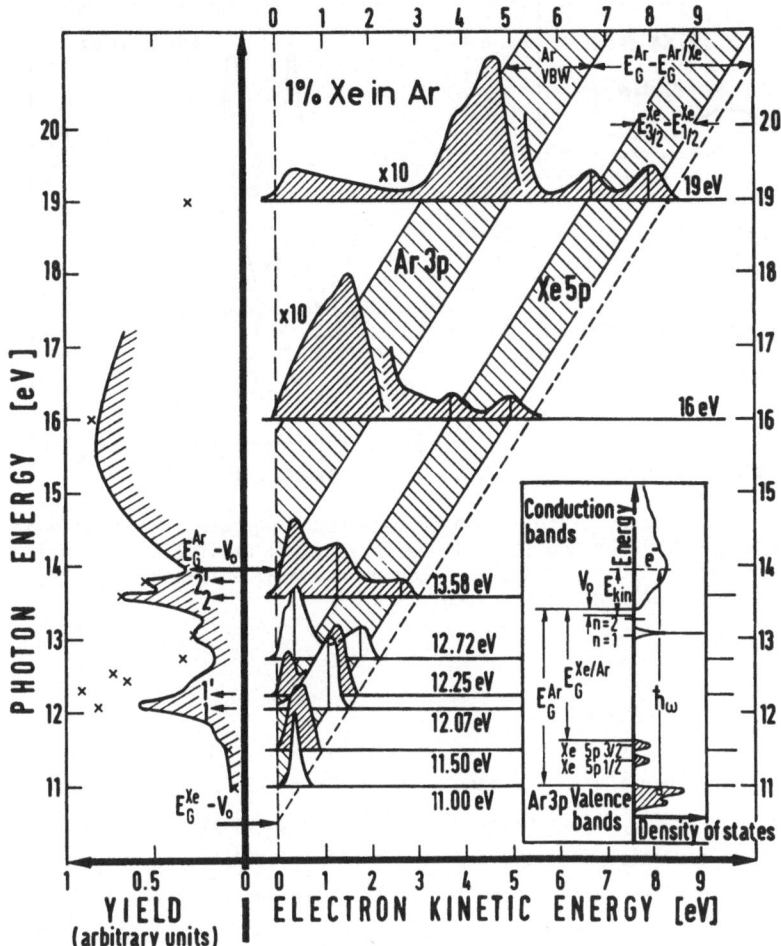

Fig. 6.40. *Right part:* photoelectron energy-distribution curves (EDC's) (counting rates versus kinetic energy) from films of 1 at.% Xe in Ar for a spectrum of photon energies. The film thickness was 50 Å. For convenience the relevant energy levels are shown in the insert. *Left part* (yield): The crosses represent the total number of emitted electrons from the EDC's. For comparison the yield spectrum of 1 at.% Xe in Ar of a 60 Å thick film is shown (———). The two sets of data were adjusted at hν = 11 and 11.5 eV (gold substrate). The energies of the n = 1, 1' and 2' exciton states are marked (from Schwentner and Koch /1976/)

n = 1 excitons before relaxation to self-trapped excitons. The difference in kinetic energy for the spectra for n = 1' and n = 1 excitons also excludes relaxation from n = 1' to n = 1 before energy transfer. For the n = 2 and n = 2' excitons of the Ar matrix again the maxima of the Xe 5p states appear in the EDC's at kinetic energies indicating energy transfer from unrelaxed n = 2 and n = 2' excitons. There is an additional maximum near the vacuum level which is attributed to energy transfer from self-trapped n = 2 excitons. This maximum cannot be explained by transfer after re-

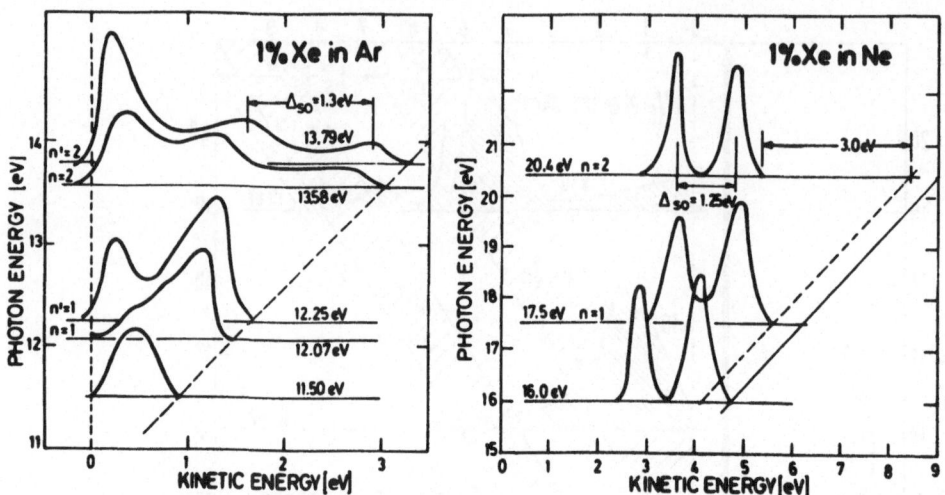

Fig. 6.41. Photoelectron energy distribution curves (counting rates versus kinetic energy) from films of 1% Xe in Ar and 1% Xe in Ne for several excitation energies (below and within the n = 1 and n = 2 host exciton bands) Δ_{SO}: spin-orbit splitting of the Xe 5p levels (after Schwentner and Koch /1976/)

laxation to n = 1 and n' = 1 excitons.

The situation is quite different for Xe in Ne. The EDC's in Fig. 6.41 are taken for photon energies corresponding to the population of Ne n = 2, n = 1 excitons and at a somewhat lower energy for direct excitation of the Xe guest atoms (hν = 16 eV). In the latter spectrum, emission from the two spin-orbit split (Δ_{SO}) Xe 5p states situated in the band gap of Ne is observed. The Xe 5p states are well reproduced when the energy is transferred from the Ne n = 1 exciton (hν = 17.5 eV) to the Xe atoms. However, there is a shift of $\simeq$ 0.7 eV of the total spectrum to lower kinetic energies indicating some relaxation either to the one-centre self-trapped exciton or to the high vibrational levels of the two-centre self-trapped exciton before energy transfer.

The striking observation is that in the EDC obtained by exciting the Ne n = 2 excitons more than 3 eV are missing. The electrons have the same energy distribution as for excitation of the n = 1 Ne excitons. This observation indicates a complete relaxation of the n = 2 excitons to n = 1 excitons and dissipation of an energy of more than 3 eV before transfer.

These data illustrate the power of EDC measurements yielding direct information about the state of relaxation before transfer. A time scale derived from these experiments for the competition of relaxation and transfer processes is given in Table 6.17.

Electron energy distribution measurements of electrons produced by energy transfer from Xe, Kr and Ar hosts to benzene molecules were conducted by Schwentner et

Table 6.17. Time hierarchy for the competition of energy transfer and relaxation. The time constants τ (in s) have the following meaning: τ_1: radiative decay of R_2^* or R^* centres of the matrix; τ_2: vibrational relaxation of R_2^* centres; τ_3: localization of excitons; $\tau_R(i \to j)$: electronic relaxation of exciton i to exciton j; τ_T: energy transfer (after Schwentner and Koch /1976/)

	time hierarchy from electron emission experiments	time hierarchy including results of Table 6.3
1% Xe in Ar		
n = 1	$\tau_1 > \tau_2 > \tau_T$	$\tau_1 \sim 10^{-5} - 10^{-9} > \tau_2 \sim 10^{-9} > \tau_T$
n = 1'	$\tau_R(1' \to 1) > \tau_T$	
n = 2	$\tau_1 > \tau_R(2 \to 1) > \tau_T \sim \tau_3$	$\tau_1 \sim 10^{-5} - 10^{-9} > \tau_R(2 \to 1) > \tau_T \sim \tau_3 \sim 10^{-12}$
1% Xe in Ne		
n = 1	$\tau_2 > \tau_1 > \tau_T > \tau_3$	$\tau_2 \sim 10^{-5} > \tau_1 \sim 10^{-6} - 10^{-8} > \tau_T > \tau_3 \sim 10^{-12}$
n = 2	$\tau_2 > \tau_1 > \begin{cases} \tau_R(2 \to 1) \\ \tau_3 \end{cases}$	$\tau_2 \sim 10^{-5} > \tau_1 \sim 10^{-6} - 10^{-8} > \tau_T > \begin{cases} \tau_R(2 \to 1) \\ \tau_3 \sim 10^{-12} \end{cases}$

al. /1978/. The spectrum of kinetic energies shows that energy transfer from unrelaxed host excitons in all three matrices is an efficient process.

6.13.4 Energy Transfer Between Localized States

Generally, in optical emission spectra from dilute rare-gas alloys (see Sect. 6.9, 11) the information concerning the primarily transferred energy cannot be recovered, since non-radiative relaxation processes are fast on the time scale of the radiative decay. Furthermore, free and localized excitons can, in principle, contribute to the emission since the threshold energies discriminating between them in EDC measurements are not relevant for these processes. The energy transfer from localized R_2^* centres of the host to the impurity occurs via the Förster-Dexter mechanism /Förster, 1948; Dexter, 1953/. It is determined by the strength of donor-acceptor dipole-dipole coupling and by the spectral overlap between the donor emission and the acceptor absorption. The transition probability for the electronic energy transfer process at the donor (D)-acceptor (A)-separation R is

$$W(R) = \alpha \, \mu_A^2 \, \mu_D^2 (F/R^6) \tag{6.49}$$

where α is a numerical constant and μ_A and μ_D are the electronic transition moments for the acceptor absorption and the donor emission, respectively, while F corresponds to the spectral overlap function

$$F = \int f_{Aa}(E) \, f_{De}(E) \, dE \tag{6.50}$$

where $f_{Aa}(E)$ and $f_{De}(E)$ represent the normalized acceptor absorption and donor emission lineshapes, respectively. The efficiency of the energy transfer process can be specified in terms of the transfer radius, R_q, where the probability, i.e., the reciprocal lifetime τ_0^{-1} for the decay of the donor state (in the absence of the acceptor), is equal to the energy transfer probability, so that

$$R_q = (\alpha \mu_A^2 \mu_D^2 F \tau_0)^{1/6} \quad . \tag{6.51}$$

Now, provided that the major contribution to τ_0^{-1} originates from radiative decay, as is the case for the R_2^* centres, the transfer radius is

$$R_q = (\lambda^3 \mu_A^2 F / 8\pi^2)^{1/6} \tag{6.52}$$

where λ is the (mean) wavelength for the donor emission. Thus R_q is essentially determined by the spectral overlap function.

For atomic impurity states in solid Ar and in solid Kr efficient spectral overlap between the R_2^* emission and the acceptor absorption is effective for Xe impurities as is evident from Fig. 6.42.

Quantitative emission studies of these electronic energy transfer processes /Cheshnovsky et al., 1972b , 1973a; Gedanken et al., 1973a/ were conducted and ana-

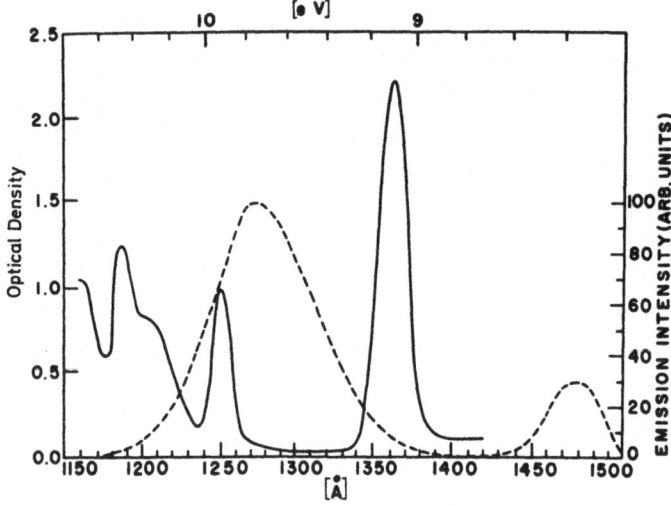

Fig. 6.42. Superposition of the absorption spectrum of solid Xe in Ar (1 ppm Xe) (———) and the emission spectrum of pure solid Ar (– – –) at 80 K (from Cheshnovsky et al. /1972b/)

lysed according to Förster's /1948/ formula

$$\frac{\eta^0 - \eta}{\eta^0} = \sqrt{\pi}\, q \exp(q^2)[1 - \text{erf}(q)] \tag{6.53}$$

$$q = (\sqrt{\pi}/2)\cdot C/C^* \tag{6.54}$$

$$C^* = 3/4\pi R_q^3 \tag{6.55}$$

where η is the donor emission quantum yield and η^0 is η at the acceptor concentration $C = 0$. A typical plot of η for the Xe/Kr system is portrayed in Fig. 6.43. The R_q values thus derived are summarized in Table 6.18 together with some theoretical data based on (6.52). The temperature dependence of R_q in the Xe/Kr system originates from the temperature dependence of the position and of the width of the Xe atomic impurity band in solid Kr.

Electronic energy transfer to molecular impurities in RGS was studied utilizing the benzene molecule as a prototype /Ophir et al., 1974; Schwentner et al., 1978; Ackermann, 1976; Hasnain et al., 1977a-d/. The relevant energy levels and transitions for the free molecule and C_6H_6 in Ar, Kr and Xe matrices are depicted in Fig. 6.44. Hasnain et al. /1977a/ measured the dependence of the luminescence intensity in the benzene emission bands on the excitation energy. The intensity is much higher for excitation of host excitons than for direct excitation of benzene molecules in the

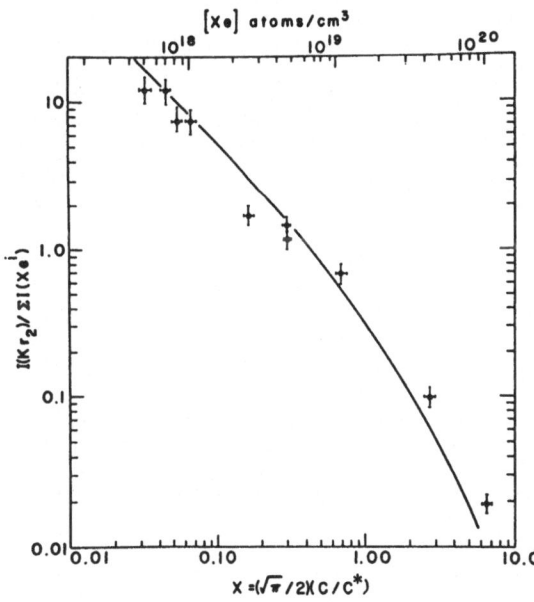

Fig. 6.43. Dependence of the ratio of Kr donor emission to Xe acceptor emission upon the acceptor concentration ($\bullet\;\bullet\;\bullet$) experimental points; (———) fit of Förster's theoretical curve (from Cheshnovsky et al. /1973a/)

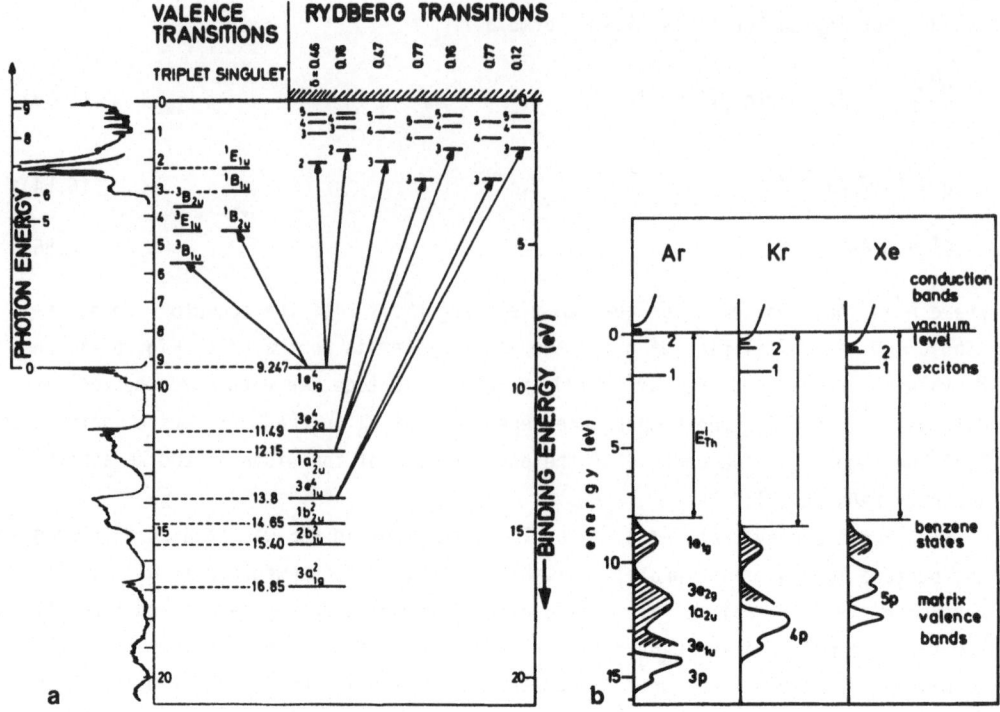

Fig. 6.44a,b. (a) One electron molecular orbital scheme for benzene. A simple presentation with one fixed vacuum level for all initial states as frequently used in solid state physics and molecular orbital pictures is shown. In such a picture neither the superposition of excitations from strongly bound levels with the continuum transitions from weakly bound levels nor correlation effects can be depicted but it indicates immediately where the final states are expected relative to the host levels when benzene is doped into a rare-gas matrix. In the left part (i) the photoelectron energy distribution curves for pure C_6H_6 (from Turner et al. /1970/) indicating the initial states and (ii) the absorption coefficients for gaseous benzene indicating allowed transitions to final states are sketched. (b) Energy level diagram for benzene in solid Xe; Kr and Ar matrices as derived from photoelectron energy distribution measurements (after Schwentner et al. /1978/)

transparent region of the host, as is demonstrated for benzene in a Kr matrix (Fig. 6.45). Evidently, energy transfer from the host to the guest molecules takes place, an observation which has been made for all rare-gas matrices.

Ackermann /1976/ established that the ratio of host emission intensity to guest emission intensity (Fig. 6.46) is quite insensitive to changes in the surface quenching rate. A careful study of the concentration and thickness dependence of host and guest emission efficiencies in the excitonic region of the rare-gas host indicated that for benzene in Xe and Kr matrices there is an important contribution of self-trapped excitons to the energy transfer for all host exciton states yielding a critical Förster-Dexter radius of 28 - 29 Å in a Xe matrix and 21 - 22 Å in a Kr matrix.

Table 6.18. Critical transfer radii R_q (in Å) for electronic energy to guest atoms and boundaries in solid (s) and liquid (l) rare-gas matrices

Host	Temperature (K)	Guest			Boundary
		Species	R_q(exp.)	R_q(cal.)	R_q(exp.)
Ar (s)	6 - 20	Xe	18 a)		
		Kr	6 b)		
Kr (s)	60	Xe	17 a)	10 a)	
(s)	110	Xe	25 a)	15 a)	
(l)	120	Xe	24 a)	21 a)	
(s)	5	C_6H_6	21 - 22 c)		
(s)					25 - 29 c)
(s)					22 d)
Xe (s)	5 - 15	C_6H_6	24 - 29 c)		
(s)					40 c)

a) Cheshnovsky et al. /1973a/
b) Gedanken, Raz and Jortner /1973a-d/
c) Ackermann /1976/
d) Ackermann et al. /1976b/

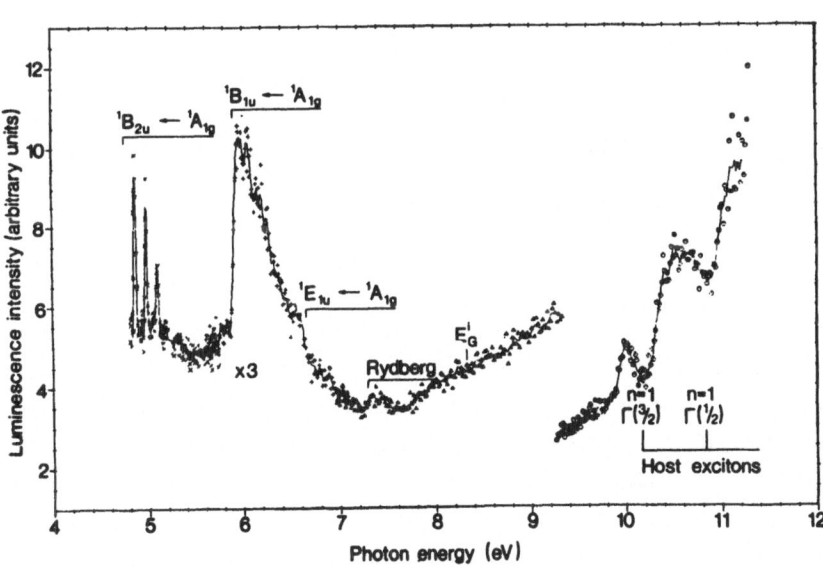

Fig. 6.45. Luminescence excitation spectrum of a benzene-doped Kr film. The film is totally absorbing above 6 eV. The position of the benzene absorption bands /Katz et al., 1969; Gedanken, Raz and Jortner, 1973a-c/ are indicated (from Hasnain et al. /1977a/)

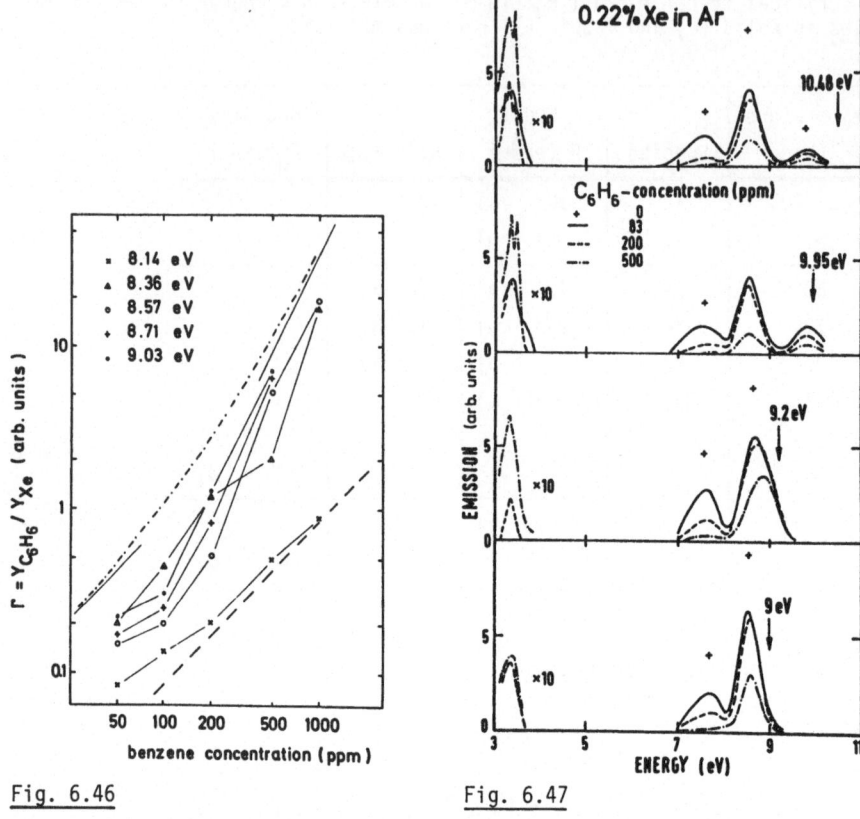

Fig. 6.46

Fig. 6.47

Fig. 6.46. Dependence of the ratio of guest emission intensity to host emission intensity in benzene doped Xe on the C_6H_6 concentration for several photon energies. (– – –) describes the result expected for the diffusion model, (–·–·) describes the result for the dipole model (from Ackermann /1976/)

Fig. 6.47. Emission spectra from benzene/Xe doubly doped Ar matrices for several excitation energies $\hbar\nu$ = 10.48; 9.95; 9.2 and 9.0 eV, respectively. The concentration of Xe was kept fixed at 0.22% while the C_6H_6 concentration has been varied from 0 to 500 ppm (from Ackermann /1976/)

In these studies it was difficult to differentiate between the contribution of electronic energy transfer from free exciton states of the host and that from trapped excitons. Energy migration from a localized donor to a molecular acceptor in RGS was explored in the study of energy transfer from Xe atoms to benzene molecules when both partners had been embedded in an Ar matrix /Ackermann, 1976/. Both centres are fixed in the Ar lattice and exciton diffusion is excluded. The transfer mechanism is exclusively of the Förster-Dexter type. Again, the intensity in the emission bands of both Xe and C_6H_6 has been measured (Fig. 6.47) after excitation of Xe exciton states. More details of the lineshape show up in the excitation spectra of the three emission bands at 1640 Å, 1460 Å and 1250 Å of Xe atoms in an Ar matrix. The analysis results in a Förster-Dexter radius of 24 Å. For the relaxation from n = 2 → n = 1, a rate constant of $1.2 \cdot 10^{10}$ s^{-1} was obtained.

Finally, we mention briefly electronic energy transfer between localized states in liquid rare gases. Electronic energy transfer between the Ar$_2^*$ and Kr$_2^*$ donor and the Xe acceptor in liquid rare gases was studied by Cheshnovsky et al. /1973a/. In the liquid phase mass diffusion has to be considered. The relative donor emission yield of Kr$_2^*$ for the liquid Xe/Kr system exhibits a linear dependence on the Xe impurity concentration. This behaviour is different from that of the Xe/Kr solid alloys (Fig. 6.43) and can be accounted for in terms of a simple kinetic scheme with a rate constant $k_{ET}\tau_0 = 6 \cdot 10^{-19}$ cm^3 which, with the value $\tau_0 \approx 10^{-6}$ s^{-1} for the radiative decay of the Kr$_2^*$ state, results in $k_{ET} = 6 \cdot 10^{-13}$ cm^{3}s^{-1}. The approximate theoretical treatment of Yokota and Tanimoto /1967/ demonstrated that the energy transfer process in the liquid can be expressed in this case by a bimolecular rate constant. Then the rate constant k_{ET} for electronic energy transfer is

$$k_{ET} = 0.51 \cdot 4\pi (R_q^6 \, \tau_0^{-1})^{1/4} \, (D^*)^{3/4} \quad . \tag{6.56}$$

With the data $\tau_0 = 10^{-6}$ s^{-1} and the relative diffusion coefficient D^* of the donor acceptor pair $D^* = 10^{-6}$ cm^{2}s^{-1} /Cheshnovsky et al., 1973a/ together with the experimental value for k_{ET}, a transfer radius $R_q = 24$ Å in the liquid was calculated. This value is in agreement with the value $R_q = 21$ Å from (6.49). Thus the energy transfer process between localized states in the liquid involves a long-range electronic energy transfer process coupled with diffusive motion.

7. Electron Transport and Electron-Hole Pair Creation Processes

Electronic transport properties have been treated by Spear and LeComber /1977/.
They discussed the results of electron and hole mobility measurements in solid
and liquid rare gases. In mobility experiments electrons or holes are accelerated
in an external electric field. Due to the strong elastic and inelastic scattering
cross sections, which allow only mean free paths of the order of some Å, the mean
kinetic energies of the electrons in the conduction band are restricted to energies below $\approx$ 1 eV (see Table 7.1). In this chapter we report on results for elec-

Table 7.1. Calculated mean free path and mean kinetic energies of electrons in
mobility experiments for liquid Ar at 84 K after Leckner /1967/

electric field (V/cm)	drift velocity (cm/µs)	mean kinetic energy (eV)	mean free path for	
			energy transfer Λ_0 (Å)	momentum transfer Λ_1 (Å)
10	0.0039	0.011	6.6	136
10^2	0.037	0.011	6.6	136
10^3	0.21	0.028	6.7	140
10^4	0.47	0.51	8.7	140
10^5	(0.25)	(2.4)	(9.0)	(60)

trons with kinetic energies of some eV's up to MeV's obtained by different techniques.

It is well known that electrons will interact with the crystal lattice by creation or absorption of phonons. The energies of the acoustic phonons ($\lesssim$ 10 meV)
(Table 1.2) in RGS are small compared to electron energies of some electron volts.
In experiments reported here changes in the electron energy by single phonon scattering events are not resolved. Therefore, electron-phonon scattering is described
by an escape depth. The escape depth ℓ_p is given by the mean distance after which

the electron is scattered below the vacuum level of the sample or is trapped and can no longer escape from the sample into the vacuum.

Inelastic interaction of the probing electron with electrons of the crystal is described by an electron-electron scattering length L. In an inelastic electron-electron scattering event a secondary electron has to be excited in an allowed state above the occupied valence bands. Due to the large band gap, the minimal energy losses will be of the order of some eV's. Unscattered electrons have been separated from scattered electrons in energy distribution measurements of photoelectrons. The derived scattering length L represents the mean free path between inelastic electron-electron scattering events. The relevant scattering process for different energy ranges following photoexcitation are sketched in Fig. 7.1.

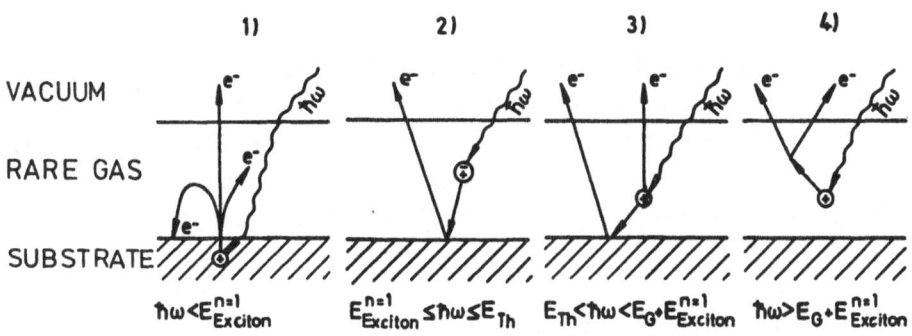

$$\hbar\omega < E^{n=1}_{Exciton} \qquad E^{n=1}_{Exciton} \le \hbar\omega \le E_{Th} \qquad E_{Th} < \hbar\omega < E_G + E^{n=1}_{Exciton} \qquad \hbar\omega > E_G + E^{n=1}_{Exciton}$$

<u>Fig. 7.1.</u> Scheme for important electron and exciton scattering processes at different energy regions ($\hbar\omega$) of the exciting photons (from Schwentner /1974/)

The energy of high energetic electrons will be diminished by subsequent electron-electron and electron-phonon energy losses. For each electron-electron energy loss a secondary electron is excited from the valence band to an excited state and either a free electron-hole pair or an exciton, i.e., a bound electron-hole pair, has been created. The accumulation of holes and excited electrons is described by a mean energy W which corresponds to a mean value $\bar{N}$ of free electron-hole pairs for a primary energy E_0:

$$\bar{N} = E_0/W \quad . \tag{7.1}$$

For practical purposes, when the charge accumulation is used for detectors one is also interested in the r.m.s. fluctuation of the number $\bar{N}$ of pairs produced, which is characterized by $(F\bar{N})^{1/2}$ with the Fano factor F /Fano, 1947/.

The study of these processes in RGS can provide new aspects; for example, concerning the so called "universal curve" for the electron-electron scattering length

L (see, for example, /Lindau and Spicer, 1974/). The threshold energies and the energy dependence near the threshold of electron-electron scattering can be a model case for insulators in view of the large band gaps and the prominent excitonic structure in RGS. Further, because only acoustic phonons with small energies and no dipole moment are present in RGS of the fcc structure, both the energy loss per electron-phonon scattering event and the cross-section are expected to be small. Therefore, ℓ_p may reach much larger values than commonly observed in other solids.

Finally, electron energy loss processes are essential for a microscopic description of light and particle excitation in RGS. In many investigations (see, for example, Chaps. 2 , 6), high energetic particles, such as X rays, α particles and electrons are used as excitation sources. Most of the particle energy is converted to highly excited electrons. The slowing down of these electrons and the final partitioning of energy into phonons, luminescence light and slow electrons shall be described here. In this context, we also mention the differences observed in branching ratios for electron and α-particle excitation, which have been attributed to interaction of excited states due to the high density in the α-particle tracks /Kubota et al., 1977; Carvalho and Klein, 1978/. A fraction of the deposited energy can be reemitted in the form of X rays. This conversion efficiency shows a monotonic increase with the atomic number Z (e.g. /Fink and Venugopala, 1974/), but considering the high density of the solid and liquid phase, reabsorption has to be taken into account.

For application purposes in high energy particle detectors liquid rare gases may be even more suitable than solid rare gases, because they combine the advantages of the electron scattering mechanisms in RGS, which are similar in liquids with less preparation problems (e.g., the trapping of charge at grain-boundaries). For particle detection either the luminescence light output (Chap. 6, /Kubota et al., 1980/) or the amount of charge collected by an external field can be used as a signal. The quantum efficiencies for both processes are expected to be high because of weak non-radiative quenching. The scattering in the mean number of charge per deposited energy should be small (i.e., small Fano factor). Finally, due to the very high mobility of electrons in rare-gas solids and liquids, the response is expected to be extremely fast. Some of the basic results fundamental for the development of fast high energy particle detectors with high efficiency and high energy resolution using rare-gas liquids will be discussed here.

After a short compilation of basic concepts describing electron-electron scattering (Sect. 7.1), experiments providing information about electron-phonon scattering ℓ_p are discussed (Sect. 7.2). The threshold behaviour of electron-electron scattering (Sect. 7.3) is basic for the charge multiplication processes (Sect. 7.5). Further, results for the energy dependence of the electron-electron scattering length L are presented in Sect. 7.4.

7.1 Concepts for the Description of Electron Scattering

Theories of the inelastic scattering of electrons with atoms go as far back as to Bohr /1913,1915/ and Bethe /1930/. They were first modified for the solid state by Fermi /1939/. Here we will only summarize the background for understanding experimental results obtained for solid and liquid rare gases and refer the reader to the many reviews on this subject.

The theoretical framework is quite elaborate for energies of primary electrons being large compared to the energy losses, i.e., when the first Born approximation holds. The differential scattering cross section per atom for an energy loss $\hbar\omega$ and momentum transfer q can be expressed by:

$$\frac{d^2\sigma}{d\omega\,dq} = \frac{2e^2}{\pi N \hbar v^2} \frac{1}{q} \, \text{Im}\left\{\frac{1}{\varepsilon(\omega,q)}\right\} \tag{7.2}$$

where N is the density of atoms, v is the velocity of the incident electron and $\varepsilon(\omega,q)$ is the complex dielectric constant. The connection of optical experiments and electron energy loss experiments is given by $\text{Im}\{1/\varepsilon(\omega,q)\}$ in the "optical limit" of zero momentum transfer q. The electron energy loss for RGS has been discussed by Sonntag /1977/. Following the treatment of Powell /1974/, the attenuation length $d^2\sigma/d\omega dq$ can be integrated for a limited energy loss range, corresponding to a particular excitation process n with mean energy ΔE_n by introducing the generalized oscillator strength $f_n(q)$. The integration concerning the momentum transfer q requires, in principle, the q dependence of $f_n(q)$ which is difficult to obtain. Therefore, $f_n(q)$ is replaced by $f_n(0)$ and a constant C_n accounts for the functional form of $f_n(q)$:

$$\sigma_n = \frac{\pi e^4}{E_0} \frac{f_n(0)}{\Delta E_n} \ln\frac{4E_0 C_n}{\Delta E_n} \quad . \tag{7.3}$$

For plasmon excitation, for example, C_n will be unity. Using the cross section σ_n for specific excitations n and taking $C \approx 1$, an approximate expression for the total inelastic scattering cross section σ_T can be written in the form:

$$\sigma_T = \frac{\pi e^4}{E_0} \sum_n \frac{f_n(0)}{\Delta E_n} \ln\frac{4E_0}{\Delta E_n} \quad . \tag{7.4}$$

If the spectrum of characteristic excitation processes, which would for example consist of the K, L, M, ... shell excitations and plasmons, and the corresponding oscillator strengths $f_n(0)$, are not available, the energy dependence of the scattering cross sections can be estimated from

$$\sigma_T = \frac{\pi e^4}{E_0} \sum_i \frac{N_i}{\Delta E_i} \ln\frac{4E_0}{\Delta E_i} \quad . \tag{7.5}$$

N_i is the number of electrons in the ith shell available for excitation and ΔE_i is an average excitation energy which will always be greater than the corresponding binding energy. In this equation, differences between the gas-phase cross section and σ_T in the solid phase can only be accounted for by different average binding energies, ΔE_i, and typical solid-state effects like plasmons do not appear explicitly.

Battye et al. /1974,1976/ have shown for typical insulators, like alkali-halides and Al_2O_3, that the atomic ionization cross section describes the experimental results within the limits of the experimental accuracy quite well. In general, empirical formulas are applied which follow the calculations by Bethe /1930/ and, for example, Rudge and Schwartz /1966/. They used the following ansatz for single ionization of the ith subshell

$$\sigma_i = \frac{a_i \, N_i}{E_0 \, P_i} \, \ln \frac{E_0}{P_i} \qquad \text{for } E_0 \geqslant P_i \tag{7.6}$$

with an empirical constant a_i and the ionization potential P_i. Values for the constants and tabulated stopping powers from 1 keV to 100 keV are given by Brown /1974/. The total atomic ionization cross section follows from

$$\sigma_T = \sum_i \sigma_i \quad . \tag{7.7}$$

These formulas are valid only for electron energies being large compared to the ionization potentials. Lotz /1967/ has presented a modified formula with three free parameters a_i, b_i and c_i which describes satisfactorily the atomic ionization cross section also near the threshold and up to some keV:

$$\sigma_T = \sum_i \frac{a_i \, N_i}{E_i \, P_i} \, \ln \frac{E_0}{P_i} \cdot \left\{ 1 - b_i \, \exp[-\tau_i(E_0/P_i - 1)] \right\} \qquad \text{for } E_0 \geqslant P_i \tag{7.8}$$

with σ_T in cm^2. The constants are listed in Table 7.2.

The mean free path L for electron-electron scattering in the solid phase is calculated from the total atomic ionization cross section by

$$L(E_0) = \frac{1}{\sigma_T(E_0)} \cdot \frac{A}{L \cdot \rho} \, \text{[cm]} \tag{7.9}$$

with the atomic weight A, the density ρ and the Loschmidt number L.

In Fig. 7.2 the calculated energy dependence of L(E) for Ar is shown in the region of 10 eV to 1000 eV using (7.8 , 9), the gas-phase ionization energy P_1 and the constants listed in Table 7.2. In the solid phase the threshold energies are lower as in the gas phase. Therefore, L(E) has also been calculated for Xe, Kr, Ar and Ne (Fig. 7.2) by replacing P_1 by P_1'. P_1' (Table 7.2) is the energy of the lowest exciton (n = 1) which, as will be discussed later, is the most probable threshold for scattering /Schwentner et al., 1973; Schwentner, 1974,1976b/.

Table 7.2. Parameters a, b, c for the calculation of atomic ionization cross sections /Lotz, 1967/ and of the mean free path for electron-electron scattering (a in 10^{-14} cm^2(eV)^2). N_i and P_i: number and binding energy of electrons in shell i. A: atomic weight; ρ: density (g/cm^3)

	P_1	N_1	P_2	N_2	P_3	N_3	a	b	c	P_1' *	A	ρ
Ne	21.6	6	48.5	2			2.8	0.92	0.19	17.6	20.2	1.504
Ar	15.8	6	29.2	2			4.0	0.62	0.40	12.08	40	1.771
Kr	14.0	6	27.5	2	93	10	4.0	0.71	0.76	10.23	83.8	3.094
Xe	12.1	6	23.4	2	67	10	4.0	0.54	0.64	8.4	131.3	3.782

* P_1': threshold energy in the solid state (eV)

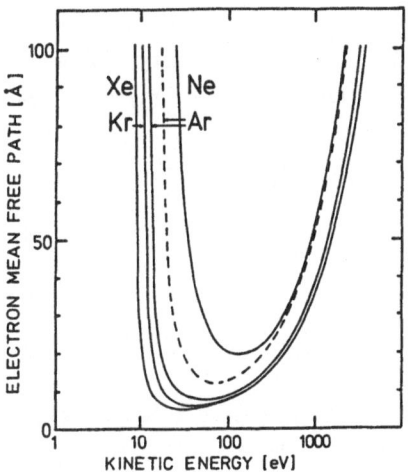

Fig. 7.2. Mean free path for electron-electron scattering in solid Xe, Kr, Ar and Ne according to (7.8, 9); (– – –) P_1 from Table 7.2; (———) P_1' from Table 7.2

In this approximation the influence of the particular properties of the solid phase on the scattering process is neglected. Differences between the solid-state cross section and the atomic cross section are expected to show up strongly near the threshold. For this region a different approximation has been proposed. In the solid-state description a primary electron with momentum k and energy E will be scattered to the momentum k' and E' by exciting a second electron from the valence band with k" and E" to k''' and E'''. Momentum conservation requires

$$k - k' = k'' - k''' + k_0 \tag{7.10}$$

and energy conservation requires

$$E - E' = E'' - E''' \tag{7.11}$$

where k_0 is a principal lattice vector (which may be zero). This problem has been formulated by Berglund and Spicer /1964/ and an extended calculation for silicon has been given by Kane /1967/. The matrix elements are obtained from a screened Coulomb interaction with a frequency and momentum dependent dielectric function. The density in k space for all relevant k vectors follows from the band structure. Kane confirmed by his calculation that the complex scattering problem can be strongly simplified without changing the numerical results significantly by ignoring the restrictions due to momentum conservation. This simplification works because of the averaging effect of the large variety of possible scattering events contributing to L(E). The transition probability $P_S(E,E')$ for scattering a primary electron at E to a lower energy E' is then given by an integration over several densities of states:

$$P_S(E,E') = \frac{2\pi}{\hbar^2} \int_{VB} |M|^2 \, \rho_{CB}(E') \, \rho_{VB}(E'') \, \rho_{CB}(E''') \, dE'' \tag{7.12}$$

where $\rho_{CB}(E')$, $\rho_{CB}(E''')$ are the final densities of states of the primary and secondary electrons, respectively, and $\rho_{VB}(E'')$ is the initial density of states of the secondary electron and M is the matrix element.

The inverse lifetime or total scattering probability P(E) follows as:

$$P(E) = \int P_S(E,E') \, dE' \tag{7.13}$$

and the mean free path L(E):

$$L(E) = v_g(E) / P(E) \tag{7.14}$$

where v_g is the electron group velocity. Schwentner /1976/ proposed a further simplified evaluation for rare-gas solids in the region near the threshold. The matrix element has been taken as constant in the limited energy range of $\approx$ 10 eV above the threshold, the width of the valence bands has been neglected for the integration of E" and parabolic conduction bands have been used,

$$\rho_{CB} \propto (E - E_G)^{1/2} \tag{7.15}$$

where E_G is the gap energy. Then the integrations can be solved analytically yielding

$$P(E) = C(E - 2E_G)^2 \tag{7.16}$$

and

$$L(E) = C(E - E_G)^{1/2} (E - 2E_G)^{-2} \tag{7.17}$$

with a free parameter C. The assumption of parabolic conduction bands is reasonable for the lower s-type conduction bands. The structure due to the higher lying d bands (see Chap. 3) may be smeared out. The strong excitonic structures will change both the density of states and the scattering threshold, as will be discussed in Sect. 7.4.

The scattering of electrons for mean kinetic energies smaller than ≈ 1 eV has been treated by Spear and LeComber /1977/ using the results of mobility measurements /LeComber et al., 1976/. The cross sections in the solid and liquid phase are nearly identical; therefore, it is sufficient to discuss here the results for the liquid phase. For the mean free path the density change has to be considered which causes differences in the mobilities for solid and liquid phases. Electrons in the conduction bands can be treated as free electrons. For low electric fields the experimental results have been predicted excellently by the theory of Leckner /1967/. This theory gives the mean free path for both the energy loss Λ_0, which corresponds in our case to inelastic electron-phonon scattering, and for the momentum transfer Λ_1, which means elastic scattering /Cohen and Leckner, 1967/ (see Table 7.1). At higher fields the mobility saturates. This has been ascribed to changes in the effective mass of the electrons caused by structures in the conduction bands. The limiting average kinetic energies of about 1 eV are reached for electric field strengths of 10^5 V/cm (see, for example /LeComber et al., 1976; Yoshino et al., 1976/).

The transient behaviour of the conductivity after pulsed excitation in mobility experiments has been analysed in terms of a thermalization time of electrons /Sowada et al., 1982/. Values for the thermalization time of the order of 0.5 ns (Ar) up to 4 ns (Xe) have been found in agreement with expectations based on the deformation potential scattering theory /Conwell, 1967/. The thermalization times are much longer as for example in semiconductors since optical phonons are not involved in RGS.

Electron phonon scattering is the dominant process at least up to the onset of inelastic electron-electron scattering around 10 eV (Fig. 7.1 and Table 7.2). Above about 1 eV, theoretical cross sections for electron phonon scattering are not available for RGS. For other insulators, such as alkali halides, the energy dependence of the electron-phonon mean free path has been calculated for optical phonons (e.g. /Glacer and Garwin, 1969/), yielding an approximately linear increase of the mean free path with electron kinetic energy. Typical experimental values for alkali halides scatter between 10 Å and 100 Å. For the interaction of electrons with the acoustic phonons of RGS's similar calculations are not available. Therefore, only the experimental data presented in Sect. 7.2 can be discussed.

For larger energies electron-electron scattering becomes an additional and important mechanism. In order to explain the dissipation of the primary energy by successive electron-electron and electron-phonon scattering events, the energy and space distribution of the avalanche of primary and secondary electrons has to be evaluated. This is a problem encountered in all photoemission experiments (see, for example, /Cardona and Ley, 1978/). Several attempts have been made to analyse the photoemission data (e.g. /Berglund and Spicer, 1964; Kane, 1966/) using empirical models, Monte Carlo calculations, and others. Furthermore, a variety of empirical formulas have been developed to describe the deposition rate of energy versus penetration depth in cathode luminescence work (e.g. /Fano, 1940; Kingsley and Prener, 1972/).

A quantity which is relevant for energy deposition is the average energy W necessary to create an electron-hole pair (7.1). In semiconductors an empirical relation between W and the band gap E_G is known /Alig and Bloom, 1975/ which has been explained in the free particle approximation using conservation of energy /Shockley, 1961/ and momentum /Klein, 1968/. It is assumed that as long as energetically possible the electron energy is used to create electrons and holes. Finally, below a threshold energy E_M the energy is insufficient for exciting another electron-hole pair. Then the remaining excess energy above E_G is lost by phonon emission. Since W must be the sum of the band gap energy and the mean excess energy of electrons and holes we get, with the density of states proportional to $\sqrt{E}$ in the free particle limit,

$$W = E_G + 2 \int_0^{E_M} E\sqrt{E}\, dE \bigg/ \int_0^{E_M} \sqrt{E}\, dE = E_G + 6E_M/5 \quad . \tag{7.18}$$

When both energy and momentum conservations are fully fulfilled

$$E_M = 3E_G/2 \tag{7.19}$$

and the general relation for semiconductors follows as:

$$W = 2.8\, E_G \quad . \tag{7.20}$$

This relation has been applied for solid and liquid rare gases (see /Spear and LeComber, 1977; Howe et al., 1968/ and Table 7.3).

Table 7.3. Mean energies W (in eV) for the production of free electron hole pairs

	W calculated		W experimental		
	$2.8\, E_G$	Doke et al. /1976/	solid	liquid	liquid
	(7.20)	(7.21)	Spear and LeComber /1977/		Doke et al. /1976/
Xe	26	15.4	24 ± 1	33 ± 2	15.6 ± 0.3
Kr	32	19.5	25 ± 1	25 ± 1	
Ar	40	23.3	27 ± 1	27 ± 1	23.6 ± 0.3

In a series of careful investigations, the evaluation of Shockley's formula has been modified for liquid rare gases /Doke et al, 1976/. In this case exciton excitation (bound electron-hole pairs) is an energy loss process competitive to free electron-hole pair creation and has to be incorporated in the mean energy W /Platzman, 1961/:

$$W = E_G + \bar{E}_{ex}(\bar{N}_{ex}/\bar{N}_{e-h}) + \int_0^{E_1} E(dN/dE) \, dE \bigg/ \int_0^{E_1} (dN/dE) \, dE \quad . \tag{7.21}$$

In addition to the exciton losses with a mean exciton energy $\bar{E}_{ex}$ and the ratio of the number of excitons to the number of electron-hole pairs $(\bar{N}_{ex}/\bar{N}_{e-h})$, both derived from optical data, further essential changes have to be noted. The hole kinetic energy has been accounted for by adding half of the valence band width to E_G which explains the missing factor of two. The free particle density of states $\sqrt{E}$ has been replaced by that derived from the band structure. More importantly, the mean excess energy E_M has been replaced by E_1, i.e., the energy of the lowest exciton, because momentum conservation has been ignored. These calculated values (Table 7.2) are approximately a factor of two lower than those derived from the empirical law for semiconductors. We shall return to this point when discussing the experimental data. We note that Doke et al. /1976/ have also calculated the Fano-factors (/Fano, 1947/; see (7.1)) for rare-gas liquids.

7.2 Electron-Phonon Scattering

Electron-phonon scattering in RGS has been studied for electron energies below the threshold of electron-electron scattering, i.e., for electrons up to ≈ 10 eV above the bottom of the conduction band. The escape depth ℓ_p has been derived from three types of photoelectron yield experiments.

In the first type of experiment (Process 1 in Fig. 7.1), photoelectrons from a metal substrate (usually Au) are injected into an RGS overlayer of known thickness d. The electrons are excited by photons with energies smaller than the first absorption line of the RGS film but larger than the work function of the substrate. The yield of electrons penetrating through the RGS film is measured. From the thickness dependence of the yield the escape depth ℓ_p is calculated. In Fig. 7.3 experimental results are compiled. The thickness dependence can be divided into three parts. For small film thicknesses (0 - 50 Å) the yield is sensitive to the cleanliness and hence to the work function of the substrate. An increase as well as a decrease with the film thickness, both up to a factor of two, have been observed. This behaviour has been attributed to changes in the work function of the layer system. For larger thicknesses of 50 Å $\leq$ d $\leq$ 300 Å a steep decrease in the yield is observed for Ar,

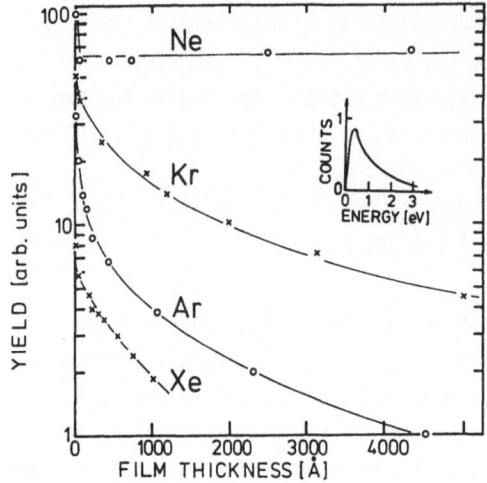

Fig. 7.3. Hot electron current from an Au substrate versus thickness of the RGS overlayer; (Ne) hν = 16 eV /Pudewill et al., 1976/; (Ar) hν = 10 eV /Schwentner et al., 1980/; (Kr) hν = 9 eV /Schwentner et al., 1980/; (Xe) hν = 7.9 eV /Ophir et al., 1975a; Schwentner, 1976/. In the insert the energy distribution of photoelectrons from the Au substrate is shown

Kr and Xe. Finally, for d > 300 Å the slope of the decrease on a logarithmic scale becomes gradually smaller. The second and third range can be used to derive the escape depth ℓ_p. Usually, an experimental escape probability for the yield Y of the form

$$Y(d) = Y_0 \, e^{-e/\ell_p} \qquad (7.22)$$

has been assumed. Evidently, the results in Fig. 7.3 cannot be explained by a straight line with the slope ℓ_p. Equation (7.22) would be correct, if the electrons could be lost for detection after one scattering process. Of course, this model is too simple for a description of the yield experiments and a number of microscopic processes have to be considered; electrons can be scattered below the vacuum level in the case of Xe and Kr by successive phonon emission, and phonon absorption also plays a role.

In Ar and Ne the vacuum level lies below the bottom of the conduction band (see Sect. 3) and all free electrons should be able to escape as far as their energy is concerned. The decrease of the yield indicates trapping of electrons at grain boundaries, at dislocations and at impurities. An important loss process diminishing the yield is elastic and inelastic backscattering of electrons to the substrate. This loss can be larger near the substrate, i.e., at small film thicknesses, rather than at larger separations from the substrate. Further, when the electron is near the substrate a positive image charge will be formed. Due to inelastic scattering processes the electron energy can be reduced below the ionization limit for the electron plus image charge system and the electron will be recaptured by the substrate.

Quantitative model calculations to explain the thickness dependence have not been published. Nevertheless, in order to obtain a rough estimate for the escape depth, the curves in Fig. 7.2 have been approximated by straight lines in the region of

Table 7.4. Escape depth ℓ_p of RGS (ℓ_p and film thickness d in Å)

	ℓ_p from Fig. 6.33		ℓ_p from Fig. 6.34	ℓ_p from (7.22)
	$50 \leqslant d \leqslant 200$	$500 \leqslant d \leqslant 5000$		
Xe	250	950		
Kr	250	4000		2900
Ar	200	3000	1000	1400 - 1700
Ne		5000	3500	

50 - 200 Å and from 200 Å up to the larger thicknesses (Table 7.4). A mean free path for inelastic electron phonon scattering can be estimated from (7.23) /Baraff, 1964/

$$\Lambda_0 = 3\ell_p(\bar{E}_{ph}/3\bar{E}_0)^{1/2} \quad . \tag{7.23}$$

With a mean electron kinetic energy $\bar{E}_0$ of $\approx$ 1 eV (see insert in Fig. 7.3), a mean phonon energy $\bar{E}_{ph} \approx 5$ meV and $\ell_p = 1000$ Å, a mean free path of 150 Å was obtained /Schwentner, 1976/. It should be kept in mind that this value can only be regarded as a crude estimate in view of the various processes contributing to ℓ_p.

In a second type of yield experiments (Processes 3, 4 in Fig. 7.1) the photoelectrons are created by photoabsorption in the rare-gas film (Fig. 7.4). The yield depends on the absorption coefficient k and the escape depth. For isotropic excitation of electrons and for and exponential escape probability the following thickness dependence is expected (see, for example, /Schwentner, 1976/ and references therein)

$$Y = \frac{1}{2} \frac{k\ell_p}{1 + k\ell_p} \{1 - \exp\left[-[d(k + 1/\ell_p)]\right]\} \quad . \tag{7.24}$$

Interference effects and reflectivity losses have been neglected. The calculated yield using several escape depths as a parameter is compared with the experimental result for pure Ne in Fig. 7.4 suggesting an escape depth of $\ell_p \approx 3500$ Å.

Due to a factor of one half, which results from the assumption that 50% of the electrons will run towards the substrate and 50% towards the surface, the efficiency is restricted to values below 0.5 electrons per photon in this simple model. For Ar (Fig. 7.4) an increase of the yield above 0.6 electrons per photon has been measured. This high yield can be explained by either including secondary electrons from the substrate or by taking into account the increased density of electrons at the surface due to the absorption process. Electrons reaching the metal substrate can produce secondary electrons quite efficiently /Petry, 1926/. These secondary electrons have been detected in energy distribution measurements of photoelectrons (see Sect.

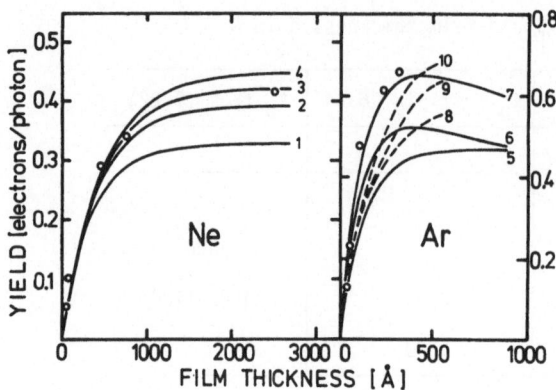

Fig. 7.4. Photoelectron yield from films of solid Ne and Ar. (o o o) represent the experimental electron yield for Ne at hν = 25 eV and Ar at hν = 18 eV. For Ne (*left*) the lines are calculated from (7.24) for several escape depths ℓ_p. (1) ℓ_p = 1000 Å; (2) ℓ_p = 2000 Å; (3) ℓ_p = 3000 Å; (4) ℓ_p = 5000 Å (from Pudewill et al. /1976/). For Ar (*right*) the solid lines are calculated from (7.24) including a contribution due to secondary electrons with the secondary electron coefficient β; (5) ℓ_p = 2000 Å, β = 0; (6) ℓ_p = 1000 Å, β = 0.2; (7) ℓ_p = 2000 Å, β = 0.3; the dashed lines are calculated using a random walk model with an electron-electron scattering length L = 10 000 Å and several electron-phonon mean free paths Λ_0; (8) Λ_0 = 400 Å; (9) Λ_0 = 100 Å; (10) Λ_0 = 5 Å; (after Schwentner /1976/)

7.5). The solid curves in Fig. 7.4 show fits for Ar for two escape depths ℓ_p including secondary electrons with different emission coefficients β. Further, the enhancement of the yield caused by the non uniform distribution of electrons follows from a random walk model calculation with three different values for Λ_0 and β = 0 (dashed lines in Fig. 7.4). For thick films, the random walk contribution will be more important than secondary electrons. A combination of both processes with ℓ_p between 1000 - 2000 Å, β between 0.2 to 0.3 and Λ_0 between 5 - 500 Å fits the experiments.

Escape depths have been determined in a third way by Hasnain et al. /1977e/. In doped RGS photoelectron emission is observed below the threshold for host emission due to energy transfer from host excitons to impurity centres and escape of an electron (see Sect. 6.13). Using a similar type of analysis, as described above, Hasnain et al. /1977e/ derived escape depths from the experimental thickness dependence of the yield in the excitonic region of Kr and Ar matrices doped with 0.01% C_6H_6. Within the experimental uncertainties they fit with the other results (Table 7.4).

Recently, the penetration of low energetic electrons through thin films of solid Xe to a metal substrate has been measured /Bader et al., 1982/. The results are more specific concerning the penetration depth because the incident electron energy is well defined in comparison to photoelectron injection from a substrate. Besides the onset for electron-electron scattering around 8.2 eV (Sect. 7.3), detailed information on the escape depth due to electron-phonon scattering has been obtained. The mean free path for electron-phonon scattering has been extracted, and its energy dependence is shown in Fig. 7.5. Inspection of Fig. 7.5 shows that this mean free path is high, as is expected at electron energies below 2 eV. The structures between 0 - 8 eV can be understood in terms of a structure factor, which allows one to deduce a mean electronic effective mass /Bader et al., 1982/.

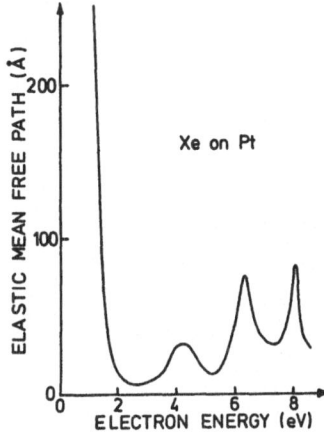

Fig. 7.5. Mean free path for electron-phonon scattering from electron transmission spectra (after Bader et al. /1982/)

Summarizing, we note that the escape depth for all RGS is unusually large in the order of 1000 Å for electrons below the electron-electron scattering threshold. These large values are mainly attributed, as has been mentioned before, to the small cross section and small energy losses for emission of acoustic phonons and to the small electron affinities which are even negative in Ar and Ne. These escape depths exceed by far all values reported for other materials.

7.3 Threshold for Electron–Electron Scattering

The threshold for electron-electron scattering E_{sc} can be referred either to the bottom of the conduction band, or to the vacuum level or to the top of the valence bands. In order to get a consistent picture, including the results of photoluminescence and photoelectron emission measurements, where electron excitation energies from the valence bands are essential, we use the top of the valence bands as a common reference energy for E_{sc}.

From electron energy loss data one obtains immediately the smallest energy losses possible. The pertinent experimental results have been collected by Sonntag /1977/

together with excitation energies from optical spectra. Despite different selection rules causing changes in the overall shape of the spectra, the lowest energy losses (for both high and small primary electron energies) and the lowest absorption lines correspond to the n = 1 excitons (Table 7.5). Of course, the details of the onset

<u>Table 7.5.</u> Threshold energies E_{sc} for electron-electron scattering. E_1: n = 1 exciton; E_G: band gap (in eV)

	$2E_1$	$E_G + E_1$	$2E_G$	a)	b)	E_{sc} c) threshold extrapolation	
Xe	16.74	17.70	18.66	17.6		18	17.0
Kr	20.34	21.78	23.22	21.5		21.56	20.5
Ar	24.12	26.22	28.32		25.5 ± 0.5	26.10	24.5
Ne	34.72	38.86	43.00				

a) Photoelectron yield   /Schwentner, Skibowski, Steinmann, 1973/
b) Luminescence yield   /Möller et al., 1976/
c) Photoelectron energy distribution   /Schwentner, 1976/

would require an extrapolation of the low energy tail of the n = 1 excitons. Neglecting this latter point, the scattering onset E_{sc} is given by

$$E_{sc} = E_G + E_1 \qquad (7.25)$$

when an excited valence electron in the conduction band is inelastically scattered to the bottom of the conduction band by exciting another valence electron to the n = 1 exciton state.

The threshold energy has been studied in photoelectron emission and luminescence experiments. In these experiments the primary electron is created together with a hole by photoabsorption, whereas in energy loss experiments an excess electron is injected and the corresponding hole is missing. Consequently, a difference in the threshold could be expected from a process corresponding to a recapture of a primary electron by one of the holes and the subsequent decay into two excitons in photoabsorption experiments. In this case E_{sc} would be smaller

$$E_{sc} = 2E_1 \quad . \qquad (7.26)$$

Other processes which deserve attention in this context are double excitations, since they could mask the onset of scattering processes. Double excitation processes as well as scattering processes result in two low energy electrons or excitations where the sum of the energies is equal to $\hbar\omega$. Therefore, they cannot be distinguished immediately. If their respective threshold energies are different, they can be identified easily. More detailed criteria could be obtained from the thickness dependence of the ratio of unscattered and scattered electrons.

A special type of double excitation, i.e. the formation of an "electronic polaron complex" has been discussed by Devreese et al. /1972/ and Kunz et al. /1972/ and applied to alkali halides. The optical transition leads to an electron-hole pair dressed by an electronic polarisation cloud and an additional exciton. Two different types of electronic polaron complexes have been distinguished: in the bound complex the dressed electron is coupled to the dressed hole via Coulomb interaction, in the free complex this Coulomb interaction is neglected. The cross-sections for the polaron complex in rare gas solids should be similar to those for alkali halides, because the coupling constants are of the same order. The threshold for the free complex lies at $E_G + E_1$ and for the bound complex near $2E_1$. On the basis of this model Möller et al. /1976/ have estimated the cross sections for both processes in Ar.

In the photoelectron yield curves /Schwentner et al., 1973/ and the more recent results by Schwentner /1974/ a decrease at a photon energy E_X is observed for solid Kr and Xe, while it is absent in the yield spectra for solid Ar. E_X can be identified with the threshold energy E_{SC} for electron-electron scattering (Table 7.5). For solid Kr and Xe just above threshold part of the primary electrons will be scattered below the vacuum level. Also near threshold the secondary electrons can only be excited to an exciton state or to the bottom of the conduction band below the vacuum level. Thus, both scattering partners are lost for the yield. In solid Ar the vacuum level lies below the bottom of the conduction band. Therefore, the primary electron can escape in any case and no significant decrease is observed at $E_X = E_{SC}$.

A luminescence yield spectrum up to $\hbar\omega = 55$ eV has been reported by Möller et al. /1976/ (Fig. 7.6). In luminescence any electron-electron scattering will cause an increase of the yield. The luminescence channel due to the self-trapped excitons is fed by excitons and conduction band states, i.e., from any possible state of the primary and secondary electrons. The first pronounced increase is observed at $\hbar\omega = 25.2 \pm 0.5$ eV (m = 1 in Fig. 7.6) and lies within the experimental accuracy at $E_{SC} = E_1 + E_G$. At these excitation energies only a doubling of the holes due to electron-electron scattering is possible. Möller et al. were able to explain the yield spectrum (Fig. 7.6) in detail: The shape of the onset at around 27 eV is due to the energy dependent scattering cross-section as derived by Schwentner /1976/; the decrease of the yield by more than a factor of two is due to the higher efficiency of the low energy, once scattered, electrons. At higher photon energies two additional pronounced thresholds are observed, denoted by m = 2 and m = 3. The experimental values of $E_{m=2} = 37.5 \pm 1$ eV and $E_{m=3} = 50 \pm 2$ eV correspond nicely to $E_G + 2E_1 = 38.3$ eV and to $E_G + 3E_1 = 50.3$ eV. Therefore, the final states at these thresholds are pre-

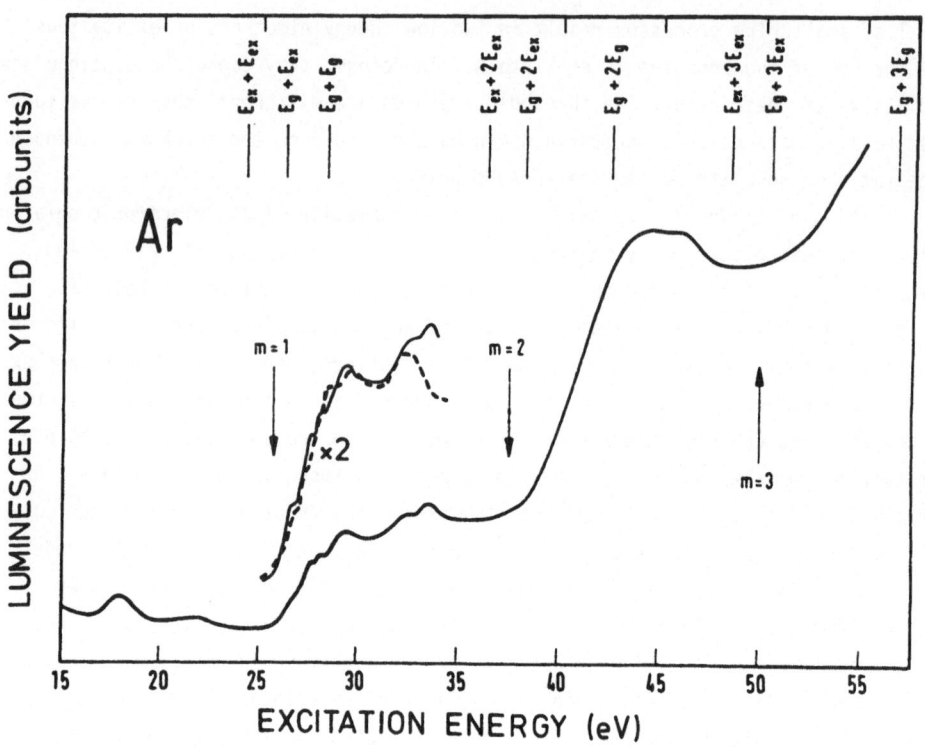

Fig. 7.6. Excitation spectrum of the R_2^* luminescence band of solid Ar. The thresholds for electron-electron scattering are indicated by m = 1,2,3. (———) shows a fit to the experimental spectrum (see text) (after Möller /1976/)

dominantly one electron hole pair plus m excitons per primary electron.

The onset of electron-electron scattering has also been identified in photoelectron energy distribution curves. These curves will be discussed in more detail in Sect. 7.4 where the energy dependence of the electron-electron scattering length is derived. With increasing excitation energy the kinetic energy of the unscattered electron increases (Fig. 7.7). The onset of electron-electron scattering manifests itself in two features: a sudden decrease in the high-energy part of unscattered electrons (maximum A) and a strong increase in the dashed region due to scattered electrons which are concentrated near zero kinetic energy. The corresponding photon energies have been collected in Table 7.5. Further, the onset can be taken from a fit of the energy dependence of the scattering length (see below). These values are approximately 1 eV lower than those obtained from Fig. 7.7. Both sets of E_{sc} values are in agreement if one notes that the fit corresponds to an extrapolation to L → ∞, whereas the former set corresponds to the well observable onset.

Summarising the results for the threshold values for electron exciton scattering E_{sc} one observes a general agreement with the value expected from electron energy

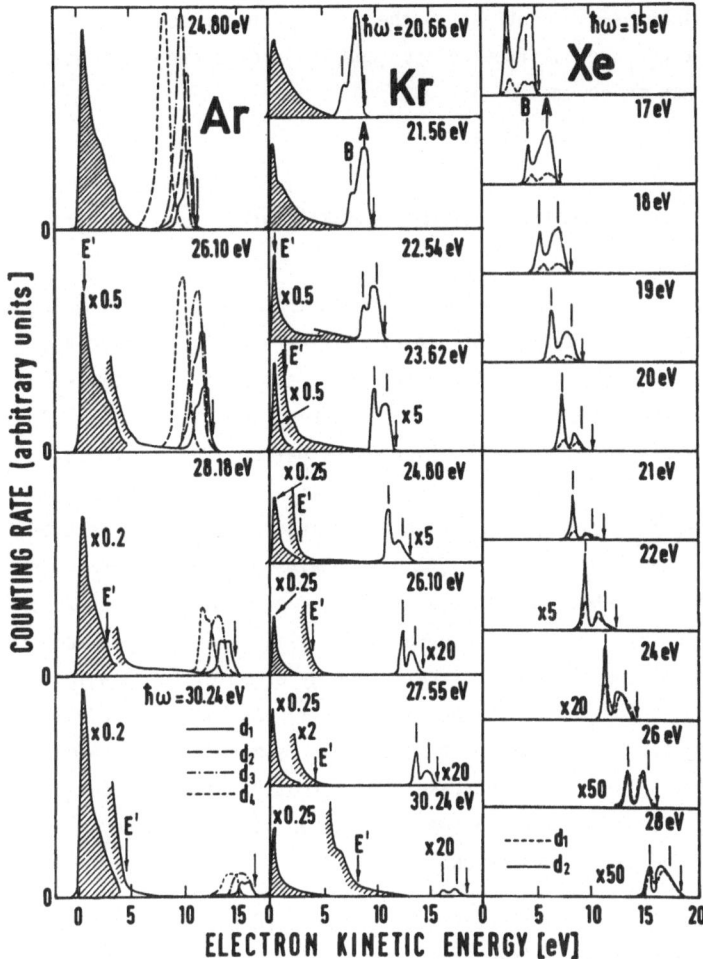

Fig. 7.7. Photoelectron energy distribution curves from solid Ar, Kr and Xe films for a spectrum of photon energies. The film thicknesses have been $d_1 = 14$ Å and $d_2 = 228$ Å for Xe (the region of scattered electrons is not shown), $d = 50$ Å for Kr and $d_1 = 30$ Å, $d_2 = 84$ Å, $d_3 = 140$ Å and $d_4 = 300$ Å for Ar. The spectra have been multiplied by the attached factors (from Schwentner /1976/)

loss experiments. This shows that scattering of primary electrons to the bottom of the conduction band and excitation of an m = 1 exciton is the dominant process, whereas the bound polaron complex with threshold energy $2E_1$, can be excluded.

The free polaron complex cannot be excluded on the basis of its threshold energy. The thickness dependence has been studied only in the case of electron energy distribution curves for Ar /Schwentner, 1974/. The maximum cross section for the polaron complex is expected at $E_G + 1.2E_1 = 28.6$ eV /Devreese et al., 1972/. In this energy region one observes an increase of the ratio of scattered to unscattered

211

electrons with sample thickness, in contrast to the prediction for double excitation. Thus, although one cannot exclude the free polaron complex from these rough estimates, its contribution has to be small.

7.4 Energy Dependence of Electron-Electron Scattering Mean Free Path

Electron energy distribution curves in the region of electron-electron scattering have been reported by Schwentner /1974,1976/. In Fig. 7.7 the EDC's of solid Ar, Kr and Xe are shown for photon energies between an energy somewhat below twice the band-gap energy and 30 eV where electron-electron scattering is important. For Ar and Xe, EDC's have been measured for several thicknesses between 10 - 300 Å. Each spectrum can be divided into two parts. The first part with high kinetic energies (A, B) is due to unscattered electrons which have been directly excited from the valence bands of the rare gas. The structure (A, B) contains information about the band structure as discussed in Chap. 3. After an inelastic electron-electron scattering event an electron excited from the valence bands will appear in the second part (the hatched region) of the EDC's. The kinetic energy of an electron which has been excited from the top of the valence band (right arrow) and has suffered an energy loss corresponding to the excitation of an exciton is marked by E'. A maximum appears below E' showing the strong increase of the scattering cross section with photon energy. Also, at photon energies below E_{sc} there are low energy electrons observed which are due to two processes:

i)   Hot electrons from the Au substrate excited by transmitted photons and penetrating through the rare-gas film (see Process 1 in Fig. 7.1 and Sect. 7.2). The intensity and shape of EDC's from the substrate support this explanation.

ii) Approximately 50% of the electrons excited in the rare-gas film reach the Au substrate and produce secondary electrons with an efficiency increasing with electron energy from 20% to 40%. The secondaries will partly leave the sample yielding an energy distribution similar to an EDC from the Au substrate (Process 3 in Fig. 7.1).

The energy dependence of the electron-electron scattering length L(E) has been derived /Schwentner, 1976/ in several ways from the EDC's (Fig. 7.7).

1)   From the decrease of the counting rate of unscattered electrons associated with a distinct initial energy when the electron kinetic energy is varied by means of varying $\hbar\omega$ (Fig. 7.8, left part).

2)   From the thickness dependence of the counting rate of unscattered electrons for the different photon energies (Fig. 7.8, right part).

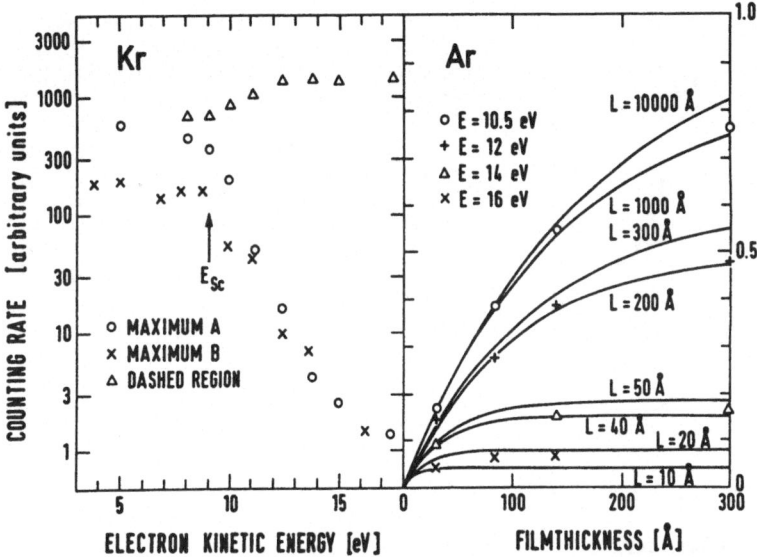

Fig. 7.8. (*Left part*) Dependence of the intensity of maximum A and B of Fig. 7.7 on the electron kinetic energy for Kr. Zero corresponds to the vacuum level of Kr. (*Right part*) The points show the thickness dependence of the intensity of unscattered electrons from Fig. 7.7 and the solid lines are calculated curves (from Schwentner /1976/)

3) From the ratio of counting rates within the part of unscattered electrons in one EDC (for example, the ratio of maximum A to maximum B in Fig. 7.7) for different photon energies.

4) The consistency of the evaluation has been checked by a comparison of counting rates at equal kinetic energies in spectra obtained with different photon energies.

The results for L(E) for solid Ar, Kr and Xe are shown in Fig. 7.9 /Schwentner, 1976/. L(E) decreases from very large values near the scattering onset within 2 eV to values of about 10 Å and within 10 eV down to 1 - 5 Å. The main sources for errors are uncertainties in the absorption constant. The independent ways of calculating L allow for averaging. The influence of electron phonon scattering on L(E) is considered to be small. Due to the small phonon energy, the phonon-scattered electron would be still in the unscattered region of the EDC. An electron-phonon scattering event would only increase the path of the electron to the surface in a random walk process.

The experimental results have been compared with the models discussed in Sect. 7.2. The full lines in Fig. 7.9 have been calculated for the best fit from the following equation:

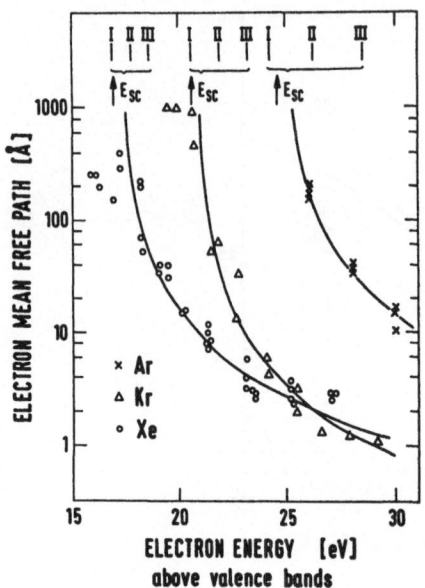

Fig. 7.9. Electron mean free path versus electron energy measured from the top of the valence bands for Ar, Kr and Xe. The points show the experimental results. The solid curves represent a fit according to (7.27). E_{SC} is the electron-electron scattering onset determined by the use of (7.27). I corresponds to twice the $n = 1$ exciton energy, II to the sum of band gap and $n = 1$ exciton energy, and III to twice the band gap for Ar, Kr and Xe, respectively (from Schwentner /1976/)

$$L(E) = C(E - E_G)^{1/2} (E - E_{SC})^{-2} \qquad\qquad (7.27)$$

with the constants C and E_{SC} as parameters. In this relation, E_G in the second term (7.17) has been replaced by E_{SC} to account for the energy losses due to excitons. The calculated curves follow the experimental values with the E_{SC} values from Table 7.5. A calculation according to the gas-phase ionization cross sections (7.8) as well as an integration of the loss function, as proposed by Powell /1974/ (7.5) do not fit the experimental scattering length data equally well /Schwentner, 1976/.

A comparison of L(E) for rare-gas solids with some other materials has been made by Schwentner /1976/. The striking, though not unexpected point, is the large energy range where electron-electron scattering does not occur as well as the very steep decrease of L(E) near the threshold. Some eV's above the threshold L(E) follows the so called "universal curve" (see, for example, /Lindau and Spicer, 1974/).

7.5 Generation of Electron-Hole Pairs by High Energy Particles

As mentioned in the introduction to this section, liquid rare gases have found wide application as detector media in high energy particle detectors. Thus, the processes underlying the occurrence of electron avalanches in liquid rare gases have been investigated in several papers /Kubota et al., 1976,1980; Takahashi et al., 1975; Doke et al., 1976/ and references cited therein. Electrons with energies of 0.48 MeV,

0.55 MeV, 0.976 MeV and 1.05 MeV from radioactive materials have been used to excite liquid Xe, Ar, Xe-doped Ar and Kr-doped Ar. A fast ionization chamber for electron collection has been filled with carefully purified liquid rare gases. Positive
ions have been shielded by grids. The voltage pulses at the collector for each exciting particle have been amplified and the distribution of pulse heights was stored
in a multichannel analyser. From the position of the maxima in the pulse-height
spectrum it could be shown that the number of electrons (i.e., pulse height) is proportional to the energy of the exciting particles. Therefore, the spectra (Fig.
7.10) represent the energy distribution of the exciting particles. The scaling factor of the energy scale (i.e., the proportional constant) is the mean electron-hole
pair creation energy W. Evidently W is smaller for Xe-doped Ar than for pure Ar and
therefore the energy scale is enlarged in the doped liquid (Fig. 7.10). From a comparison with the known pair-creation energy of a gas chamber filled with Ar (95%) +
CH_4 (5%) (W_{gas} = 26.09 ± 0.13 eV), the pair-creation energies for liquid Ar and Kr
have been derived. The experimental values agree excellently with the calculated
value of Doke et al. /1976/ (Table 7.3). These values seem to be more reliable than

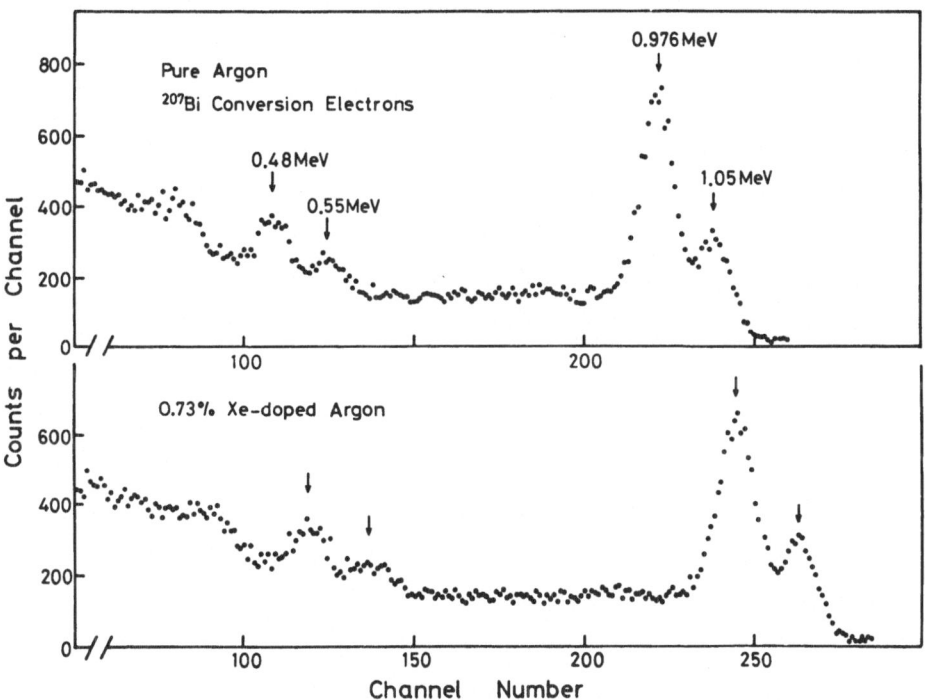

Fig. 7.10. Pulse-height spectrum of [207]Bi conversion electrons measured with the
liquid Ar-ionization chamber with and without Xe at an electric field of 13.7 kV/cm
(from Kubota et al. /1976/)

the larger ones measured in the course of mobility experiments (Table 7.3).

The experimental support for the model calculation of Doke et al. /1976/ shows that this model is appropriate for insulators whereas the generally accepted rule used for semiconductors fails. Further, the assumptions made in the calculation [see Sect. 7.1 and (7.21)] are justified. The assumed loss in efficiency due to exciton excitations has been verified independently in the experiment on Xe-doped Ar. In this case, the exciton energy of Ar is sufficient to ionize the Xe dopand (see also Chap. 6). Through this process the loss due to Ar excitons can be eliminated. Indeed, the increase in ionization yield observed for Xe-doped liquid Ar (Fig. 7.10) corresponds to the calculated $\bar{N}_{ex}/\bar{N}_{e-h}$ ratio (7.21). With the W values also the calculated Fano factors have been corroborated.

The interest in liquid Xe as a detector medium is motivated by the large electron mobility and the large atomic number. In addition, the efficiency due to the low W values and the energy resolution due to the small Fano factors in the liquid phase are better than in the gas phase. For 1 MeV gamma rays a resolution of 3 keV FWHM is expected which is near the actual energy resolution of 1.5 keV of conventional Ge(Li) detectors with large volume /Doke et al., 1976/.

8. Concluding Remarks

The understanding of the electron structure of the ground and excited states in condensed rare gases is fairly complete for pure RGS. In this book we have attempted to shift the focus of interest to structurally and compositionally disordered rare gases, including solid rare-gas alloys, metal rare-gas solid mixtures, liquid rare gases and liquid rare-gas alloys. The basic experimental results concerning the electronic states of these materials have been surveyed. The theory of the electronic structure of disordered systems does not provide general unified concepts comparable to the Bloch theorem for ideal crystalline solids. We have not attempted to wrench the diverse systems of disordered condensed rare gases from their natural settings. The current theoretical interpretation relies in many cases on conventional solid state concepts, e.g., band structure, densities of states and Wannier and Frenkel exciton states. Impurity states in alloys behave more as impurity states in crystalline materials. Excitons in liquids were considered within the framework of solid state approach, while metal rare-gas solids were discussed in terms of metal-nonmetal transitions from a heavily doped disordered semiconductor to an amorphous metallic material.

We have encountered some rather exotic systems. Medium relaxed electronically excited states and, in particular, bubble states in liquid helium are examples for novel phenomena in isolated centres. Metal rare-gas solid mixtures provide a nice demonstration for the effects of microscopic clustering and metal-nonmetal transitions in "expanded" low-temperature metals. Even pure metallic xenon at 330 kilobars and 32 K has been prepared /Nelson and Ruoff, 1979/.

Complementary and supplementary to the information regarding electronic structure is the new and diverse information concerning excited-state dynamics in condensed rare gases. During the last few years this field has made an impressive experimental progress. While until recently all the information concerning excited-state relaxation phenomena originated from optical emission and electron emission studies under steady-state conditions, new experiments in this field advanced photoselective excitation and time-resolved interrogation methods on a time scale of ≈ 100 ps. Further experimental progress in time-resolved studies will undoubtedly unveil the nature of a variety of interesting non-radiative relaxation phenomena involving exciton selftrapping, vibrational relaxation, electronic relaxation and energy transfer in these materials.

It has often been stated that the conceptually simple intermolecular interactions in RGS render them to be useful prototypes for the description of insulators and molecular crystals. These materials will also serve as prototypes for the understanding of radiationless processes. The description of relaxation processes in excited states is relatively simple, in view of the simplicity of the phonon states in these materials, where interconversion of electronic energy into vibrational energy involves only acoustic phonons. However, the achievements of the theory of electronic relaxation in insulators are still fragmentary and phenomenological. Further experimental and theoretical progress is required to establish models and concepts for a more complete and unified description of excited-state dynamics in insulating materials.

The present survey of electronic excitations in condensed rare gases illustrates the interplay between molecular and solid-state concepts. Solid-state theoretical concepts were extensively utilized for the description of band structure and electronic excitations in pure RGS as well as in alloys and liquids. Nevertheless, even in that domain a tight-binding type approach resting on the "molecular" properties of the constituents proved a useful starting point. The analysis of medium-relaxed excited states requires the molecular approach both for the self-trapped excitons in the pure materials as well as excited impurity states in doped solid and liquid rare gases. The experimental and theoretical studies of rare-gas clusters may lead to a merging between the molecular and solid state points of view. Studies of molecular clusters in supersonic beams would extend this interesting research area to finite systems.

References

Abeles, B., Ping Sheng, M.D. Coutts, Y. Arie /1975/: Adv. Phys. *24*, 407
Abeles, B., H.L. Pinch, J.I. Gittleman /1975/: Phys. Rev. Lett. *35*, 247
Abouaf, R., B.A. Huber, P.C. Cosby, R.P. Saxon, J.T. Moseley /1978/: J. Chem. Phys. *68*, 2406
Ackermann, Ch. /1976/: Thesis, Universität Hamburg and DESY Internal Report F41 - 76/04
Ackermann, Ch., R. Brodmann, G. Tolkiehn, G. Zimmerer, R. Haensel, U. Hahn /1976a/: J. Luminescence *12/13*, 315
Ackermann, Ch., R. Brodmann, U. Hahn, A. Suzuki, G. Zimmerer /1976b/: phys. stat. sol. (b) *74*, 579
Alig, R.C., S. Bloom /1975/: Phys. Rev. Lett. *35*, 1522
Altarelli, M., W. Andreoni, F. Bassani /1975/: Solid State Comm. *16*, 143
Ammeter, J.H., D.C. Schlosnagle /1973/: J. Chem. Phys. *59*, 4784
Anderson, P.W. /1958/: Phys. Rev. *109*, 1492
Andreoni, W., M. Altarelli, F. Bassani /1975/: Phys. Rev. *B11*, 2352
Andreoni, W., F. Perrot, F. Bassani /1976/: Phys. Rev. *B14*, 3589
Andreoni, W., M. De Crescenzi, E. Tosatti /1978/: Solid State Comm. *26*, 425
Andrews, L., G.C. Pimentel /1967/: J. Chem. Phys. *47*, 2905
Antoniewicz, P.R. /1977/: Phys. Rev. Lett. *38*, 374
Armstrong, S., R. Grinter, J. McCombie /1981/: J. Chem. Soc. Faraday Trans. 2 *77*, 123
Arrighini, C.P., F. Biondi, C. Guidotte /1974/: Phys. Lett. *48*, 385
Asaf, U., I.T. Steinberger /1971/: Phys. Lett. *34A*, 207
Asaf, U., I.T. Steinberger /1974/: Phys. Rev. *B10*, 4464
Atzmon, R., O. Cheshnovsky, B. Raz, J. Jortner /1974/: Chem. Phys. Lett. *29*, 310
Auzel, F. /1978/: in *Luminescence of Inorganic Solids*, ed. by DiBartolo (Plenum Press, New York) p. 67
Avci, R., C.P. Flynn /1976/: Phys. Rev. Lett. *37*, 864
Avci, R., C.P. Flynn /1978/: Phys. Rev. Lett. *41*. 428

Bader, G., G. Perluzzo, L.G. Caron, L. Sanche /1982/: Abstracts of the "Tenth Molecular Crystal Symposium", Quebec, Canada p.14
Bakale, G., U. Sowada, W.F. Schmidt /1976/: J. Phys. Chem. *80*, 2556
Baldini, G. /1962/: Phys. Rev. *128*, 1562
Baldini, G., R.S. Knox /1963/: Phys. Rev. Lett. *11*, 127
Baldini, G. /1965/: Phys. Rev. *137A*, 508
Balling, L.C., M.D. Havey, J.E. Dawson /1978/: J. Chem. Phys. *69*, 1670
Baraff, G.A. /1964/: Phys. Rev. *135A*, 528
Barker, I.A. /1976/: in *Rare Gas Solids*, ed. by M.K. Klein and J.A. Venables, Vol. I (Academic Press, London) p. 212
Barnes, A.J., W.J. Orville-Thomas, A. Müller, R. Gaufrès (eds.) /1981/: *Matrix Isolation Spectroscopy* (Reidel Publishing Company, Dordrecht)
Baroni, S., G. Grosso, L. Martinelli, G. Pastori Parravicini /1979/: Phys. Rev. *B20*, 1713

Baroni, S., G. Grosso, G. Pastori-Parravicini /1980/:   Phys. Rev. *B22*, 6440
Baroni, S., G. Grosso, G. Pastori-Parravicini /1981/:   Phys. Rev. *B23*, 6441
Basak, S., M.H. Cohen /1979/:   Phys. Rev. *B20*, 3404
Basov, N.G., O.V. Bogdankevich, V.A. Danilychev, A.G. Devyatkov, G.N. Kashnikov,
     N.P. Lantsov /1968/:   JETP Lett. *7*, 317
Basov, N.G., E.M. Balashov, O.V. Bogdankevich, V.A. Danilychev, G.N. Kashnikov,
     N.P. Lantzov, D.D. Khodkevitch /1970a/:   J. Luminescence *1/2*, 834
Basov, N.G., V.A. Danilychev, Yu.M. Popov, D.D. Khodkevich /1970b/:   JETP Lett.
     12, 329
Basov, N.G., V.A. Danilychev, A.G. Molchanov, Yu.M. Popov, D.D. Khodkevich /1973/:
     Isv. Ak. Nauk. SSR, Ser. Fiz. *37*, 494
Battye, F.L., J.G. Jenkin, J. Liesegang, R.C.G. Lechey /1974/:   Phys. Rev. *B9*, 2887
Battye, F.L., J. Liesegang, R.C.G. Lechey /1976/:   Phys. Rev. *B13*, 2646
Beaglehole, D. /1965/:   Phys. Rev. Lett. *15*, 551
Belov, A.G. I.Ya. Fugol, E.V. Savchenko /1973/:   Solid State Comm. *12*, 1
Belyaeva, A.A., Y.B. Predtechenskii, L.D. Shcherpa /1969/:   Bull. Acad. Sci. USSR,
     Phys. Ser. *33*, 825
Berglund, C.N., W.E. Spicer /1964/:   Phys. Rev. *136*, A 1030; A 1044
Bernstorff, S., P. Laporte, R. Reininger, V. Saile, I.T. Steinberger, J.L. Subtil
     /1983/:   Annales of the Israel Physical Society, Vol. *6*, p. 270, and to be
     published
Berreman, D.W. /1967/:   Phys. Rev. *163*, 855
Bethe, H. /1930/:   Ann. Phys. (5) *5*, 525
Blair, M., D. Pooley, D. Smith /1972/:   J. Phys. C *5*, 1537
Boehmer, W., R. Haensel, N. Schwentner, E. Boursey /1980/:   Chem. Phys. *49*, 225
Bohr, N. /1913/:   Phil. Mag. *25*, 10
Bohr, N. /1915/:   Phil. Mag. *30*, 581
Bonifield, T.D., F.H.K. Rambow, G.K. Walters, M.V. McCuskers, D.C. Lorents,
     R.A. Gutcheck /1980/:   Chem. Phys. Lett. *69*, 290 and J. Chem. Phys. *72*, 2914
Bonnot, A., A.M. Bonnot, F. Coletti, J.M. Debever, J. Hanus /1974/:   J. Phys.
     (France) C3 *35*, 49
Boschi, R.A., D.R. Salahub /1972/:   Mol. Phys. *24*, 289
Böttcher, E.H., W.F. Schmidt /1984/:   phys. stat. sol. (b) *126*, K165
Boursey, E., J.Y. Roncin, M. Damaney /1970/:   Phys. Rev. Lett. *25*, 1279
Boursey, E., M.-C. Castex, V. Chandrasekharan /1977/:   Phys. Rev. *B16*, 2858
Breithaupt, B., J.E. Hulse, D.M. Kolb, H.H. Rotermund, W. Schröder, W. Schritten-
     lacher /1983/:   Chem. Phys. Lett. *95*, 513
Briant, C.L., J.J. Burton /1975/:   J. Chem. Phys. *63*, 2045
Brodmann, R., R. Haensel, U. Hahn, U. Nielsen, G. Zimmerer /1974a/:   Chem. Phys.
     Lett. *29*, 250
Brodmann, R., R. Haensel, U. Hahn, U. Nielsen, G. Zimmerer /1974b/:   in *Vacuum
     Ultraviolet Radiation Physics*, ed. by E.E. Koch, R. Haensel, and C. Kunz
     (Vieweg-Pergamon, Braunschweig) p. 344
Brodmann, R. /1976/:   Thesis, Universität Hamburg and DESY Internal Report
     F41 - 77/02
Brodmann, R., G. Tolkiehn, G. Zimmerer /1976/:   phys. stat. sol. (b) *73*, K99
Brodmann, R., G. Zimmerer /1977/:   J. Phys. B: Atom. Molec. Phys. *10*, 3395
Brodmann, R., G. Zimmerer /1978/:   Chem. Phys. Lett. *56*, 434
Brown, D.B. /1974/:   in *Handbook of Spectroscopy*, ed. by J.W. Robinson, Vol. I
     (CRC Press, Cleveland) p. 248 ff.
Børgesen, P., J. Schou, H. Sørensen, C. Classen /1982/:   Appl. Phys. *A29*, 57

Calvani, P., B. Maraviglia, C. Messana /1972/:   Phys. Lett. *39A*, 123
Calvani, P., C. De Simone, B. Maraviglia /1973/:   Phys. Lett. *44A*, 5
Cardona, M., L. Ley (eds.) /1978,1979/:   *Photoemission in Solids, Vol. I and Vol. II*,
     Topics in Applied Physics, Vol. 26, 27 (Springer, Berlin, Heidelberg, New York)
Carvalho, M.J., G. Klein /1978/:   J. Luminescence *18/19*, 487
Castex, M.C. /1974/:   Chem. Phys. *5*, 448
Castex, M.C. /1977/:   in Extended Abstracts, 5th Intern. Conf. on VUV Radiation
     Physics, Montpellier, Vol. I, p. 123
Cate, R.C., J.G. Wright, N.E. Cusack /1970/:   Phys. Lett. *32A*, 467

Chance, R.R., A. Prock, R. Silvey /1976/: J. Chem. Phys. *65*, 2527
Chandrasekharan, V., E. Boursey /1978/: Ber. Bunsenges. Phys. Chem. *82*, 49
Chandrasekharan, V., E. Boursey /1979/: Phys. Rev. B *19*, 3299
Cheshnovsky, O., B. Raz, J. Jortner /1972a/: Chem. Phys. Lett. *15*, 475
Cheshnovsky, O., B. Raz, J. Jortner /1972b/: J. Chem. Phys. *57*, 4628
Cheshnovsky, O., B. Raz, J. Jortner /1973a/: J. Chem. Phys. *59*, 5554
Cheshnovsky, O., A. Gedanken, B. Raz, J. Jortner /1973b/: Chem. Phys. Lett. *22*, 23
Cheshnovsky, O., A. Gedanken, B. Raz, J. Jortner /1973c/: Solid State Comm. *13*, 639
Cheshnovsky, O., U. Even, J. Jortner /1977/: Solid State Comm. *22*, 745
Cheshnovsky, O., U. Even, J. Jortner /1979/: Phys. Lett. *71A*, 255
Cheshnovsky, O., U. Even, J. Jortenr /1982/: Phys. Rev. *B25*, 3350
Chiang, T.C., D.E. Eastman, F.J. Himpsel, G. Kaindl, M. Aono /1980a/: Phys. Rev. Lett. *45*, 1846
Chiang, T.C., G. Kaindl, D.E. Eastman /1980b/: Solid State Comm. *36*, 25
Chiang, T.C., G. Kaindl, D.E. Eastman /1982/: Solid State Comm. *41*, 661
Cho, K., Y. Toyozawa /1969/: J. Chem. Soc. Japan *26*, 71
Cho, K., Y. Toyozawa /1971/: J. Chem. Soc. Japan *30*, 1555
Cohen, J.S., B. Schneider /1974/: J. Chem. Phys. *61*, 3230
Cohen, M.H., J. Lekner /1967/: Phys. Rev. *158*, 305
Cohen, M.H., H. Fritzche, S.R. Ovshinsky /1969/: Phys. Rev. Lett. *22*, 1065
Cohen, M.H., J. Jortner /1973/: J. de Physique *35*, C4-345
Cohen, M.H., J. Jortner /1974/: Phys. Rev. A *10*, 978
Cohen, M.H., J. Jortner, I. Webman /1978/: in "1th Conference on the Electrical Transport and Optical Properties of Inhomogeneous Media" ed. by J.C. Garland and D.B. Tanner (Am. Inst. Phys., New York) Conf. Proc. No. 40, 63
Cohen, M.H., J. Jortner, I. Webman /1978/: Phys. Rev. B *17*, 4555
Cohen, R.W., G.D. Cody, M.D. Coutts, B. Abeles /1973/: Phys. Rev. *B8*, 3689
Cole, M.W. /1974/: Review of Modern Physics *46*, 451
Coletti, F., J. Hanus /1977/: Extended Abstracts, 5th Intern. Conf. on VUV Radiation Physics, Montpellier, Vol. I, p. 78
Coletti, F., A.M. Bonnot /1978/: Chem. Phys. Lett. *55*, 92
Coletti, F., J.M. Debever, G. Zimmerer /1984/: J. Physique Lett. *45*, L-467
Coletti, F., J.M. Debever, G. Zimmerer /1985/: to be published
Conwell, E.M. /1967/: in *Solid State Physics*, ed. by F. Seitz, Suppl. Vol. 9 (Adademic Press, New York)
Cook, G.A. /1961/: *Argon, Helium and the Rare Gases* (Interscience, New York)
Coufal, H.J., U. Nagel, M. Bürger, E. Lüscher /1974/: Phys. Lett. *A47*, 327
Coufal, H.J., U. Nagel, E. Lüscher /1978/: Ber. Bunsenges. Phys. Chem. *82*, 133
Coulson, C.A. /1951/: *Valence* (Oxford University Press, Oxford)
Creuzburg, M., K. Teegarden /1968/: Phys. Rev. Lett. *20*, 593
Creuzburg, M. /1971/: Solid State Comm. *9*, 665
Creuzburg, M., G. Völkl /1977/: Extended Abstracts, 5th Intern. Conf. on VUV Radiation Physics, Montpellier, Vol. I, p. 74

Dagens, L., F. Perrot /1972/: Phys. Rev. *B5*, 641
Danilychev, V.A., G.N. Kashnikov, Yu.M. Popov /1970/: Preprint No. 136, Lebedev Institute, Moscow
Danor, R., O. Cheshnovsky, Y. Even, J. Jortner /1979/: Phil. Mag. *B39*, 99
Das, T.P., A.N. Jette, R.S. Knox /1964/: Phys. Rev. *134*, A 1079
Dash, J.G. /1975/: *Films on solid surfaces* (Academic Press, New York)
Davis, H.T., S.A. Rice, L. Meyer /1962/: Phys. Rev. Lett. *9*, 81
Davydov, A.S. /1971/: *Theory of Molecular Excitons* (Plenum Press, New York)
Debever, J.M., A. Bonnot, A.M. Bonnot, F. Coletti, J. Hanus /1974/: Solid State Comm. *14*, 989
Dehmer, P.M., J.L. Dehmer /1978a/: J. Chem. Phys. *68*, 342
Dehmer, P.M., J.L. Dehmer /1978b/: J. Chem. Phys. *69*, 125
Dehmer, P.M., S.T. Pratt /1982/: J. Chem. Phys. *76*, 843
Dennis, W.S., E. Durbin, W.A. Fitzsimmons, O. Heybey, G.K. Walters /1969/: Phys. Rev. Lett. *23*, 1083
Devreese, J.T., A.B. Kunz, T.C. Collins /1972/: Solid State Comm. *11*, 673

Dexter, D.L. /1953/: J. Chem. Phys. *21*, 836
Dexter, D.L. /1958/: Solid State Phys. *6*, 353
Dobbs, E.R., G.O. Jones /1957/: Rep. Prog. Phys. *20*, 516
Dössel, O., H. Nahme, R. Haensel, N. Schwentner /1983/: J. Chem. Phys. *79*, 665
Doke, T., A. Hitachi, S. Kubota, A. Nakamoto, T. Takahashi /1976/: Nucl. Instr.
   Methods *134*, 353
Dressler, K. /1970/: Memories de la Societé Royale des Sciences de Liège, Vol. *20*,
   357
Druger, S.D., R.S. Knox /1969/: J. Chem. Phys. *50*, 3143
Dubost, H., R. Charneau /1976/: Chem. Phys. *12*, 407
Dunning, F.B., R.D. Rundel, R.F. Stebbing /1975/: Rev. Sci. Instrum. *46*, 697

Eatah, A.I., N.E. Cusack, J.G. Wright /1975/: Phys. Lett. *51A*, 149
Economou, E.N., M.H. Cohen, K.F. Freed, E.S. Kirkpatrick /1974/: in *Amorphous and
   Liquid Semiconductors*, ed. by J. Tauc (Plenum Press, London) p. 101
Edwards, S. /1965/: Proc. Phys. Soc. *85*, 1
Endo, H., A.I. Eatah, J.G. Wright, N.E. Cusack /1973/: J. Phys. Soc. Japan *34*, 666
Englman, R., J. Jortner /1970/: Mol. Phys. *18*, 145
Englman, R. /1972/: *The Jahn Teller Effect in Molecules and Molecular Crystals*
   (Wiley, New York)
Ermler, W.C., Y.S. Lee, K.S. Pitzer, N.W. Winter /1978/: J. Chem. Phys. *69*, 976
Even, U., J. Jortner /1972/: Phil. Mag. *25*, 715

Fano, U. /1940/: Phys. Rev. *58*, 544
Fano, U. /1947/: Phys. Rev. *72*, 26
Farrell, H.H., M. Strougin, J.M. Dickey /1972a/: Phys. Rev. *B6*, 4703
Farrell, H.H., M. Strougin /1972b/: Phys. Rev. *B6*, 4711
Fermi, E. /1939/: Phys. Rev. *56*, 1242
Filinski, I. /1972/: phys. stat. sol. (b) *49*, 577
Fink, R.W., P. Venugopala /1974/: in *Handbook of Spectroscopy*, Vol. I, ed. by
   J.W. Robinson (CRC Press, Cleveland) p. 219
Fischbach, J.U., D. Fröhlich, M.N. Kabler /1973/: J. Luminescence *6*, 29
Fischbach, M.R., H.A. Roberts, F.L. Hereford /1969/: Phys. Rev. Lett. *23*, 462
Fischer, S., S.A. Rice /1968/: Phys. Rev. *176*, 409
Fitzsimmons, W.A. /1983/: in *Atomic Physics 3*, ed. by J. Smith and K. Walters
   (Plenum, New York) p. 477
Flynn, C.P. /1976/: Phys. Rev. *B14*, 5294
Förster, Th. /1948/: Ann. Phys. *2*, 55
Forstmann, F., D.M. Kolb, D. Leutloff, W. Schulze /1977/: J. Chem. Phys. *66*, 2806
Forstmann, F., D.M. Kolb /1978/: Ber. Bunsenges. Phys. Chem. *82*, 30
Forstmann, F., S. Ossicini /1980/: J. Chem. Phys. *73*, 5997
Fowler, W.B. /1963/: Phys. Rev. *132*, 1591
Fowler, W.B. (ed.) /1968/: *Physics of Color Centers* (Academic Press, London)
Frenkel, J. /1931a/: Phys. Rev. *37*, 17
Frenkel, J. /1931b/: Phys. Rev. *37*, 1276
Frenkel, J. /1936/: Physik. Z. Sowjetunion *9*, 158
Friedman, L.R., D.P. Tunstall (eds.) /1978/: *The Metal–Non-Metal Transition in
   Disordered Systems*, Proceedings of the 19th Scottish Universities Summer School
   in Physics (SUSSP Publication, Edinburgh)
Fugol, I.Ya., E.V. Savchenko, A.G. Belov /1972/: ZhETF Pis. Red. *16*, 245
Fugol, I.Ya., A.G. Belov, J.V. Savchenko, Yu.B. Poltoratskii /1974/: Solid State
   Comm. *15*, 525
Fugol, I.Ya., A.G. Belov /1975/: Solid State Comm. *17*, 1125
Fugol, I.Ya., A.G. Belov, E.V. Savchenko, Yu.B. Poltoratski /1975/: Fiz.nizh.temp.
   USSR *1,2*, 203, Sov. J. Low Temp. Phys. *1,2*, 98
Fugol, I.Ya., A.G. Belov, Yu.B. Poltoratski, E.V. Savchenko /1976/: Fiz.nizh.temp.
   USSR *2*, 400
Fugol, I.Ya., E.I. Tarasova /1977/: Fiz.nizh.temp. USSR *3*, 366
Fugol, I.Ya. /1978/: Advances in Physics *27*, 1
Fugol, I.Ya., O.N. Grigorashchenko, E.V. Savchenko /1982/: phys. stat. sol. (b)
   111, 397

Gaethke, R., P. Gürtler, R. Kink, E. Roick, G. Zimmerer /1984/: phys. stat. sol. (b)
    124, 335
Gedanken, A., B. Raz, J. Jortner /1972a/: Chem. Phys. Lett. *14*, 172
Gedanken, A., B. Raz, J. Jortner /1972b/: Chem. Phys. Lett. *14*, 326
Gedanken, A., B. Raz, J. Jortner /1973a/: J. Chem. Phys. *58*, 1178
Gedanken, A., B. Raz, J. Jortner /1973b/: J. Chem. Phys. *59*, 1630
Gedanken, A., B. Raz, J. Jortner /1973c/: J. Chem. Phys. *59*, 2752
Gedanken, A., B. Raz, J. Jortner /1973d/: J. Chem. Phys. *59*, 5471
Gedanken, A., Z. Karsch, B. Raz, J. Jortner /1973e/: Chem. Phys. Lett. *20*, 163
Gerick, U. /1977/: Diplomarbeit, Universität Hamburg
Gilbert, T.L., A.C. Wahl /1971/: J. Chem. Phys. *55*, 5247
Gillen, K.T., R.P. Saxon, D.C. Lorentz, G.E. Ice, R.E. Olson /1976/: J. Chem. Phys.
    64, 1925
Gillis, N.S., N.R. Werthamer, T.R. Koehler /1968/: Phys. Rev. *165*, 951
Ginter, M.L., R. Battino /1970/: J. Chem. Phys. *52*, 4469
Gittleman, J.I., B. Abeles /1977/: Phys. Rev. *B15*, 3273
Gleason, R.E., T.D. Bonfield, J.W. Keto, G.K. Walters /1977/: J. Chem. Phys. *66*,
    1589
Gold, A. /1961/: J. Phys. Chem. Solids *18*, 218
Granier, R., M.-C. Castex, J. Granier, M.J. Romand /1967/: Compt. Rend. Acad. Sci.
    (Paris) *264B*, 778
Granier, R. /1969/: Ann. Phys. *4*, 383
Grobman, W.D. /1975/: Comments on Solid State Phys. *7*, 27
Grosso, G., L. Martinelli, G. Pastori-Parravicini /1978a/: Solid State Comm. *25*,
    435
Grosso, G., L. Martinelli, G. Pastori-Parravicini /1978b/: Solid State Comm. *25*,
    835
Grosso, G., L. Martinelli, G. Pastori-Parravicini /1982/: Solid State Comm. *44*, 1317
Gruen, D.M., S.L. Gaudioso, R.L. McBeth, J.L. Lerner /1974/: J. Chem. Phys. *60*, 89
Gruen, D.M. /1976/: *Cryochemistry*, ed. by M. Moskovitz and G.A. Ozin (Wiley, New
    York) Chap. 10
Gruzdey, P.F., A.V. Longinov /1975/: Opt. Spectrosc. *38*, 611
Grubermann, S.L., W.A. Goddard /1975/: Phys. Rev. *A12*, 1203
Gürtler, P., E.E. Koch /1980/: J. Molec. Structure *60*, 259
Gürtler, P., E. Roick, G. Zimmerer, M. Pouey /1983/: Nucl. Instr. and Methods in
    Phys. Res. *208*, 835
Guse, M.P., A.B. Kunz /1975/: phys. stat. sol. (b) *71*, 631

Haensel, R., C. Kunz /1967/: Z. Angew. Physik *23*, 276
Haensel, R., G. Keitel, P. Schreiber, C. Kunz /1969a/: Phys. Rev. Lett. *22*, 398
Haensel, R., G. Keitel, P. Schreiber, C. Kunz /1969b/: Phys. Rev. *188*, 1375
Haensel, R., G. Keitel, E.E. Koch, M. Skibowski, P. Schreiber /1969c/: Phys. Rev.
    Lett. *23*, 116
Haensel, R., G. Keitel, C. Kunz, P. Schreiber, B. Sonntag /1970a/: Mem. Soc. Roy.
    Sci. Lg. 5 série 20, 169
Haensel, R., G. Keitel, C. Kunz, P. Schreiber /1970b/: Phys. Rev. Lett. *25*, 208
Haensel, R., G. Keitel, E.E. Koch, M. Skibowski, P. Schreiber /1970c/: Opt. Comm.
    2, 59
Haensel, R., G. Keitel, E.E. Koch, N. Kosuch, M. Skibowski /1970d/: Phys. Rev.
    Lett. *25*, 1281
Haensel, R., G. Keitel, N. Kosuch, U. Nielsen, P. Schreiber /1971/: J. Phys.
    (France) *32*, C4 - 236
Haensel, R., N. Kosuch, U. Nielsen, B. Sonntag, U. Rössler /1973/: Phys. Rev. *B7*,
    1577
Hagena, O.F., W. Obert /1972/: J. Chem. Phys. *56*, 1793
Hahn, U., N. Schwentner, G. Zimmerer /1977/: Opt. Comm. *21*, 237
Hahn, U. /1978/: Thesis, Universität Hamburg
Hahn, U., N. Schwentner, G. Zimmerer /1978/: Nucl. Instr. Methods *152*, 261
Hahn, U., N. Schwentner /1979/: J. Luminescence *18/19*, 23
Hahn, U., N. Schwentner /1980/: Chem. Phys. *48*, 53
Hahn, U., R. Haensel, N. Schwentner /1982/: phys. stat. sol. (b) *109*, 233
Halpern, B., R. Gomer /1965/: J. Chem. Phys. *43*, 1069

Hansen, J.P., E.L. Pollock /1972/: Phys. Rev. *A5*, 2214

Hanus, J., F. Coletti, A.M. Bonnot, J.M. Debever /1974/: *Vacuum Ultraviolet Radiation Physics*, ed. by E.E. Koch, R. Haensel and C. Kunz (Vieweg, Pergamon, Braunschweig, New York) p. 341

Harmsen, A., E.E. Koch, V. Saile, N. Schwentner, M. Skibowski /1974/: *Vacuum Ultraviolet Radiation Physics*, ed. by E.E. Koch, R. Haensel and C. Kunz (Vieweg, Pergamon, Braunschweig, New York) p. 339

Harmsen, A. /1975/: Diplomarbeit, Universität Hamburg

Hasnain, S.S., T.D.S. Hamilton, I.H. Munro, E. Pantos, I.T. Steinberger /1977a/: Phil. Mag. *35*, 1299

Hasnain, S.S., P. Brint, T.D.S. Hamilton, I.H. Munro /1977b/: Phil. Mag. *36*, 629

Hasnain, S.S., T.D.S. Hamilton, I.H. Munro /1977c/: Nuovo Cimento *39*, 500

Hasnain, S.S., I.H. Munro, T.D.S. Hamilton /1977d/: J. Phys. C: Solid State Phys. *10*, 1097

Hasnain, S.S., T.D.S. Hamilton, I.H. Munro /1978a/: J. Phys. *C11*, L 261

Hasnain, S.S., P. Brint, T.D.S. Hamilton, I.H. Munro /1978b/: Chem. Phys. Lett. *56*, 134

Hasnain, S.S., P. Brint, T.D.S. Hamilton, I.H. Munro /1978c/: J. Molec. Spectroscopy

Hasnain, S.S., T.D.S. Hamilton, I.H. Munro, E. Pantos /1978d/: J. Molec. Spectroscopy

Hasnain, S.S., T.D.S. Hamilton, I.H. Munro, P. Brint /1978e/: J. Luminescence *18/19*, 429

Herbst, J.F. /1977/: Phys. Rev. *B15*, 3720

Hermann, K., J. Noffke, K. Horn /1980/: Phys. Rev. *B22*, 1022

Hermanson, J. /1966/: Phys. Rev. *150*, 660

Hermanson, J., J.C. Phillips /1967/: Phys. Rev. *150*, 652

Herzberg, G. /1950/: *Molecular Spectra and Molecular Structure*, Vol. I, II, III (Van Nostrand Reinhold Company, New York)

Heumüller, R., M. Creuzburg /1978/: Optics Comm. *26*, 363

Heumüller, R. /1978/: Thesis, Universität Regensburg

Hickmann, A.P., N.F. Lane /1971/: Phys. Rev. Lett. *26*, 1216

Hickmann, A.P., W. Steets, N.F. Lane /1975/: Phys. Rev. *B12*, 3705

Hilder, G., N.F. Cusack /1977/: Phys. Lett. *62A*, 163

Hill, J.L., O. Heybey, G.K. Walter /1971/: Phys. Rev. Lett. *26*, 1213

Himpsel, F.J. /1983/: Advances in Physics *32*, 1

Hingsammer, J., E. Lüscher /1968/: Helvetia Physica Acta *41*, 914

Holstein, T. /1959/: Ann. Phys. (N.Y.) *8*, 343

Hormes, J., B. Karrasch /1982/: Chem. Phys. *70*, 29

Hormes, J., R. Grinter, B. Breithaupt, D.M. Kolb /1983/: J. Chem. Phys. *78*, 158

Horn, K., M. Scheffler, A.M. Bradshaw /1978/: Phys. Rev. Lett. *41*, 822

Horn, K., A.M. Bradshaw /1979/: Solid State Comm. *30*, 545

Horn, K., C. Mariani, L. Cramer /1982/: Surface Science *117*, 376

Hornbeck, J.A., J.P. Molnar /1951/: Phys. Rev. *84*, 621

Horton, G.K. /1968/: Am. J. Phys. *36*, 93

Hoshen, J., J. Jortner /1972/: J. Chem. Phys. *56*, 933, 4138, 5550

Howe, S., R.G. LeComber, W.E. Spear /1968/: unpublished

Hsu, Y.P., P.M. Johnson /1973/: J. Chem. Phys. *59*, 136

Huang, K., A. Rhys /1950/: Proc. Roy. Soc. *A204*, 406

Huang, S.S.S., G.R. Freeman /1978/: J. Chem. Phys. *68*, 1355

Huang, S.S.S., G.R. Freeman /1981/: Phys. Rev. *A24*, 714

Huber, E.E., D.A. Emmons, R.M. Lerner /1974/: Opt. Comm. *11*, 155

Hulse, J., J. Küppers, K. Wandelt, G. Ertl /1980/: Applications of Surface Science *6*, 453.

Hunderi, O., R. Ryberg /1975/: Phys. Lett. *51A*, 167

Jacobi, K., D. Schmeisser, D.M. Kolb /1980/: Chem. Phys. Lett. *69*, 113

Jacobi, K., Y. Hsu, H.M. Rotermund /1982/: Surface Science *114*, 683

Jahnke, J.A., N.A.W. Holzwarth, S.A. Rice /1972/: Phys. Rev. *A5*, 463

Jansen, L. /1964/: Phys. Rev. *135*, A1292

Jen, C.K., V.A. Bowers, E.L. Cochran, S.N. Foner /1962/: Phys. Rev. *126*, 1749

Johnson, R.E., M. Inokuti /1983/: Nucl. Instr. and Methods *206*, 289

Jordan, B. /1978/: diploma work, Universität Hamburg

Jortner, J., J.L. Katz, S.I. Choi, S.A. Rice /1964/: J. Chem. Phys. *42*, 309

Jortner, J., L. Meyer, S.A. Rice, E.G. Wilson /1965/: J. Chem. Phys. *42*, 4250

Jortner, J. /1968/: Phys. Rev. Lett. *20*, 244

Jortner, J. /1974/: *Vacuum Ultraviolet Radiation Physics*, ed. by E.E. Koch,
    R. Haensel and C. Kunz (Vieweg Pergamon, Braunschweig, New York) p. 263

Jortner, J. /1976/: Mol. Phys. *32*, 379

Jortner, J., M.H. Cohen /1976/: Phys. Rev. B *13*, 1548

Jortner, J., A. Gaathon /1977/: Can. J. Chem. *55*, 1801

Jortner, J., S. Leach /1980/: J. de Chimie Physique *77* (1), 7

Jortner, J., E.E. Koch, N. Schwentner /1984/: in *Photophysics and Photochemistry
    in the Vacuum Ultraviolet*, ed. by S. McGlynn, G. Findley and R. Huebner (Reidel
    Publ. Co. Dordrecht) p. 515

Kaindl, G., T.-C. Chiang, D.E. Eastman, F.J. Himpsel /1980a/: Phys. Rev. Lett. *45*,
    1808

Kaindl, G., T.-C. Chiang, D.E. Eastman, F.J. Himpsel /1980b/: in *Ordering in Two
    Dimensions*, ed. by S.K. Sinha (North Holland, Amsterdam) p. 99

Kane, E.O. /1966/: Phys. Rev. *147*, 335

Kane, E.O. /1967/: Phys. Rev. *159*, 624

Kashnikov, G.N. /1971/: Thesis, Lebedev Institute, Moskow

Katz, B., M. Brith, B. Sharf, J. Jortner /1969/: J. Chem. Phys. *50*, 5195

Kepler, R.G. /1976/: *Treatise on Solid State Chemistry*, Vol. 3, ed. by N.B. Hannay
    (Plenum Publishing Corp., New York) p. 615

Kessler, T., R. Markus, H._Nahme, N. Schwentner /1985a/: submitted to phys. stat.
    sol.

Kessler, T., H. Nahme, N. Schwentner /1985b/: accepted Optics Comm.

Kessler, T., H. Nahme, N. Schwentner /1985/: to be published

Keto, J.W., M. Stockton, W.A. Fitzsimmons /1972/: Phys. Rev. Lett. *28*, 792

Keto, J.W., R.E. Gleason, G.K. Walters /1974a/: Phys. Rev. Lett. *33*, 1365

Keto, J.W., F. Soley, H. Stockton, W.A. Fitzsimmons /1974b/: Phys. Rev. *10*, 872,
    887

Keto, J.W., R.E. Gleason, T.D. Bonfield, G.K. Walters, F.K. Soley /1976/: Chem.
    Phys. Lett. *42*, 125

Keto, J.W., R.E. Gleason, F.K. Soley /1979/: J. Chem. Phys. *71*, 2676

Kimura, T., G.R. Freeman /1974/: J. Chem. Phys. *60*, 4081

Kingsley, J.D., J.S. Prener /1972/: J. Appl. Phys. *43*, 3073

Kink, R., A. Lohmus, M. Selg, T. Soovik /1977/: phys. stat. sol. (b) *84*, K61

Kink, R., M. Selg /1979/: phys. stat. sol. (b) *96*, 101

Kink, R., A. Lohmus, M. Selg /1981/: phys. stat. sol. (b) *107*, 479

Kirkpatrick, S. /1971/: Phys. Rev. *27*, 1722

Kittel, C. /1971/: *Introduction to Solid State Physics* (Wiley, New York)

Klaberer, J.B., R.D. Etters /1977/: J. Chem. Phys. *66*, 3233

Klein, C.A. /1968/: J. Appl. Phys. *39*, 2029

Klein, M.K., J.A. Venables (eds.) /1976,1977/: *Rare Gas Solids*, Vol. I and Vol. II
    (Academic Press, London, New York)

Knox, R.S. /1959/: Phys. Chem. Solids *9*, 265

Knox, R.S., F. Bassani /1961/: Phys. Rev. *124*, 652

Knox, R.S. /1963/: *Theory of Excitons* (Academic Press, New York)

Koch, E.E., B. Raz, V. Saile, N. Schwentner, M. Skibowski, W. Steinmann /1974a/:
    Japan. J. Appl. Phys. Suppl. 2, Pt 2, 775

Koch, E.E., V. Saile, N. Schwentner, M. Skibowski /1974b/: Chem. Phys. Lett. *28*,
    562

Koch, E.E., R. Nürnberger, N. Schwentner /1978/: Ber. Bunsenges. Phys. Chem.
    82, 110

Koch, E.E. (ed.) /1983/: *Handbook on Synchrotron Radiation*, Vol. 1a,b (North
    Holland Publ. Company, Amsterdam)

Kolb, D.M. /1981/: in *Matrix Isolation Spectroscopy*, ed. by A.J. Barnes, W.J.
    Orville-Thomas, A. Müller and R. Gaufrès (Reidel Publ. Company, Dordrecht) p. 447

Kolb, D.M., F. Forstmann /1981/: in *Matrix Isolation Spectroscopy*, ed. by A.J.
    Barnes, W.J. Orville-Thomas, A. Müller and R. Gaufrès (Reidel Publ. Company,
    Dordrecht) p. 347

Kotani, A., Y. Toyozawa /1979/: in *Synchrotron Radiation Techniques and Applications*, ed. by C. Kunz (Springer Verlag, Berlin, Heidelberg, New York) p. 169
Kovalenko, S.I., E.J. Indan, A.A. Khudoteplaya /1972/: phys. stat. sol. (a) *13*, 235
Kovalenko, S.I., E.J. Indan, A.A. Solodovnik, I.N. Krupski /1975/: Soviet. Phys. Low Temp. *1*, 1027
Kramer, B. /1976/: in *Physics of Structurally Disordered Solids*, Proceedings of an advanced NATO Study Institute, ed. by S.S. Mitra (Plenum Press, New York, London) p. 291
Kreitman, M., E.E. Barnett /1965/: J. Chem. Phys. *43*, 364
Kubo, R., Y. Toyozawa /1955/: Prog. Theoret. Phys. *13*, 160
Kubo, R.J. /1962/: J. Phys. Soc. Japan *17*, 975
Kubota, S., A. Nakamoto, T. Tokahashi, S. Konno, T. Hamada, M. Miyajima, A. Hitachi, E. Shibamura, T. Doke /1976/: Phys. Rev. *B13*, 1649
Kubota, S., M. Hishida, J. Raun /1978a/: J. Phys. C, Solid State Phys. *11*, 2645
Kubota, S. A. Nakamoto, T. Tabahashi, T. Hamada, E. Shibamura, M. Miyajima, K. Masuda, T. Doke /1978b/: Phys. Rev. *B17*, 2762
Kubota, S., M. Hishida, K. Gess /1979/: unpublished
Kunsch, P.L., F. Coletti /1979/: J. Chem. Phys. *70*, 726
Kunz, A.B., J.T. Devreese, T.C. Collins /1972/: J. Phys. C: Solid State *5*, 3259
Kunz, A.B., D. Mickish /1973/: Phys. Rev. *B8*, 779
Kunz, A.B., T.C. Collins, D. Esterling, D.C. Licciardello, D.J. Mickish /1975a/: Phys. Rev. *B11*, 3210
Kunz, A.B., D.J. Mickish, S.K.V. Mirmira, T. Shima, F.-J. Himpsel, V. Saile, N. Schwentner, E.E. Koch /1975b/: Solid State Comm. *17*, 761
Kunz, C. (ed.) /1979/: *Synchrotron Radiation, Techniques and Applications* (Springer, Berlin, Heidelberg, New York)
Kupferman, S.L., F.M. Pipkin /1965/: Phys. Rev. *166*, 207
Kusmartsev, F.V., E.I. Rashba /1982/: Czech. J. Phys. *B32*, 54
Kunkel, W.B., M.N. Rosenbluth /1969/: in *Plasmaphysics in Theory and Application*, ed. by W.B. Kunkel (Mc-Graw-Hill, New York) p. 2

Laporte, P., I.T. Steinberger /1977/: Phys. Rev. *A15*, 2538
Leckner, J. /1967/: Phys. Rev. *158*, 130
Leckner, J., B. Halpern, S.A. Rice, R. Gomer /1967/: Phys. Rev. *156*, 351
LeComber, P.G., J.B. Wilson, R.J. Loveland /1976/: Solid State Comm. *18*, 377
Leichner, P.I. /1973/: Phys. Rev. *A 6*, 815
Leung, C.H., L. Emery, K.S. Song /1983/: Phys. Rev. *B28*, 3474
Leutloff, D., D.M. Kolb /1979/: Ber. Bunsenges. Phys. Chem. *83*, 666
Lilly, R.A. /1976/: J. Opt. Soc. Am. *66*, 245
Lindau, I., W.E. Spicer /1974/: J. Electron. Spectrosc. *3*, 409
Lipari, N.O. /1970/: phys. stat. sol. (b) *40*, 691
Lipari, N.O., W.B. Fowler /1970/: Phys. Rev. *B2*, 3354
Lipari, N.O. /1972/: Phys. Rev. *B6*, 4071
Llacer, J., E.L. Garwin /1969/: J. Appl. Phys. *40*, 2766, 2776
Lorentz, D.C., R.E. Olson /1972/: Semiannual Technical Report, No. 1 (Stanford Res. Inst., Menlo Park, Calif.)
Lotz, W. /1967a/: Zeitschrift für Physik *206*, 205
Lotz, W. /1967b/: J. Opt. Soc. Am. *57*, 873
Lotz, W. /1968/: J. Opt. Soc. Am. *58*, 236, 915
Loveland, R.J., P.G. LeComber, W.E, Spear /1972/: Phys. Lett. *39A*, 225
Luchner, K., H. Micklitz /1978/: J. Luminescence *18/19*, 882
Lumb, K. /1978/: *Luminescence Spectroscopy* (Academic Press, New York)

Maecker, H., T. Peters /1954/: Z. Physik *139*, 448
Malzfeld, W., W. Niemann, P. Rabe, N. Schwentner /1983/: in *EXAFS and Near Edge Structure*, ed. by A. Bianconi, L. Incoccia and S. Stipcich (Springer, Berlin, Heidelberg, New York) p. 203
Mandel, T., G. Kaindl, K. Horn, M. Iwan, H.U. Middelmann, C. Mariani /1983/. Solid State Comm. *46*, 713
Mann, B., A. Behrens /1978/: Ber. Bunsenges. Phys. Chem. *82*, 136
Maradudin, A.A. /1966/: Solid State Physics *18*, 273

Mariani, C., K. Horn, A.M. Bradshaw /1982/: Phys. Rev. *B25*, 7798
Markham, J.J. /1959/: Review of Modern Physics *31*, 956
Martin, M., S.A. Rice /1970/: Chem. Phys. Lett. *7*, 94
Martin, M. /1971/: J. Chem. Phys. *54*, 3289
Martin, T.P. /1977/: Phys. Rev. *B15*, 4071
Martin, T.P. /1978/: J. Chem. Phys. *69*, 2036
Martinelli, L., G. Pastori-Parravicini /1977/: J. Phys. C: Solid State Phys. *10*, L687
Mattheis, L.F. /1964/: Phys. Rev. *A 133*, 1399
Matthew, J.A.D., M.G. Devey /1976/: J. Phys. *C 9*, L 413
Matthias, E., R.A. Rosenberg, E.D. Poliakoff, H.G. White, S.T. Lee, D.A. Shirley /1977/: Chem. Phys. Lett. *52*, 239
Maxwell-Garnett, J.C. /1904/: Phil. Trans. Roy. Soc. London *203*, 385
McCarty, M., G.W. Robinson /1959/: Molecular Physics *2*, 415
McNeal, N.A., A.M. Goldman /1977/: Phys. Lett. *61A*, 268
Messing, I., B. Raz, J. Jortner /1977a/: Chem. Phys. *23*, 23
Messing, I., B. Raz, J. Jortner /1977b/: Chem. Phys. *23*, 351
Messing, I., J. Jortner /1977c/: Chem. Phys. *24*, 183
Messing, I., B. Raz, J. Jortner /1977d/: Chem. Phys. *25*, 55
Messing, I., B. Raz, J. Jortner /1977e/: J. Chem. Phys. *66*, 2239
Messing, I., B. Raz, J. Jortner /1977f/: J. Chem. Phys. *66*, 4577
Meyer, B. /1965/: J. Chem. Phys. *43*, 2986
Meyer, B. /1971/: *Low Temperature Spectroscopy* (Elsevier, New York)
Meyer, B. /1978/: Ber. Bunsenges. Phys. Chem. *82*, 24
Meyer, L. /1969/: Adv. Chem. Phys. *16*, 343
Miller, J.C., B.S. Ault, L. Andrews /1977/: J. Chem. Phys. *67*, 2478
Miller, J.C., R.L. Nowery, E.R. Kransz, S.M. Jacobs, H.W. Kim, P.N. Schatz, L. Andrews /1981/: J. Chem. Phys. *74*, 6349
Miller, L.S., S. Howe, W.E. Spear /1968/: Phys. Rev. *166*, 871
Millet, P., A. Birot, H. Brunet, J. Galy, B. Pons-Germain, J.L. Teyssier /1978/: J. Chem. Phys. *69*, 92
Miranda, R., E.V. Albano, S. Daiser, G. Ertl, K. Wandelt /1983/: Phys. Rev. Lett. *51*, 782
Miron, E., B. Raz, J. Jortner /1972/: J. Chem. Phys. *56*, 5265
Mitchell, R.P., G.W. Rayfield /1971/: Phys. Lett. *37*, 231
Miyakawa, T., D.L. Dexter /1969/: Phys. Rev. *184*, 166
Möller, H. /1976/: Diplomarbeit Universität Hamburg and DESY Internal Report F41 - 76/14
Möller, H., R. Brodmann, G. Zimmerer, U. Hahn /1976/: Solid State Comm. *20*, 401
Molchanov, A.G. /1972/: Fiz. Tverd. Tela *4*, 9
Monahan, K., V. Rehn, E. Matthias, E. Poliakoff /1977/: J. Chem. Phys. *67*, 1784
Monahan, K., V. Rehn /1978/: Nucl. Instr. Meth. *152*, 255
Moore, C.E. /1958/: *Atomic Energy Levels*, Washington, NBS Circular 467
Moskovits, M., G.A. Ozin (eds.) /1976/: *Cryochemistry* (Wiley, New York)
Moskovits, M., J.E. Hulse /1977/: J. Chem. Phys. *67*, 4271
Moss, E.E., F.L. Hereford /1963/: Phys. Rev. Lett. *11*, 63
Mott, N.F. /1949/: Proc. Phys. Soc. *A62*, 416
Mott, N.F. /1966/: Phil. Mag. *13*, 989
Mott, N.F. /1967/: Adv. in Phys. *16*, 49
Mott, N.F., E.A. Davis /1971/: *Electronic Processes in Non-Crystalline Solids* (Oxford University Press, Oxford)
Mott, N.F. /1972/: Phil. Mag. *26*, 1249
Mott, N.F. /1974/: *Metal-Insulator Transitions* (Taylor and Francis, London)
Mott, N.F. /1975/: Phil. Mag. *31*, 217
Mott, N.F. /1978/: Phil. Mag. B *37*, 377
Muller, D.F., M.W. Wilson, M. Rothschild, C.K. Rhodes /1982/: Phys. Rev. *A25*, 1004
Mulliken, R.S. /1964/: Phys. Rev. *136*, 962
Mulliken, R.S. /1970/: J. Chem. Phys. *52*, 5170
Mulliken, R.S. /1974/: Radiation Research *59*, 357
Munro, I.H., N. Schwentner /1983/: Nucl. Instr. and Methods in Phys. Res. *208*, 819

Nagasawa, N., T. Karasawa, N. Miura, T. Nanba /1972/: J. Phys. Soc. Japan *32*, 1155
Nagasawa, N., T. Nanba /1974/: Optics Comm. *11*, 152
Nagel, D. /1976/: Diplomarbeit Universität Hamburg
Nagel, D., B. Sonntag /1978/: Ber. Bunsenges. Phys. Chem. *82*, 38
Nanba, T., N. Miura, N. Nagasawa /1974/: J. Phys. Soc. Japan *36*, 158
Nanba, T., N. Nagasawa /1974a/: J. Phys. Soc. Japan *36*, 1216
Nanba, T., N. Nagasawa, M. Ueta /1974b/: J. Phys. Soc. Japan *37*, 1031
Nasu, K., Y. Toyozawa /1981/: J. Phys. Soc. Jap. *50*, 235
Natanson, G., F. Amar, R.S. Berry /1983/: J. Chem. Phys. *78*, 399
Nelson, D.A., A.L. Rouff /1979/: Phys. Rev. Lett. *42*, 383
Ng, C.Y., D.J. Trevor, B.H. Mahan, Y.T. Lee /1977/: J. Chem. Phys. *66*, 446
Nicolis, G., S.A. Rice /1967/: J. Chem. Phys. *46*, 4445
Nitzan, A., S. Mukamel, J. Jortner /1975/: J. Chem. Phys. *63*, 200
Nowak, G., J. Fricke /1985/: J. Phys. B: At. Mol. Phys. 1355
Nürnberger, R., F.-J. Himpsel, E.E. Koch, N. Schwentner /1977/: phys. stat. sol.
   (b) *81*, 503

O'Brian, J.F., K.J. Teegarden /1966/: Phys. Rev. Lett. *17*, 919
Oka, T., K.V.S.R. Rao, J.L. Redpath, R.F. Firestone /1974/: J. Chem. Phys. *61*,
   4740
Onodera, Y., Y. Toyozawa /1968/: J. Phys. Soc. Jap. *24*, 341
Ophir, Z. /1970/: Master Thesis, Tel-Aviv University
Ophir, Z., B. Raz, J. Jortner /1974/: Phys. Rev. Lett. *33*, 415
Ophir, Z., N. Schwentner, B. Raz, M. Skibowski, J. Jortner /1975a/: J. Chem. Phys.
   63, 1072
Ophir, Z., B. Raz, J. Jortner, V. Saile, N. Schwentner, E.E. Koch, M. Skibowski,
   W. Steinmann /1975b/: J. Chem. Phys. *62*, 650
Ophir, Z. /1976/: Thesis, Tel-Aviv University
Ossicini, S., F. Forstmann /1981/: J. Chem. Phys. *75*, 2076
Ozin, G.A. /1977/: Catal. Review Science Eng. *16*, 191
Ozin, G.A., H. Huber /1978/: Inorg. Chem. *17*, 155
Ozin, G.A., H. Huber /1979/: Inorg. Chem. *18*, 1402
Ozin, G.A. /1983/: Angewandte Chemie *95*, 706

Paccioni, G., D. Playsic, J. Koutecky /1983/: Ber. Bunsenges. Phys. Chem., in press
Packard, R.E., F. Reif, C.M. Surko /1970/: Phys. Rev. Lett. *25*, 1435
Pantos, E., S.S. Hasnain, I.T. Steinberger /1977/: Chem. Phys. Lett. *46*, 395
Parinello, M., E. Tossatti, N.H. March, M.P. Tosi /1977/: Lettre al Nuovo Cimento
   18, 341
Peierls, R.E. /1932/: Ann. Physik {5} *13*, 905
Pendry, J.B. (ed.) /1974/: *Low Energy Electron Diffraction* (Academic Press, London)
Perlin, Y.E. /1963/: Sov. Phys. Uspekhi *6*, 542
Person, W.B. /1958/: J. Chem. Phys. *28*, 319
Petry, R.L. /1926/: Phys. Rev. *28*, 362
Phelps, D.J., R. Avici, C.P. Flynn /1975/: Phys. Rev. Lett. *34*, 23
Phelps, D.J., C.P. Flynn /1976a/: Phys. Rev. *B14*, 5279
Phelps, D.J., R.A. Tilton, C.P. Flynn /1976b/: Phys. Rev. *B14*, 5254
Phillips, J.C. /1966/: Solid State Physics *18*, 55
Pimentel, G.C. /1958/: J. Am. Chem. Soc. *62*, 80
Pimentel, G.C. /1978/: Ber. Bunsenges. Phys. Chem. *82*, 2
Platzman, R.L. /1961/: Int. J. Appl. Rad. Isotopes *10*, 116
Pollak, G.L. /1964/: Rev. Mod. Phys. *36*, 749
Popielawski, J., S.A. Rice /1967/: J. Chem. Phys. *47*, 2292
Posener, D.W. /1959/: Australian J. Phys. *12*, 184
Potts, W.A., H.J. Lempka, D.G. Streets, W.C. Price /1970/: Phil. Trans. Roy. Soc.
   268A, 59
Powell, B.M., G. Dolling /1977/: in *Rare Gas Solids*, ed. by M.L. Klein and J.A.
   Venables, Vol. II (Academic Press, London) p. 921
Powell, C.J. /1974/: Surface Science *44*, 29
Powell, R.C., Z.G. Soos /1975/: J. Luminescence *11*, 1

Powell, R.C. /1978/: in *Luminescence of Inorganic Solids*, ed. by B. DiBartolo (Plenum, New York) p. 547
Price, W.C. /1936/: J. Chem. Phys. *4*, 547
Prigogine, I. /1957/: *The Molecular Theory of Solutions* (North Holland, Amsterdam) Chap. 9
Pryce, M.H.L. /1966/: in *Phonons in Perfect Lattices and in Lattices with Point Imperfections*, ed. by R.W.H. Stevenson (Oliver and Bryd, Edinburgh, London) p. 403
Pudewill, D., F.-J. Himpsel, V. Saile, N. Schwentner, M. Skibowski, E.E. Koch /1976a/: phys. stat. sol. (b) *74*, 485
Pudewill, D., F.-J. Himpsel, V. Saile, N. Schwentner, M. Skibowski, E.E. Koch, J. Jortner /1976b/: J. Chem. Phys. *65*, 5226
Purdum, H., P.A. Montano, G.K. Shenoy, T. Morrison /1982/: Phys. Rev. *B25*, 4412

Quinn, J.J., J.G. Wright /1977/: *Liquid Metals 1976*, Inst. Phys. Conf. Ser. No. 30 (Bristol and London) p. 430

Raether, H. /1965/: *Solid State Excitations by Electrons* in Springer Tracts in Modern Physics *38* (Springer, Berlin, Heidelberg, New York) p. 85
Raether, H. /1974/: in *Vacuum Ultraviolet Radiation Physics*, ed. by E.E. Koch, R. Haensel and C. Kunz (Vieweg Pergamon, Braunschweig, New York) p. 591
Raether, H. /1980/: *Excitation of Plasmons and Interband Transitions by Electrons* in Springer Tracts in Modern Physics *88* (Springer, Berlin, Heidelberg, New York)
Rashba, E.I. /1976/: Izv. AN SSR, Ser. Fiz. Vol. *40*, 535
Rashba, E.I. /1981/: in *Defects in Insulating Crystals*, ed. by V.M. Tuchkevich and K.K. Shvarts,
Rashba, E.I., M.D. Sturge (eds.) /1982/: *Excitons* (North Holland Publ. Company, Amsterdam, New York, Oxford)
Rashba, E.I. /1982/: in *Excitons*, ed. by E.I. Rashba and M.P. Sturge (North Holland) p. 543
Raz, B., J. Jortner /1969/: Chem. Phys. Letters *4*, 155
Raz, B., J. Jortner /1970a/: Proc. Roy. Soc. London *A 317*, 113
Raz, B., J. Jortner /1970b/: Chem. Phys. Lett. *4*, 511
Raz, B., J. Jortner /1971/: Chem. Phys. Lett. *9*, 222
Raz, B., A. Gedanken, U. Even, J. Jortner /1972/: Phys. Rev. Lett. *28*, 1643
Raz, B., O. Cheshnovsky, J. Jortner /1976/: in *Molecular Energy Transfer*, ed. by R.D. Levine and J. Jortner (J. Wiley and Sons, New York) p. 237
Rehn, V. /1980/: Nucl. Instr. Methods *177*, 193
Reilly, M.H. /1967/: J. Phys. Chem. Solids *28*, 2067
Reimann, C.T., R.E. Johnson, W.C. Brown /1984/: Phys. Rev. Lett. *53*, 600
Reiniger, R., S. Bernstorff, P. Laporte, V. Saile, I.T. Steinberger /1983a/: Annals of the Israel Physical Society, Vol. *6*, p. 273
Reiniger, R., U. Asaf, I.T. Steinberger, P. Laporte, S. Bernstorff, V. Saile /1983b/: Annals of the Israel Physical Society, Vol. *6*, p. 282
Reiniger, R., U. Asaf, I.T. Steinberger, S. Basak /1983c/: Phys. Rev. *B28*, 4426
Reiniger, R., U. Asaf, I.T. Steinberger, V. Saile, P. Laporte /1983d/: Phys. Rev. *B28*, 3193
Reiniger, R., I.T. Steinberger, S. Bernstorff, V. Saile, P. Laporte /1984/: Chem. Phys. *86*, 189
Resca, L., R. Resta, S. Rodriguez /1978a/: Phys. Rev. B *18*, 696 and *18*, 702
Resca, L., R. Resta, S. Rodriguez /1978b/: Sol. State Comm. *26*, 849
Resca, L., S. Rodriguez /1978c/: Phys. Rev. B *17*, 3334
Resca, L., R. Resta /1979/: Phys. Rev. B *19*, 1683
Resca, L., R. Resta /1980/: Phys. Rev. B *21*, 4889
Resta, R. /1978/: phys. stat. sol. (b) *86*, 627
Rhodes, Ch.K. /1984/: in *Topics in Applied Physics*, Vol. 30, 2nd ed. (Springer, Berlin, Heidelberg, New York)
Rice, S.A., J. Jortner /1966/: J. Chem. Phys. *44*, 4470
Rice, S.A., J. Jortner /1967/: in *Physics and Chemistry of the Organic Solid State*, Vol. III, ed. by F. Fox, M. Labes and A. Weissberger (Interscience Publishers, New York) Chap. 4

Rice, S.A., G. Nicolis, J. Jortner /1968/: J. Chem. Phys. *48*, 2484
Rimbey, P.R. /1977/: J. Chem. Phys. *67*, 698
Riseberg, L.A., H.W. Moos /1968/: Phys. Rev. *174*, 429
Roberts, H.A., F.L. Hereford /1973/: Phys. Rev. *A7*, 284
Roberts, I., E.G. Wilson /1973/: J. Phys. C, Sol. State Phys. *6*, 2169
Robin, J., R. Bergeon, L. Galatry, B. Vodar /1956/: Discussions Faraday Soc. *22*, 30
Robin, M., H. Basch, N.A. Kuebler, B.E. Kaplan, J. Meinwald /1968/: J. Chem. Phys. *48*, 5037
Robin, M.B., N.A. Kuebler /1970/: J. Mol. Spectrosc. *33*, 247
Robin, M.B. /1974/: *Higher Excited States of Polyatomic Molecules*, Vol. I, II (Academic Press, New York)
Roick, E., R. Gaethke, G. Zimmerer, P. Gürtler /1983/: Sol. State Comm. *47*, 333
Roick, E. /1984/: Thesis, Universität Hamburg
Roick, E., R. Gaethke, P. Gürtler, T.O. Woodruff, G. Zimmerer /1984/: J. Phys. C: Solid State Phys. *17*, 945
Rößler, U. /1970/: phys. stat. sol. (b) *42*, 345
Rößler, U. /1971/: phys. stat. sol. (b) *45*, 483
Rößler, U., O. Schütz /1973/: phys. stat. sol. (b) *56*, 483
Rößler, U. /1976/: in *Rare Gas Solids*, Vol. I, ed. by M.K. Klein and J.A. Venables (Academic Press, London) p. 505
Rudge, M.R.H., S.B. Schwartz /1966/: Proc. Phys. Soc. (London) *88*, 563
Rudnick, W. /1983/: Diplomarbeit, Universität Kiel
Rupin, J.M., M. Morlais, S. Robin /1967/: Compt. Rend. Acad. Sci. (Paris) *265B*, 1177
Ryberg, R., O. Hunderi /1977/: J. Phys. C., Solid State Phys. *10*, 3559

Saile, V. /1976/: Thesis, Universität München
Saile, V., N. Schwentner, E.E. Koch, M. Skibowski, W. Steinmann, Z. Ophir, B. Raz, J. Jortner /1974/: in *Vacuum Ultraviolet Radiation Physics* ed. by E.E. Koch, R. Haensel and C. Kunz (Pergamon-Vieweg, Braunschweig) p. 352
Saile, V., M. Skibowski, W. Steinmann, P. Gürtler, E.E. Koch, A. Kozevnikov /1976a/: Applied Optics *15*, 2559
Saile, V., M. Skibowski, W. Steinmann, P. Gürtler, E.E. Koch, A. Kozevnikov /1976b/: Phys. Rev. Lett. *37*, 305
Saile, V., W. Steinmann, E.E. Koch /1977/: Extended Abstracts, 5th Intern. Conf. on VUV Radiation Physics, Montpellier, Vol. I, p. 74
Saile, V., H.W. Wolff /1977/: Proc. 7th Intern. Vac. Congr. and 3rd Intern. Conf. Solid Surfaces, Vienna, p. 391
Saile, V., E.E. Koch /1979/: Phys. Rev. *B20*, 784
Saile, V. /1980/: Applied Optics *19*, 4115
Saile, V., E.E. Koch /1980a/: Phys. Rev. *B21*, 4892
Saile, V., E.E. Koch /1980b/: unpublished
Saile, V., D. Rieger, W. Steinmann, T. Wegehaupt /1980/: Phys. Lett. *A79*, 221
Samson, J.A.R. /1967/: *Techniques of Vacuum Ultraviolet Spectroscopy* (Wiley, New York)
Sando, K.M. /1971/: Mol. Phys. *21*, 439
Saxon, R.P., B. Liu /1976/: J. Chem. Phys. *64*, 3291
Saxton, M.J., J.M. Deutsch /1974/: J. Chem. Phys. *60*, 2800
Scharber, R.S., S.E. Webber /1971/: J. Chem. Phys. *55*, 3985
Scheffler, M., K. Horn, A.M. Bradshaw, K. Kambe /1978/: Surface Science *80*, 69
Scher, H., R. Zallen /1970/: J. Chem. Phys. *53*, 3959
Schmidt, W.F. /1974/: in *Electron Solvent Interactions*, ed. by L. Kevon (Elsevier Publ., Amsterdam)
Schneider, B., J.S. Cohen /1974/: J. Chem. Phys. *61*, 3240
Schnepp, O., K. Dressler /1960/: J. Chem. Phys. *33*, 49
Schnyders, H., S.A. Rice, L. Meyer /1966/: Phys. Rev. *150*, 127
Schörner, R. /1977/: Diplomarbeit, Universität Regensburg
Schörner, R., E. Schuberth, M. Creuzburg /1977/: Extended Abstracts, 5th Intern. Conf. on VUV Radiation Physics, Montpellier, Vol. I, p. 181
Schou, J., H. Sørensen, P. Børgensen /1984/: Nucl. Instr. Methods in Phys. Research *B5*, 44

Schuberth, E., M. Creuzburg /1975/: phys. stat. sol. (b) *71*, 797
Schuberth, E. /1976/: Thesis, Universität Regensburg
Schuberth, E., M. Creuzburg, W. Müller-Lierheim /1976/: phys. stat. sol. (b) *76*, 301
Schuberth, E. /1977/: phys. stat. sol. (b) *84*, K91
Schuberth, E., M. Creuzburg /1978/: phys. stat. sol. (b) *90*, 189
Schulze, W., D.M. Kolb /1974a/: J. Chem. Soc. Faraday Transactions II *70*, 1098
Schulze, W., D.M. Kolb, G. Klipping /1974b/: 5th Intern. Cryogenic Engineering Conf., Kyoto, p. 268
Schulze, W., D.M. Kolb, H. Gerischer /1975/: J. Chem. Soc. Faraday Transactions II, *71*, 1763
Schulze, W., H.U. Becker, H. Abe /1978/: Ber. Bunsenges. Phys. Chem. *82*, 138
Schulze, W., H.U. Becker, H. Abe /1978/: Chem. Phys. *35*, 177
Schwarz, K.W. /1975/: in *Advances in Chemical Physics XXXIII*, ed. by I. Prigogine and S.A. Rice (Wiley, New York)
Schwentner, N., M. Skibowski, W. Steinmann /1973/: Phys. Rev. *B8*, 2965
Schwentner, N. /1974/: Thesis, Universität München
Schwentner, N., F.J. Himpsel, E.E. Koch, V. Saile, M. Skibowski /1974a/: in *Vacuum Ultraviolet Radiation Physics*, ed. by E.E. Koch, R. Haensel and C. Kunz (Vieweg, Pergamon, Braunschweig, New York) p. 355
Schwentner, N., E.E. Koch, V. Saile, M. Skibowski, A. Harmsen /1974b/: in *Vacuum Ultraviolet Radiation Physics*, ed. by E.E. Koch, R. Haensel and C. Kunz (Vieweg, Pergamon, Braunschweig, New York) p. 792
Schwentner, N., F.J. Himpsel, V. Saile, M. Skibowski, W. Steinmann, E.E. Koch /1975/: Phys. Rev. Lett. *34*, 528
Schwentner, N., E.E. Koch /1976a/: Phys. Rev. *B14*, 4687
Schwentner, N. /1976b/: Phys. Rev. *B14*, 5490
Schwentner, N. /1978/: in *Luminescence of Inorganic Solids*, ed. by B. DiBartolo and D. Pacheco (Plenum Press, New York) p. 645
Schwentner, N., E.E. Koch, Z. Ophir, J. Jortner /1978/: Chem. Phys. *34*, 281
Schwentner, N., U. Hahn, D. Einfeld, G. Mühlhaupt /1979/: Nuclear Instr. and Meth. *167*, 499
Schwentner, N. /1980/: Applied Optics *19*, 4104
Schwentner, N., H.W. Rudolf, G. Martens /1981/: phys. stat. sol. (b) *106*, 183
Schwentner, N., O. Dössel, H. Nahme /1982/: in *Laser Techniques for Extreme Ultraviolet Spectroscopy*, AIP Conference Proceedings 90, ed. by McIlrath and Freeman (American Institute of Physics, New York)
Schwentner, N., E.E. Koch, J. Jortner /1984/: in *Energy Transfer Processes in Condensed Matter*, ed. by B. Di Bartolo (Plenum, New York) p. 417
Shanfield, Z., P.A. Montano, P.H. Barett /1975/: Phys. Rev. Lett. *35*, 1789
Shante, V.K.S., S. Kirkpatrick /1971/: Adv. Phys. *20*, 325
Shockley, W. /1961/: Solid State Electron. *2*, 35
Siegbahn, K., C. Nordling, A. Fahlman, R. Nordberg, K. Hamrin, J. Hedman, G. Johansson, T. Bergmark, S.E. Karlsson, I. Lindgren, B. Lindberg /1967/: Nova Acta Regiae Soc. Sci. Uppsal., Ser. IV, Vo. 20, Uppsala
Siegbahn, K., C. Nordling, G. Johansson, J. Hedman, P.F. Hedén, K. Hamrin, U. Gelius, T. Bergmark, L.O. Werme, R. Manne, Y. Bear /1969/: *ESCA Applied to Free Molecules* (North-Holland, Amsterdam)
Silbey, R. /1976/: Annal. Rev. Phys. Chem. *27*, 203
Slater, J.C., G.F. Koster /1954/: Phys. Rev. *94*, 1498
Smalley, P.E., D.A. Auerbach, P.S.H. Fitch, D.H. Levy, L. Wharton /1977/: J. Chem. Phys. *66*, 3778
Soley, F.J., W.A. Fitzsimmons /1974/: Phys. Rev. Lett. *32*, 988
Soley, F.J., R.K. Leach, W.A. Fitzsimmons /1975/: Phys. Lett. *55*, 49
Sommer, W.T. /1964/: Phys. Rev. Lett. *12*, 271
Sommer, K., U. Hahn, B. Jordan, N. Schwentner, G. Zimmerer /1978/: unpublished
Song, K.S. /1969/: J. Phys. Soc. Jap. *26*, 1131
Song, K.S. /1971/: Can. J. Phys. *49*, 26
Song, K.S., L.J. Lewis /1979/: Phys. Rev. *B19*, 5349
Song, K.S., C.H. Leung /1981/: J. Phys. C. Solid State Phys. *14*, L359
Song, K.S. /1982/: U.S.-Japan Joint Seminar on Recombination Induced Defect Formation (Argonne)

Sonnenblick, Y., E. Alexander, Z.H. Kalman, I.T. Steinberger /1977/: Chem. Phys. Lett. *52*, 276
Sonntag, B. /1977/: in *Rare Gas Solids*, Vol. II, ed. by M.K. Klein and J.A. Venables (Academic Press, London) p. 1021
Sowada, U., W.F. Schmidt, G. Bakale /1977/: Can. J. Chem. *55*, 1885
Sowada, U., J.M. Warman, M.P. de Haas /1982/: Phys. Rev. *B25*, 3434
Spear, W.E., P.G. LeComber /1977/: in *Rare Gas Solids*, Vol. II, ed. by M.K. Klein and J.A. Venables (Academic Press, London) p. 1119
Spiegelman, F., J.P. Malrieu /1978/: Chem. Phys. Lett. *57*, 214
Springett, B.E., J. Jortner, M.H. Cohen /1968/: J. Chem. Phys. *48*, 2720
Steets, W., A.P. Hickmann, N.F. Lane /1974/: Chem Phys. Lett. *28*, 31
Steinberger, I.T., O. Schnepp /1967/: Sol. State Comm. *5*, 417
Steinberger, I.T., C. Atluri, O. Schnepp /1970/: J. Chem. Phys. *52*, 2723
Steinberger, I.T., U. Asaf /1971/: in Proc. 2nd Intern.. Conf. on *Conduction in Low Mobility Materials*, ed. by N. Klein, D.S. Tannhauser and M. Pollak (Taylor Francis, London) p. 453
Steinberger, I.T. /1973/: J. Appl. Opt. *12*, 614
Steinberger, I.T., U. Asaf /1973/: Phys. Rev. *B8*, 914
Steinberger, I.T., I.H. Munro, E. Pantos, U. Asaf /1974/: in *Vacuum Ultraviolet Radiation Physics*, ed. by E.E. Koch, R. Haensel and C. Kunz (Vieweg, Pergamon, Braunschweig, New York); see also Phys. Lett. *47A*, p. 299
Steinberger, I.T., E. Pantos, S.S. Hasnain, I.H. Munro, T.D.S. Hamilton /1976/: in *Molecular Spectroscopy of Dense Phases* (Elsevier, Amsterdam) p. 467
Steinberger, I.T., P. Maaskant, S.E. Webber /1977/: J. Chem. Phys. *66*, 4722
Stern, E.A., S.M. Heald /1983/: in *Handbook on Synchrotron Radiation*, Vol. 1, ed. by E.E. Koch (North Holland, Amsterdam) p. 955
Stewart, T.E., G.S. Hurst, T.E. Bortner, J.E. Parks, F.W. Martin, H.L. Weidner /1970/: J. Opt. Soc. Am. *60*, 1290
Stock, M., E.W. Smith, R.E. Drullinger, M.M. Hessel, J. Pourcin /1978/: J. Chem. Phys. *68*, 1785
Stockton, M., J.W. Keto, W.A. Fitzsimmons /1970/: Phys. Rev. Lett. *24*, 654
Stockton, M., J.W. Keto, W.A. Fitzsimmons /1972/: Phys. Rev. *A5*, 372
Suemoto, T. /1977/: Thesis, University of Tokyo
Suemoto, T., Y. Kondo, H. Kanzaki /1977/: Phys. Lett. *61A*, 131
Suemoto, T., Y. Kondo, H. Kanzaki /1978/: Solid State Comm. *25*, 669
Suemoto, T., H. Kanzaki /1979/: J. Phys. Soc. Japan *46*, 1554
Suemoto, T., H. Kanzaki /1980/: Phys. Soc. Jap. *46*, 1039
Suemoto, T., H. Kanzaki /1981/: Phys. Soc. Jap. *50*, 3664
Surko, C.M., F. Reif /1968/: Phys. Rev. *175*, 229
Surko, C.M., G.J. Dick, F. Reif, W.C. Walker /1969/: Phys. Rev. Lett. *23*, 842
Surko, C.M., R.E. Packard, G.J. Dick, F. Reif /1970/: Phys. Rev. Lett. *24*, 657
Swyler, K.J., M. Creuzburg /1976/: J. Luminescence *1,2*, 842

Takahashi, T., S. Konno, T. Hamada, M. Miyajima, S. Kubota, A. Nakamoto, A. Hitachi, E. Shibamura, T. Doke /1975/: Phys. Rev. *A12*, 1771
Tauchert, W.K. /1975/: Thesis, Freie Universität Berlin
Tauchert, W.K., W.F. Schmidt /1975/: Z. Naturforsch. *30a*, 1085
Tauchert, W.K., H. Jungblut, W.F. Schmidt /1977/: Can. J. Chem. *55*, 1860
Thompson, C.E. /1965/: J. Opt. Soc. Am. *55*, 1184
Thonnard, N., G.S. Hurst /1972/: Phys. Rev. *A5*, 1110
Tilton, R.A., D.J. Phelps, C.P. Flynn /1974/: Phys. Rev. Lett. *32*, 1006
Tilton, R.A., C.P. Flynn /1975/: Phys. Rev. Lett. *34*, 20
Tilton, R.A., D.J. Phelps, C.P. Flynn /1976a/: Phys. Rev. *B14*, 5265
Tilton, R.A., C.P. Flynn /1976b/: Phys. Rev. *B14*, 5289
Tolkiehn, G. /1976/: Dimplomarbeit, Universität Hamburg
Toyozawa, Y. /1958/: Progr. Theor. Phys. *20*, 53
Toyozawa, Y. /1961/: Progr. Theor. Phys. *26*, 29
Toyozawa, Y. /1974/: in *Vacuum Ultraviolet Radiation Physics*, ed. by E.E. Koch, R. Haensel and C. Kunz (Vieweg, Pergamon, Braunschweig, New York) p. 317
Toyozawa, Y., Y. Shinozuka /1980/: J. Phys. Soc. Japan *48*, 472
Tsai, B.P., T. Bear /1974/: J. Chem. Phys. *61*, 2047

Turner, D.W., C. Baker, A.D. Baker, C.R. Brundle /1970/: *Molecular Photoelectron Spectroscopy* (Wiley, London, New York)
Ueba, H., S. Ichimura /1976/: J. Phys. Soc. Jap. *41*, 1974
Ueba, H. /1977/: J. Phys. Soc. Jap. *43*, 353
Unwin, R. K. Horn, P. Geng /1980/: Vakuumtechnik *29*, 149

Van Zee, R.J., C.A. Baumann, W. Weltner /1981/: J. Chem. Phys. *74*, 6977
Van Zee, R.J., C.A. Baumann, S.V. Bhat, W. Weltner /1982/: J. Chem. Phys. *76*, 5636
Venables, J.A., B.L. Smith /1977/: in *Rare Gas Solids*, Vol. II, ed. by M.K. Klein and J.A. Venables (Academic Press, London) p. 609

Waclawski, B.J., J.F. Herbst /1975/: Phys. Rev. Lett. *35*, 1594
Wadt, W.R., P.J. Hay, L.R. Kahn /1978/: J. Chem. Phys. *68*, 1752
Wannier, G.H. /1937/: Phys. Rev. *52*, 191
Webber, S.E., S.A. Rice, J. Jortner /1965/: J. Chem. Phys. *42*, 1907
Weber, G. /1979/: Physica Scripta *20*, 240
Webman, I., J. Jortner, M.H. Cohen /1975/: Phys. Rev. *B11*, 2885
Webman, I. /1976/: Thesis, Tel-Aviv University
Webman, I., J. Jortner, M.H. Cohen /1976a/: Phys. Rev. *B13*, 713
Webman, I., J. Jortner, M.H. Cohen /1976b/: Phys. Rev. *B14*, 4737
Webman, I., J. Jortner, M.H. Cohen /1977/: Phys. Rev. *B15*, 5712
Weinert, C.M., F. Forstmann, H. Abe, R. Grinter, D.M. Kolb /1982/: J. Chem. Phys. *77*, 3392
Welker, T. /1978/: Ber. Bunsenges. Phys. Chem. *82*, 40
Welker, T., T.P. Martin /1979/: J. Chem. Phys. *70*, 5683
Wenck, H.D. /1979/: Diplomarbeit, Universität Hamburg
Wenck, H.D., S.S. Hasnain, M.M. Nikitin, K. Sommer, G. Zimmerer, D. Haaks /1979/: Chem. Phys. Lett. *66*, 138
Weymann, W., F.M. Pipkin /1965/: Phys. Rev. *A 137*, 490
Wiese, W.L., M.W. Smith, B.M. Miles /1969/: in *Atomic Transition Probabilities*, II, *192* NS RDS - NBS 22
Wilcke, H. /1979/: Diplomarbeit, Universität Kiel
Wilcke, H., W. Böhmer, N. Schwentner /1983/: Nucl. Instr. and Methods in Phys. Res. *204*, 533
Winick, H., S. Doniach (eds.) /1980/: *Synchrotron Radiation Research* (Plenum Press, New York)
Wöste, L. /1982/: in *Impact of Cluster Physics in Material Science and Technology* (Reidel Publishing Company, Dordrecht)
Wolf, H.C. /1967/: in *Advances in Atomic and Molecular Physics*, Vol. 3, ed. by D.R. Bates (Academic Press, New York, London) p. 119
Wolff, H.W. /1977/: Extended Abstracts 5th Intern. Conf. on VUV Radiation Physics, Montpellier, p. 214
Wolff, H.W. /1979/: Thesis, Universität Hamburg
Wright, M.R., R.P. Frosch, G.W. Robinson /1960/: J. Chem. Phys. *33*, 934

Yakhot, V., M. Berkowitz, R.B. Gerber /1975/: Chem. Phys. *10*, 61
Yakhot, V. /1976/: Chem. Phys. *14*, 441
Yokota, M., O. Tanimoto /1967/: J. Phys. Soc. Jap. *22*, 779
Yoshino, K., U. Sowada, W.F. Schmidt /1976/: Phys. Rev. *A14*, 438
Younger, S.M., W.L. Wiese /1978/: Phys. Rev. *17*, 1944

Zallen, R., H. Scher /1971/: Phys. Rev. *B4*, 4471
Zimmerer, G. /1976/: in Proc. of Intern. Summer School on Synchrotron Radiation Research, Vol. I, ed. by A.N. Mancini and I.F. Quercia, Alghero, p. 453
Zimmerer, G. /1978/: in *Luminescence of Inorganic Solids*, ed. by B. DiBartolo (Plenum Press, New York) p. 627
Zimmerer, G. /1979/: J. Luminescence *18/19*, 875

Subject Index

absorption spectroscopy 7, 11, 17 ff.

   of excitons 33 ff.

   of metal impurities 86

   of molecular impurities 60, 76

Ag, matrix isolated 92, 96, 100

alloys 48 ff., 54 ff.

Ar, see binding energies, exciton, luminescence

   dielectric function 34

   luminescence spectrum in Ne 160

   luminescence spectrum liquid 135

   luminescence spectrum solid 124

   photoelectron spectrum 57 ff., 211

   structure 8

   transient absorption 122

   valence bands 26

   valence excitons 44

autoionization 108

band structure 22 ff., 200

   of alloys 57 ff.

   of monolayers 30

   two dimensional 29 ff.

benzene in rare gases, luminescence 191 ff.

   photoemission 190

bilinear collision of excimers 140 f.

binding energies 6, 8, 27, 40 f., 50, 119, 121, 125, 136, 156 ff., 181

Born approximation 197

bubble formation 123

bubbles in liquid He 145 ff.

Ca_2 matrix isolated 97

clusters 28, 81, 85, 97, 99

cluster beams 4

conductivity, metal rare gas mixtures 101 ff.

conduction bands 22 ff.

configuration coordinate model 151, 172 f.

core excitons 46 ff.

crystal field splitting 88

crystal growth 14

Cu matrix isolated 91

deformation potential 112 ff.

diffusion

   length in energy transfer 183

   of excitons 176 ff.

     photoinduced 92

disorder 84

dissociation energies, rare gas molecules 8

effective mass 40

electron affinities 27, 68, 75, 79 ff.

electron (continued)

    diffraction 85

    diffusion length 183

    energy bands 22 ff.

    energy loss cross section 197 ff.

    energy transfer 108, 175

    hole pair creation 202, 214 ff.

    hole recombination 108

    mobility 194, 201

    phonon coupling 109 ff., 152 ff.,
        165 ff.

    polaron complex 209

    relaxation 108, 165 ff., 184

    states 22 ff., 63 ff.

    thermalization times 201

    transport, see scattering

elementary excitations 5, 107, see also
    excitons

energies, see also binding energies

    of luminescence bands 119

    of luminescence from impurities
        156 ff.

    of luminescence in liquids 136

    of transient absorption 121

energy dissipation 107 ff.

    gap law 117, 130, 166 ff.

    transfer, electronic 175 ff.

    transfer radii 191

EXAFS extended X-ray absorption fine
    structure 86

excimer 91

    centres 113

excitation sources 15 ff.

excited state, see exciton

excited state dynamics Chap. 6

exciton Chap. 3

    binding energy 27, 116

    central cell correction 36

    core 46 ff., 55 f.

    diffusion 176

    dispersion 109 ff., 116

    dynamics Chap. 6

    effective mass 35

exciton (continued)

    free 126 ff., 133, 177 ff.

    Frenkel 35 ff.

    impurity states, see impurity states

    intermediate 35 ff.

    lattice distortion 110 ff.

    lifetime 119

    line shape 43, 53, 72 ff., 110 ff.

    in liquids 65 ff.

    localization 112 ff., 117 ff.

    longitudinal 42, 45

    luminescence 117 ff., 124 ff.

    motion 109 ff., 176 ff.

    phonon interaction 109 ff.

    polariton 133

    radii 36

    screening 104

    self trapping 108, 112 ff., 176

    spin orbit splitting 27 ff., 88

    surface 38 ff.

    theory 35 ff.

    tunneling 113 ff., 116

    Wannier-Mott 35 ff.

experimental techniques 11 ff.

Fano factor 195, 203, 216

fluorescence, see luminescence

Förster-Dexter mechanism 108, 176 ff.,
    187 ff.

generalized oscillator strength 197

Hg matrix-isolated 93, 95, 102 ff.

harmonic approximation, lattice deforma-
    tion 114

He, liquid luminescence spectra 135,
    143, 145

    structure 8

Huang-Rhys coupling constant 167